Livestock Biodiversity

Genetic resources for the farming of the future

Stephen J. G. Hall
MA, PhD
Department of Biological Sciences, University of Lincoln

Blackwell
Science

Editorial offices:
Blackwell Science Ltd, 9600 Garsington Road, Oxford OX4 2DQ, UK
Tel: +44 (0) 1865 776868
Blackwell Publishing Professional, 2121 State Avenue, Ames, Iowa 50014-8300, USA
Tel: +1 515 292 0140
Blackwell Science Asia Pty Ltd, 550 Swanston Street, Carlton, Victoria 3053, Australia
Tel: +61 (0)3 8359 1011

First published 2004

ISBN 0-632-05499-9

A catalogue record for this title is available from the British Library

Hall, Stephen J.G.
Livestock biodiversity: genetic resources for the farming of the future / Stephen J.G. Hall.–
1st ed. p. cm.
Includes bibliographical references (p.217).
ISBN 0-632-05499-9 (hardback : alk. paper)
1. Livestock–Germplasm resources. 2. Biological diversity conservation. I. Title.

SF105.3.H35 2004
636.08'21–dc22
2004005421

Set in 10/12.5 Times
by Kolam Information Services Pvt. Ltd, Pondicherry, India
Printed and bound in India
by Gopsons Paper Ltd, Noida.

The publisher's policy is to use permanent paper from mills that operate a sustainable forestry policy, and which has been manufactured from pulp processed using acid-free and elementary chlorine-free practices. Furthermore, the publisher ensures that the text paper and cover board used have met acceptable environmental accreditation standards.

For further information on Blackwell Publishing, visit our website:
www.blackwellpublishing.com

Contents

Preface

This book has been written in response to three stimuli. First, all over the world there is evidence that the diversity of domesticated animals is under threat and students, opinion leaders, researchers and policy makers should be aware of this and should be encouraged to do what they can to protect this component of global biodiversity. Second, it is now clear that the true functions and values of livestock are not merely the production of food and materials and the provision of draft power. Livestock have always had cultural and social significance, and they provide many services to society which are now beginning to be widely appreciated. These range from landscape conservation, to serving as a reliable investment where banking is poorly developed. These broader roles, how they relate to specific breeds, and how they change and develop, need to be better understood. Third, the conservation and sustainable development of livestock biodiversity has been seen as falling mainly within the range of interest of geneticists and animal breeders. While their insights and the new techniques they are developing are indispensable, there are other disciplines which have much to offer. The fruitful interactions of apparently rather different sciences have been seen in the emergence of conservation biology (Frankham *et al.*, 2002; Primack, 2002) in response to the crisis of floral and faunal extinctions. Similarly, livestock biodiversity needs inputs from several disciplines if it is to be conserved and developed sustainably. A recent review of the currently available scientific approaches is given by Simm *et al.* (2004).

This book aims to give an overview of the patterns and processes of evolution of livestock biodiversity, to emphasize its place in our cultural heritage and its value to the farming of today and of the future, to suggest future areas of research, to celebrate its richness and to advocate its conservation and sustainable utilization.

The study of livestock biodiversity embodies the kinds of issues that modern versatile graduates need to be able to tackle. This book is written at advanced undergraduate and junior postgraduate level for students of genetics, animal breeding and animal science, conservation biology and social and development studies. It is also intended for policy makers and as a background text for researchers in these general areas.

Stephen J.G. Hall

Acknowledgements

This book was begun while I was in Professor Don Broom's group in the Department of Clinical Veterinary Medicine, University of Cambridge. It was continued while I was at the Overseas Development Institute, London, and at De Montfort University, and has been completed at the Department of Biological Sciences, University of Lincoln.

The ideas have been developed during and after innumerable conversations with very many colleagues. I owe particular debts to Dr Roger Blench, Dr Juliet Clutton-Brock, Dr Martyn Murray and two referees. Without the interest, understanding and support of my parents, Dr and Mrs J.G. Hall, and of my wife Sue and daughters Nicola and Katie, the work would have been much more arduous.

The book is dedicated to the memory of Professor Peter Jewell, who was one of the pioneers of the conservation of rare breeds, as well as being a most valued friend, and inspiring mentor and colleague.

Introduction

Domestication of animals was a vital step in the development of human culture and many of the major issues facing the world at the beginning of the twenty-first century are related to livestock. The kinds of inter-species relationships so familiar in Nature – predation, parasitism, symbiosis, etc. – have been joined by another, domestication, under which animals have been shaped to meet human needs. In the 10 000 years since the major domestications began, thousands of distinctive taxa have emerged from a few relatively narrow genetic bases. Much of this biodiversity is now at risk.

What is the place of livestock biodiversity in the world today and why is it important for the future? The current belief is that the world should be able to feed its human population, which may total eight billion in 2020, provided fair distribution can be achieved and that intensification can continue. Increasing affluence, especially in the developing world, is expected to increase global meat demand from 209 million tonnes (in 1997) to 327 million tonnes in 2020; milk demand is expected to rise from 422 million to 648 million tonnes (de Haan *et al.*, 2001). The trends are also for a shift towards pigs and poultry in both developed and developing countries.

As crop production increases, land for pasture decreases, but the amount of straw and other residues increases. World annual crop production has been estimated at about 2750 million tonnes, and of crop residues 3750 million tonnes (Smil, 1999). The ruminant can remain confident of its food supply. In wealthy countries food quality and safety are important and, in some, animal welfare is as well. Perhaps the day will come when the animals of the factory farm are once more put out to grass but such free-range, welfare-oriented husbandry systems will only contribute a small proportion to the world's food supplies.

In many societies such as the European Union (EU), tracts of land are valued primarily for their scenic beauty and cultural significance – grazing livestock are often a key element in this maintenance (Hindmarch Pienkowski, 1998). In the Lake District of northern England, or in the parklands around historic houses, 50 years ago livestock were an important part of the food chain, but today they are often kept in order to cut the grass.

The thousands of breeds that exist were mostly developed in very different circumstances from those of today, but they are not anachronisms, they represent the genetic diversity that will enable the livestock farming of the future to respond to new challenges. These include emergent diseases, consumer dissatisfaction with current livestock systems, environmental disruption arising from climate change and shortage of grazing land in developing countries as a result of increasing crop production. This diversity is under threat as high-performing breeds are freely available in a global market. For example Canada supplies 20% of the world's internationally traded dairy cattle semen and embryos, worth $128 million annually (July 2001: www.dairyinfo.agr.ca/glance.pdf), and in 1999 importations of day-old chicks into the UK totalled 98 472 662 (www.parliament.the-stationery-office.co.uk). During the late 1990s about 1.6 million semen doses and 50 000 dairy cows were exported annually from the Netherlands (Siemes, 2001). In these circumstances local genotypes, produced in small numbers, find it difficult to compete.

Threats to wildlife biodiversity are easy to describe, explain and quantify. There is no arguing with photographs of poached rhino, or elephant, or gorillas, or torched forest, and the links between habitat loss, overexploitation and extinction are easy to explain. Threats to breeds are nowhere near as obvious. Repeated mating with another breed will change the genotype of a livestock population radically in a few generations, a very different concept of extinction from that which applies in wildlife. In many countries, traditional breeds are hardly known at all to policy makers or opinion leaders.

Agricultural policy makers will be seen by future generations as having abdicated their responsibilities if they fail to support conservation of livestock biodiversity, or find other organizations willing to do so. The Food and Agriculture Organization of the United Nations (FAO) has a 50-year record of promoting research into livestock biodiversity and of advocating its conservation, but there have been very few examples of internationally funded projects to support livestock biodiversity in the developing world and relatively few government-funded projects in richer nations. On the positive side, the FAO has helped to create a climate of awareness, which was recognized in the Convention on Biological Diversity, and this book tries to add to that achievement.

Part 1
The Nature of Livestock Biodiversity

Today's livestock biodiversity is the fruit of a two–stage process. First there was domestication, then there was breed differentiation. Both involved genetic change and are thus both evolutionary and cultural processes. Domestication of plants and animals was of fundamental importance for the development of the human cultures that are economically dominant today. Modern genetics has cast fresh light on this process: molecular genetics has been explicitly applied to the study of where and when domestication happened. Quantitative and population genetics have not – yet. How today's breeds arose is well known in some regions of the world and is almost a complete mystery in others.

Chapter 1
Biology of domestication

Introduction

What are the evolutionary processes that operate under domestication, and how do they differ from those observed in nature? What was the biology of the process of domestication, and how much of this can we infer from studies on present-day domesticated animals, to add to what is known from archaeology and from human sciences? Could the processes giving rise to new heritable variation have taken new forms under domestication?

Domestication and models of evolution

Humans probably existed for three million years as hunter-gatherers before the domestication of plants and animals began, about 10 000 years ago (Clutton-Brock, 1999). The effects of this cultural change, 400 human generations ago, on the evolution of our species would be difficult to demonstrate, but clearly the effects on the animals and plants which were then enfolded into our society were profound (Figures 1.1a, b).

Domestication and breed development are the evolutionary processes that are most visible to us and this is one reason why they have been used to help formulate evolutionary principles and to persuade people of the truth of these principles. Darwin used the existence of breeds and varieties in his arguments for evolution by natural selection. In order to convey the idea that heritable adaptation (and not merely individual developmental modification) could occur in nature, Darwin emphasized how humans could practise artificial selection to create animals and plants that were adapted to human use. There was an analogy here, suited to the public attitudes of his time, between man's activities and those of a 'benevolent, omniscient "super breeder" who, unlike human breeders, selects only for the benefit of the organism' (Richards, 1998).

This is the best known example of how knowledge of domestication helped to show how evolution could have taken place. Later, Sewall Wright's studies on Shorthorn cattle highlighted the importance of population structure in evolution. Currently, understanding of quantitative genetics, obtained from work on livestock, is being applied widely to studies on evolution in nature (for example Roff, 1997).

(a)

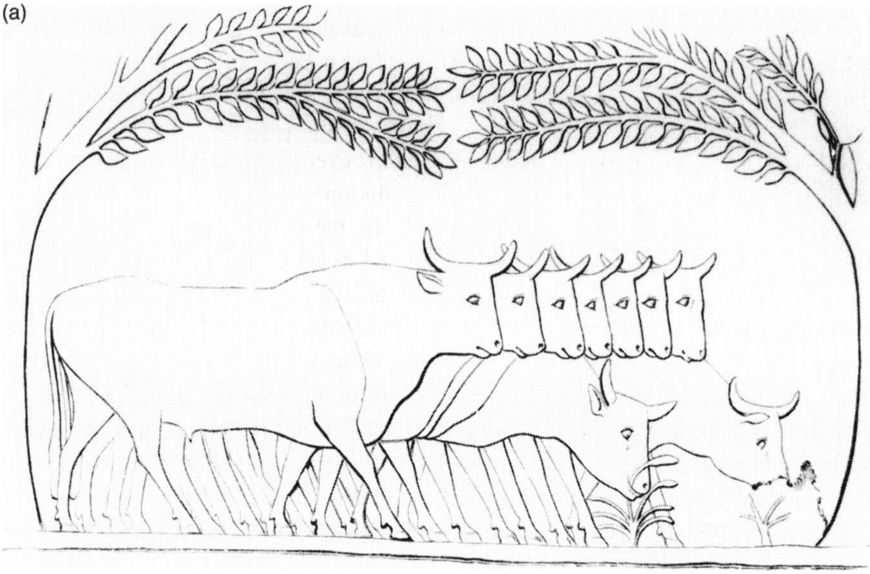

(b)

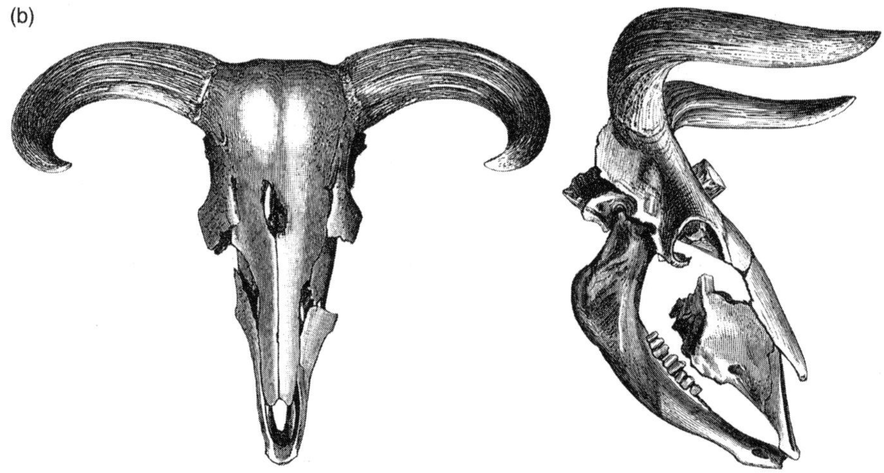

Figure 1.1 Domestication – a new relationship. This aurochs skull (b) (in the Haddon Museum, Cambridge University) still has a Neolithic stone axe embedded in it – relic of a violent encounter in what is now Burwell Fen, Cambridgeshire. The axe is depicted in the side view. In sharp contrast with the Egyptian cattle (a) in bas-relief in the temple of Hatshepsu, the themes are tranquillity, orderliness, uniformity and pride of possession. Both reproduced from McKenny Hughes (1896).

Rates of evolution under domestication

Domestication has been defined as 'the process of enfolding a species into human society and taking responsibility for its husbandry and control over its breeding' (Clutton-Brock, 1992a).

Under domestication animals are subjected to breeding plans, which can be based on artificial selection, on relationship, or on phenotypic likeness (Lush, 1943). The breeds observed today have each emerged by an interaction of one or more of these processes with the operation of founder effects and inbreeding, and probably migration and natural selection.

Rapid evolutionary change is possible, perhaps even typical, when domesticated animals are artificially selected, and rates of evolution can be compared (Figure 1.2). If a phenotypic trait changes by a factor of e (2.718, the base of natural logarithms) in one million years its rate of evolution is defined as one darwin (Gingerich, 1983; Hendry & Kinnison, 1999). In animal breeding, annual rates of change in growth rate between 1 and 3% are possible (Simm, 1998, p.80; McKay *et al.*, 2000), i.e. around 380 000 darwins. Racing performance in Thoroughbred horses has improved at a rate of about 14 000–29 000 darwins (calculated from Eckhardt *et al.*, 1988). On a geological time scale, rates are mainly between 0.1 and 400 darwins, but within natural species or populations (microevolutionary scale) rates can be much higher, for example during a colonization event or recovery from a catastrophe. Evolution of dwarfism in island Red deer proceeded at a rate of 10 000 darwins (Lister, 1989).

Technological domestication

Nowadays many animals are farmed or kept in ways which do not involve their being enfolded into human society. If they can be described as domesticated it is only in the sense that their husbandry is fully under human control rather than that their relationship with humans is of any depth. Before the twentieth century chickens were essentially farmyard scavengers, pigs were kept mainly to deal with the

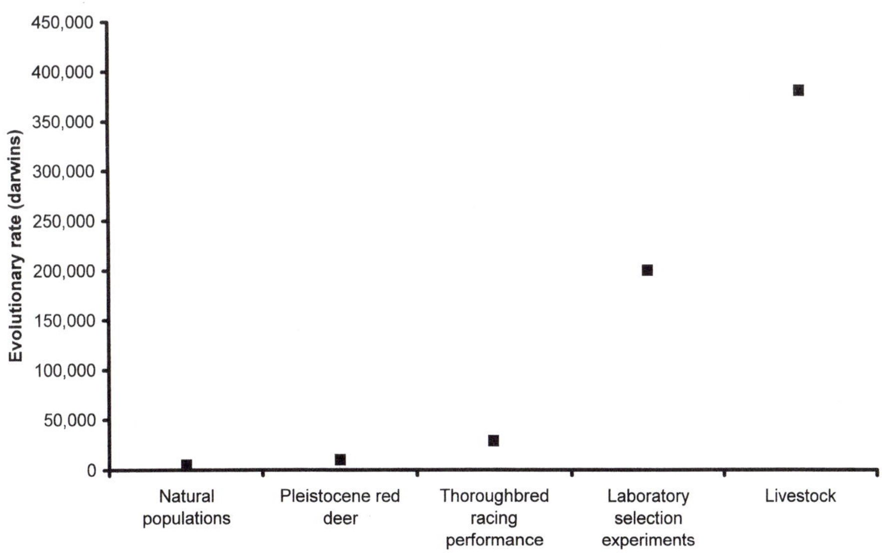

Figure 1.2 Evolutionary rates, when measured in darwins (see text), can be compared between populations.

leftovers of food production and processing and aquaculture was limited to herbivorous fish such as carp kept in ponds – all husbandry systems which involve fairly intimate human–animal interactions.

In recent years the availability of technology (housing and transport systems, feed formulation, meat processing and marketing) has enabled non-grazing livestock to be farmed intensively. More than 220 aquatic species are now farmed in a wide range of systems (Naylor *et al.*, 2000); farmed fish supplies totalled 29 million tonnes in 1997 compared with fisheries landings of about 95 million tonnes. The previously close human–animal interaction has been replaced by systems where animals are kept well apart from people. In poultry farming, certain attributes of a species which were not particularly valuable when it was being kept as a farmyard fowl have proved fundamental to the development of modern poultry industries (Table 1.1).

Molecular genetic insights into domestication

Our understanding of domestication comes from archaeology, and from knowledge of the behaviour and biology of humans and of the wild ancestors and present-day relatives of domesticated animals. Current knowledge and opinion is summarized by Hemmer (1990), Clutton-Brock (1999) and Price (2002). Providing new insights into the process, the molecular genetics of present day livestock is making it possible to test hypotheses generated by earlier work (see Chapter 7).

Table 1.1 Key features of the chicken making it highly suitable for modern intensive farming. From Hartmann (1989).

Feature	Consequence
High fecundity and short generation interval	Intense selection and rapid multiplication possible
Incubation can be mechanized	No reliance on broody hens. This is essential for commercial production, and enables selection against broodiness
Eggs remain fertile in storage	Family size can be modified for selection and increased flexibility of breeding
Depending on genotype, sex can be identified at hatching using sex-linked plumage characteristics	A relatively recent development permitting separate breeding programmes for egg and meat production. Previously all chicks had to be kept until old enough to be sexed. By then it was not economic to discard the males and they had to be reared for meat. Consequently selection programmes could not ignore meat qualities and specialist egg-producing breeds were not economically viable
Responsiveness of sexual maturation and egg laying to artificial light	Reduced seasonal variation in production and improved profitability
Apparent adaptability to novel feeding and housing conditions	Automatic feeding and high health, low labour-requirement housing possible

Early work on livestock breeds in the 1950s and 1960s, inspired by studies on human blood groups and other heritable variation in biochemical characteristics, showed how breeds could be grouped on the basis of shared descent, but mutation rates were too slow and allelic diversity too narrow to permit deductions relating to the evolutionary process. Today, mitochondrial DNA (mtDNA) lineages can be traced back in such a way that timings and locations of genetic bottlenecks suggestive of domestication can be deduced. Mitochondrial DNA is inherited in the female line; none is passed on by sperm, and there is no recombination so the only source of variation is mutation. Thus, for example, the divergence of the banteng *Bos javanicus* from the ancestral bovine stock can be dated to 2–4 million years ago (Kikkawa *et al.*, 1995). Between 610 000 and one million years ago (MacHugh *et al.*, 1997) the ancestors of today's two groups of true cattle, the zebu (humped) cattle of India and the Far East, and the taurine (humpless) cattle of the west and Near East, split. Whether or not zebu and taurine cattle truly represent distinct species (*Bos indicus* and *Bos taurus* respectively) has been thoroughly debated (Clutton-Brock, 1999); these findings clearly support the notion of separate domestications of two distinct races of aurochs to yield these distinct breed groups of cattle. Domestication of taurine cattle from their wild progenitor, the aurochs, is well attested in the Near East, and is also strongly implied in Africa (Bradley *et al.*, 1996; Mommens *et al.*, 1999). While most Chinese breeds descend from an interbreeding of Near Eastern taurine cattle with cattle derived from the Indian domestication of a distinct race of aurochs to form the zebu, there may also have been an independent domestication of zebu in China (Yu *et al.*, 1999). It is debated (Bailey *et al.*, 1996; Troy *et al.*, 2001) whether modern taurine cattle all descend from a single relatively small ancestral population of aurochs.

Microsatellites have been effective in resolving many questions on the relationships between present-day breeds (see Chapter 7) and may also shed light on the process of domestication. Present-day breeds geographically located near to where domestication took place have been claimed to still possess some of the allelic variation of the original domesticated population, which has been lost from breeds that are geographically further away. Microsatellite studies of Near Eastern cattle breeds (Loftus *et al.*, 1999) imply this. Similarly the reason why the Awassi (of Syria and nearby countries) is of all the sheep breeds examined by Byrne *et al.* (cited by Bruford, 2004) the most diverse in microsatellites, may be because it originated close to where the species was domesticated.

Incoming domesticated stocks may have been mated with resident wild conspecifics. In Europe, aurochs and domesticated cattle may have interbred, and indeed bones intermediate in size between the two types have been found in Poland (Lasota-Moskalewska & Kobryn, 1989). Davis and Payne (1993) report a bone deposit from third millennium BC England containing at least 184 skulls of domesticated cattle and one of an aurochs. This kind of immigration has been shown experimentally to have a beneficial effect on the reproductive fitness of inbred populations. Spielman and Frankham (1992) doubled reproductive fitness of small, inbred *Drosophila* populations that had declined significantly in fitness compared with an outbred control population by introduction of single immigrants. Lively *et al.* (1990) obtained similar results for fish where inbred individuals had heavier parasite burdens than control animals, but these were reduced after new individuals with greater genetic variation were introduced into the inbred population.

Some breeds are evidently the result of hybridization between species. The Diqing cattle in China carry yak-type mtDNA (Yu *et al.*, 1999). Interestingly, this breed is adapted to a high altitude, cold environment, like that of the yak. Presumably there was introgression of yak genes after domestication (female cattle–yak hybrids can be fertile). In Indonesia, Bali cattle are considered to be domesticated banteng and indeed they do have banteng mtDNA. However there is zebu mtDNA in the Malaysian population of Bali cattle, confirming interspecies hybridization (Nijman *et al.*, 2003).

The two breed groups of water buffalo, the river type and the swamp type, have different chromosome numbers ($2n = 50$ and 48, respectively). The F1 generation has $2n = 49$ and apparently produces gametes with either 24 or 25 chromosomes. In the river type, chromosomes 4 and 9 are distinct, while in the swamp type they have joined to form a large chromosome (Harisah *et al.*, 1989). The two types are interfertile but are very distinct functionally and in appearance, with the river type (relatively short, upturned horns) being primarily a dairy animal with several well-defined breeds, while the swamp type (long, swept-back horns) is used more for work and beef. The river type is native to the Indian subcontinent and spread westwards, reaching Egypt and Europe during the early Middle Ages (Clutton-Brock, 1999) while the swamp type is found throughout south-east Asia and is morphologically uniform throughout its range. They clearly diverged before domestication, which necessarily implies two domestication events, but there is not yet agreement on when this divergence happened (Flamand *et al.*, 2003).

Similar patterns can be found in other species. It is likely that there were three wild ancestors for sheep (Hiendleder *et al.*, 1998; Townsend, 2000). Current thinking is that each of these ancestors was a different lineage of the Asiatic mouflon *Ovis gmelini*. Pigs were clearly domesticated independently in western Asia and in China (Watanabe *et al.*, 1986; Huang *et al.*, 1999; Giuffra *et al.*, 2000; Kijas & Andersson, 2001). Goat mtDNA reveals a similar pattern, with three lineages implying separate domestication in west Asia and in eastern and southern Asia (Luikart *et al.*, 2001). Intercontinental transport has been important for goats and there has been such widespread mixing of lineages that today only about 10% of mtDNA can be ascribed to one continent or another; in cattle the proportion is 50% (MacHugh & Bradley, 2001).

Horses seem to have been domesticated in many different places in the grasslands of Asia (Vilà *et al.*, 2001; Jansen *et al.*, 2002). In contrast, the chicken seems to have been the result of only one domestication event (of a single subspecies of the red jungle fowl) which probably took place in Thailand (Fumihito *et al.*, 1994). There are two mtDNA lineages in rabbits, type A which is only found in wild rabbits in southern Spain, and type B which is found elsewhere in Spain and in all husbanded rabbits (Monnerot *et al.*, 1994). The two types probably diverged two million years ago. Goose breeds are derived from two species, the greylag *Anser anser* and the Chinese or swan goose *A. cygnoides*. Homologous chromosomes show differences between these lineages and some breeds show heterozygosity in this respect (Romanov, 1999).

Changes in animals following domestication

Foxes *Vulpes vulpes* and *Alopex lagopus* have been kept on fur farms on a large scale since about 1950. They have been selected for behavioural adaptations to husbandry, one of the main ones being a generalized reduction of fear responses (Nimon & Broom, 2001). Behavioural development in individual mammals and birds, in the vast majority of cases, depends upon interactions with others. This dependence on learning means that humans can act as adoptive parents, and this is the basis for the taming of young animals. Probably, any young mammal or bird can be tamed, but on reaching adulthood a tamed individual of a non-domesticated species will be much less predictable behaviourally than a tamed adult of a domesticated species (Lickliter, 1991; Clutton-Brock, 1999).

Populations of animals often become genetically adapted to captivity (especially in zoos) and the extent of this is measured by their reproductive fitness (Frankham *et al.*, 2002). What is not clear is how the tamed state of the individual evolved into the domesticated state of the population. In a comprehensive review, Price (2002) defines domestication as 'that process by which a population of animals becomes adapted to man and to the captive environment by some combination of genetic changes occurring over generations and environmentally induced developmental events recurring during each generation.' Clearly, a definition must accommodate cultural, genetic and experiential aspects and avoid merely stating that phenotype reflects the action of genotype and environment. One mechanism might be that, in domesticated animals, genes conferring 'tameability' have become fixed, but many other speculations are possible.

Skeletal changes

After domestication animals generally became smaller than their wild progenitors and this is one of the main archaeological criteria of domestication (Clutton-Brock, 1999; Davis, 1987). Size reduction was not paralleled by size changes in contemporary wild species like gazelles, implying it was not an effect of a harsher climate (Davis, 1987, Chapter 6). Circumstantial evidence suggests that small size would be advantageous under domestication and thus would be selected for. Smaller animals may have been easier to manage, would reach puberty sooner, and larger flocks and herds could be kept. Allometry of bite size and body weight (Illius & Gordon, 1987) – the different body measurements do not necessarily change in the same proportion as body mass – means that smaller grazing animals are more efficient grazers than larger ones, and this could have enabled them to cope with the lack of night-time grazing that probably accompanied domestication. Corralling animals at night is a standard way of manuring cropland and of facilitating collection of dung, and it also protects livestock from thieves and predators.

This size reduction was not permanent because throughout history body sizes have increased again (Figure 1.3).

What could be the genetic basis for this process? Body size is influenced by large numbers of genes, each having a small effect (Hastings, 1996), and in livestock, heritabilities for adult body weight are moderate to high (Falconer & Mackay, 1996). By this it is meant that a large proportion of the variation in adult body weight has a genetic cause. When populations are small and immigration is restricted, variation in the genes for body size can become depleted (for example in the Soay sheep of St. Kilda, Scotland, where heritability of body weight is only 0.06; Milner *et al.*, 1999) and body weight would then be very dependent on the environment. The archaeological evidence of increasing body weight presumably reflects improved husbandry and, probably, the restoration of genetic variation due to crossbreeding among different, previously isolated stocks.

Body conformation has changed under domestication. The wild ancestors of cattle, sheep, pigs and goats would have displayed and fought using the power of their forequarters, and indeed their conformation suggests this. Farmed livestock also fight in this way (Jensen, 2002) but, under domestication, artificial selection has taken over

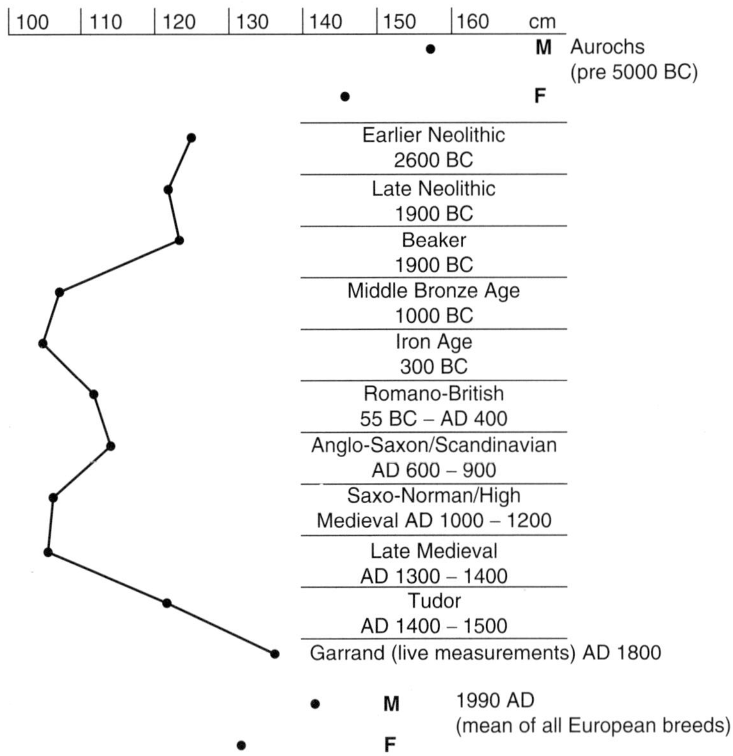

Figure 1.3 Mean withers heights (height at the shoulder) in British cattle, from several sources (Armitage, 1982; Grigson, 1982; Hall & Clutton-Brock, 1988; Simon & Buchenauer, 1993. The domesticated cattle introduced in Neolithic times are much shorter than the aurochs; heights increased during Roman times (probably reflecting evolution under better husbandry conditions), and again after the Middle Ages.

and has placed more emphasis on the hind quarters where the more palatable meat is located and where the growth constraints on muscle masses are less than on the shoulders (Figure 1.4).

Under domestication the mammalian skull shows marked changes in conformation, with the shortening of the jaws and facial region being very common, and resulting from a human preference for a flatter, more baby-like face (Clutton-Brock, 1999). It is striking how the teddy bear has also evolved in this way; people prefer them to have more baby-like features and there has been evolution of the teddy bear away from the original, more natural-featured phenotype (Hinde & Barden, 1985; Morris *et al.*, 1995). Skeletal tissue is known to be very plastic in response to subtle environmental changes. When pigs were run on treadmills, their leg bones became stronger and thicker, and so did their skulls (possibly as a result of changes in growth hormone concentrations; Gura, 2000). While there must be a genetic basis to the change of skull conformation under domestication, this has not been studied experimentally.

Many studies (Price, 2002) have suggested that captivity and domestication lead to the brain case becoming smaller, presumably reflecting adaptation to the captive environment. In 120 years of domestication the brain of mink has diminished by 20% from its original volume. The brains of the pig and the dog are said to be only 60% of their pre-domestication volume. Przewalski's horses in zoos have brains 14% smaller

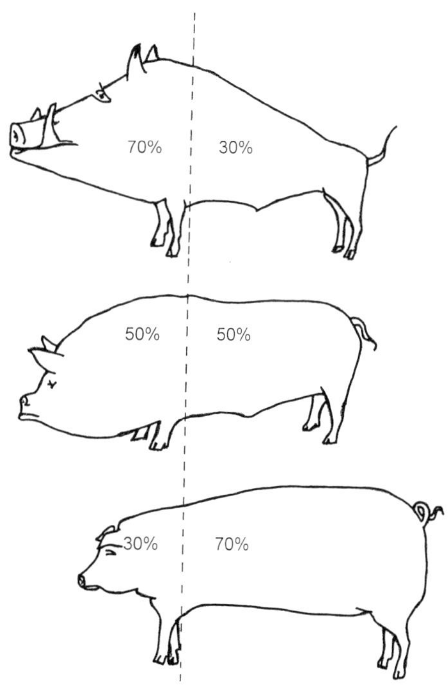

Figure 1.4 Pig conformation. In this caricature of pig conformation the distribution of body mass in the wild boar is contrasted with that in the old-fashioned pig, and in the modern meat pig (after Davidson & Coey, 1966). (Reproduced with permission from Asociación Argentina Productores de Porcinos.)

than those in the wild. However, Appleby (1999) says that if brain sizes of domestic pigs and wild boar are compared at the same age rather than at the same body weight, there is no difference. There is a need for a critical review of the effects of domestication on brain volume and skull shape, corrected for age and environmental effects.

Fat

Patterns of fat deposition may have changed as a result of domestication. Reduced predation pressure might encourage fat and this is clearly seen in birds (Gosler *et al.*, 1995), but demonstrations in mammals are lacking. With animals being prevented from migrating or seeking food elsewhere, under domestication there might have been selection for fat deposition as an aid to survival. Indeed, Ørskov (1998) suggested that in harsh environments it might be a better strategy to let ruminants store energy as fat on the body rather than for farmers to conserve feed to give to their animals later.

In western livestock breeds that have been selected for meat production, the fat is deposited either on or within the muscle, making the meat more palatable. In contrast, unimproved breeds or those which have been selected for maternal qualities rather than for meat production tend to store fat in internal depots, such as around the kidneys (Kempster, 1980; Mendizabal *et al.*, 1999). Fat-tailed and fat-rumped sheep have been selected to have exaggerated fat deposits (Figures 1.5, 1.6).

Meat with a higher fat content is better suited to cooking over an open fire than lean meat and Kyle (1987) speculated that one reason why deer were not fully

Figure 1.5 Sabi sheep. Ram of the Sabi fat-tailed sheep breed, in north-east Zimbabwe, 1997.

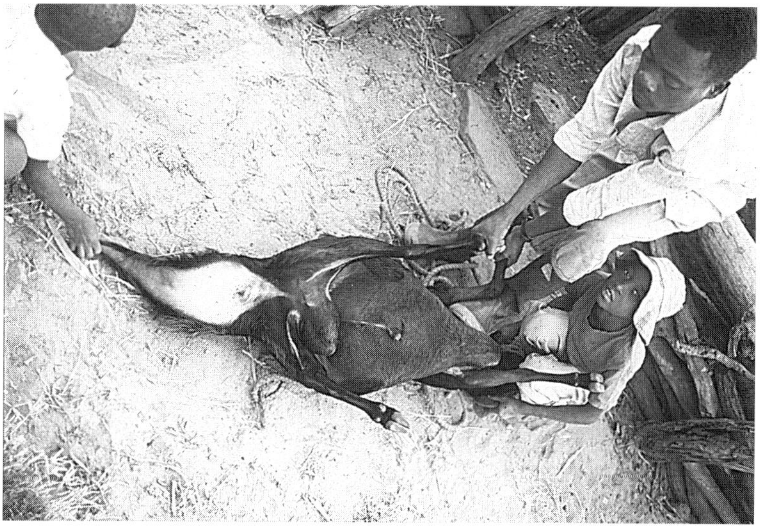

Figure 1.6 Sabi sheep. The circumference of the animal's tail was 24 cm at the base.

domesticated was because of the difficulty of producing palatable meat. Red deer only become fat when they are relatively old – while a well-grown stag may be 23% fat, at half the mature body weight young male deer are only 10% fat. At this stage of growth, lambs are 40% fat (Kay *et al.*, 1984) and horses are 13% fat (Robelin, 1986). Goats are lean and the fattest carcases usually encountered are around 14% but the record (in the Katjang goat of Malaysia and Indonesia) is 25% (Devendra & Burns, 1983). Sheep can be 45% fat as adults (Butler-Hogg, 1984) while British cattle are around 20% fat (Butler-Hogg & Wood, 1982). A 9 kg force-fed goose can be up to 16% body fat, or 28% if the liver is included (Romanov, 1999).

Muscle

Meat comprises muscle fibres and connective tissue. In cattle, taurine breeds have been found to have tougher connective tissue and zebu breeds have tougher muscle fibres. The overall result is that meat from the latter is tougher (Gazzola *et al.*, 1999); another factor is that post-slaughter tenderization is biochemically inhibited in zebu (Warriss, 2000, p.104). This could reflect a basic physiological difference between the two ancestral stocks, or possibly a longer history of selection for meat tenderness in taurine breeds.

Muscle fibres are of two main types; red (Type 1, or slow-twitch oxidative, with aerobic metabolism) and white (Type 2a fast-twitch oxidative glycolytic and Type 2b fast-twitch, mainly with anaerobic glycolytic metabolism). White fibres do not need such an extensive capillary supply as red, and they might therefore be expected to proliferate when increased muscular bulk is selected. In pigs this is indeed the case, with domestic breeds having a higher percentage of white fibres than wild boar in the longissimus dorsi muscle (Weiler *et al.*, 1995). There is also plasticity, with exercise leading to an increase in size of Type 1 fibres, though with excess growth hormone present giant Type 1 fibres are found, which indicates a degenerative condition.

Fatness and musculature are influenced by growth hormone concentration. Lean meat percentage is higher and backfat thickness lower in pigs with higher basal GH concentration, and domesticated pigs generally have higher basal GH than wild boar. The implication is that under heavy selection for muscular hypertrophy, metabolic imbalances and pathological conditions will become expressed in skeletal muscle. There are several different alleles of the gene coding for GH in pigs, as in other mammals, leading to differences in plasma GH concentrations. Knorr *et al.* (1997) made crosses between European wild boar and Piétrain, and between Meishan and Piétrain (the Meishan is a Chinese breed with the advantage of high prolificacy but also some disadvantages, including excess fat). GH variant was not related to fatness in the former, but it was in the latter cross. This implies the alleles in Meishan differ from those in Piétrain, while those in the wild boar and Piétrain are the same, and this is consistent with the separate domestication of Chinese pigs mentioned on page 8. Japanese Black cattle yield more tender beef than Holsteins; there is more intramuscular fat in the former reflecting a lower amplitude of GH secretory pulses, and a lower proportion of GH secreting cells in the adenohypophysis (Matsuzaki *et al.*, 2001).

Organ size

Effects of domestication upon organ size are very obvious in pigs. For example, the heart of the German Landrace is 0.29% of body weight, while that of the wild boar is 0.63%. The porcine heart has a poorer tolerance of exertion than that of other species and a high proportion of premature deaths of pigs is due to heart failure (Thielscher, 1987). The ability of pigs to go feral (see Chapter 6) implies that they have not, however, lost cardiac capacity irreversibly.

Lundstrom (1995) made crosses between Large White pigs and wild boar and found the best way to distinguish animals with different degrees of wild boar ancestry was by measuring the length of the small intestine. For animals with mainly wild boar ancestry this was about 16 m (about 2 m less than that of predominantly Large White pigs) and this lower length was associated with reduced growth rate and increased fat deposition.

Behaviour

Behavioural characteristics predisposed certain species to domestication (Clutton-Brock, 1999). On domestication the environment and social structure of animals changed dramatically and with it the opportunity or necessity of performing much of the behavioural repertoire (Jensen, 2002). Behaviours characteristic of the juvenile state continue to be expressed into adulthood, and this is especially obvious in dogs. There are many examples of behaviours that are important in the wild continuing to be expressed in captivity. Mason *et al.* (2001) have shown that in spite of having been 'bred in captivity for 70 generations', usually without access to water for swimming, mink show a stress response (elevated circulating cortisol) when deprived of the opportunity to swim, and that the animals will 'work' (by opening a weighted door) to gain access to a water pool. Indeed, there is only one documented example of a behaviour ceasing to be expressed under domestication (Desforges & Wood-Gush, 1976): wild mallard ducks form pairs and develop territories while under the same conditions domesticated ducks do not. But there are many quantitative differences between wild and domesticated genotypes. Gustafsson *et al.* (1999) found that sows with a wild boar sire showed the same frequency and pattern of nest building behaviour as domestic sows but there were quantitative differences in maternal behaviour.

Behavioural strategies that are less energetically costly may have been favoured under domestication. Andersson *et al.* (2001) found that bantam chickens (domestic birds) used a less costly feeding strategy than their crosses with jungle fowl (wild type). The latter moved more between two separate patches of food, without eating more. Schuurman *et al.* (1999) report similar results from comparing Landrace and Landrace x wild boar pigs.

Domesticated salmon perform less nest building than wild salmon (Fleming *et al.*, 1996) and also differ in their behavioural responses to different environments (Fleming & Einum, 1997). Farmed fish were less cautious in their behaviour and were more aggressive in a tank while the wild type were more aggressive in running water.

Anti-predator behaviours, clearly vital in the wild, can continue to be important in the domesticated state. In Colorado, guard dogs kept with sheep flocks reduce predation by black bears and mountain lions. They also reduce predation by coyotes upon lambs but have no effect on predation upon ewes (Andelt & Hopper, 2000). This implies that ewes have an ability to defend themselves against coyotes, but not against the other two predators. In Norway, sheep breeds differ in susceptibility to predation by wolverines (Landa *et al.*, 1999) while as in Colorado they are defenceless against bears.

Intriguingly, heavily parasitized sheep behave in a less timid way than uninfected sheep (Fell, 1991). Studying the interaction of cattle and wildlife in the context of transmission of tuberculosis, Sauter and Morris (1995) found a possible relationship between social dominance, positive response to the tuberculin test, and likelihood of 'investigating a sedated possum'. In general mammals carrying heavy parasite loads are more likely to be predated (Watve & Sukumar, 1995). Thus, under domestication, one of the selective forces favouring parasite resistance (differential predation) will be less effective than it was in the wild.

There exists genetic variation in various traits that are relevant to the domesticated state, for example taming, socialization to humans, and positive response to handling show much variation in lambs (Markowitz *et al.*, 1998).

Genetic processes under domestication: the origin of heritable variation

There are two ways of studying genetics. In the quantitative approach, no knowledge is assumed of the individual genes that influence a quantitative trait. The aim is to provide information and predictions which will enable the animals of highest breeding value to be identified and used (i.e. the ones which are most superior to the population mean for the character(s) under selection). Quantitative genetics is at the basis of conventional animal breeding, for which it was developed and in which it has been highly successful.

The other approach is essentially one of molecular physiology, and interest is focused on the regulation of gene expression and on the chain of molecular events by which the phenotype is encoded. Examples include pig coat colours (see Chapter 7). The breeding of animals for performance characters has not yet benefited greatly from the molecular approach, but with the development of modern genomics the two approaches are likely to converge (Goddard, 2003).

Incorporated genetic material and mutations

New genetic variation arises from mutation and there are several reasons why mutation rates might be higher under domestication.

Environmental mutagens

Human populations characterized as 'tropical zone/tribal/non-industrialized' have higher mutation rates than 'temperate zone/civilized/industrialized' populations (Neel & Rothman, 1981). This has been suggested to be due to diet-related mutagens (prevalent when food is kept and cooked under poor conditions) and to certain diseases. Domesticated animals might be more exposed to such mutagens than wild animals because they are likely to be given food that has been stored or processed in some way.

Exposure to disease

The genome of some viruses (the retroviruses) consists entirely of RNA and they propagate themselves by transcribing the RNA into DNA that then acts as a template for synthesis of RNA (Nicholas, 1996a). This DNA template can become incorporated into the host genome and have a phenotypic effect. There are many examples in pigs (Chapter 3) and in other species. One might hypothesize that domesticated species have more genetic material of retroviral origin than wild species because of greater exposure to viral pathogens. There is no evidence of incorporation of bacterial genomes in vertebrates (Stanhope *et al.*, 2001).

Stressors

Domesticated animals can be regarded as pre-adapted (or not pre-adapted) to varying degrees to specific aspects of life with humans (Fraser *et al.*, 1997). An incompleteness of pre-adaptation leads to stress. Kohane & Parsons (1988) describe how mutation and recombination are more prevalent when animals are under stress, such as is the case in domestication. Parsons (1990) discussed the possible interaction between environmental stress development and the genotype. This transmission is termed *epigenetic inheritance*. Under domestication unusual phenotypes arising in this way would be less liable to natural selection and a genetic background favouring their appearance would not be so strongly selected against as in the wild. Various molecular mechanisms could be possible, for example phenotypic variation among genetically identical individuals may be at least partly due to methylation of cytosine and to remodelling of chromatin, leading to alterations in gene expression (Rakyan *et al.*, 2001). Conventionally, these alterations have been believed not to be heritable but there have been many studies, often using experimental populations of *Drosophila*, which show that environmentally induced phenotypes can be transmitted to progeny.

Heterozygosity

In the context of a discussion on the evolutionary rate of microsatellites, Amos (1999) describes how microsatellites may have greater mutability when in the heterozygous state. Microsatellites consist of repeated sequences of DNA nucleotides of no obvious function; microsatellite alleles differ in the number of repeats and, therefore, in length. During meiosis the homologous chromosomes pair and heterozygous sites can undergo a repair process – parents whose microsatellite alleles differ considerably in length are more likely to produce offspring which carry mutants. This would create a positive feedback, with further heterozygosity being generated by these mutants, itself producing further mutation. In the case of microsatellites gains in length are more commonly found than reductions, and increased mutation rates would be expected to lead to longer microsatellite alleles.

There are some interesting species differences, with microsatellite alleles in sheep being more diverse (Byrne *et al.*, cited by Bruford, 2004) and longer than their homologues in cattle (Crawford *et al.*, 1998). Amos (1999) suggests heterozygotes would generally be more common in large populations. Microsatellite alleles are larger, with higher allele-size variances and with more genetic variation, in domestic sheep than in Rocky Mountain bighorn sheep (Forbes *et al.*, 1995) and populations of the latter are likely to be smaller than those of most breeds. Such a process need not be restricted to microsatellites. Amos (1999) also refers to hybrizymes, which are novel proteins arising in hybrid zones from alleles that are not found in either parental population. There are examples of this in small rodents (Sage *et al.*, 1993) but it remains to be seen if the population structure of livestock species is conducive to hybrizyme evolution.

Chromosomal characteristics

Chromosome number can vary within wild and domestic species. Acrocentric chromosomes (in which the centromere is at the end of the chromosome) can fuse to form a single metacentric chromosome (i.e. one in which the centromere is internal; Nicholas, 1987). In sheep there are some translocations associated with particular breeds and some Portuguese breeds of cattle effectively have $2n = 58$

instead of the normal 60 (Bunch *et al.*, 1990; Rangel-Figueiredo & Iannuzzi, 1991). These translocations can be of economic significance in that fertility can be reduced (Popescu & Tixier, 1984), more so in cattle than in sheep. In zebu cattle, the Y chromosome is acrocentric and this has been used to clarify the breed affinities of the southern African sanga and east African shorthorned zebu breeds. Though it had been widely held that the former had affinities with the zebu this is not supported and its affinities are much closer to taurines (Frisch *et al.*, 1997).

The Przewalski's horse has $2n = 66$ and the domestic horse has $2n = 64$. Equids are known to have a high rate of karyotypic evolution (Bush, 1996) and the Przewalski's horse was proposed as the relict of an ancestral horse species. The $2n = 66$ karyotype hypothetically gave rise to the $2n = 64$ karyotype of the domestic horse by fusion of two pairs of acrocentric chromosomes to form one pair of metacentrics. However, mtDNA studies showed that the Przewalski's horse is similar to modern domestic horses, especially the native Mongolian horse (Ishida *et al.*, 1995). This implied that the chromosomal fusion was quite recent – after the lineage of the Mongolian horse had separated from that of other horses. Studies on the Y chromosome (Wallner *et al.*, 2003) then confirmed that the Przewalski's horse is the descendant of a stock that split off from the domestic horse over 100 000 years before domestication, the similarity of mtDNA being explained by gene flow from domestic mares into the Przewalski's horse population. So the Przewalski's horse is a relict of an offshoot from the main domestic horse lineage.

Heritability

If the total variance of a specified quantitative character in a population is V_P, some of this (termed V_G) will be due to genetic factors, so the proportion V_G/V_P is a measure of how much of the phenotypic variation in an individual is due to its genotype. This is called 'heritability in the broad sense' or the 'degree of genetic determination'. The remaining variance is due to environmental variation, so $V_P = V_G + V_E$ (see Chapter 3). V_G can itself be partitioned three ways, into variance due to the genes received from the parents (V_A, the variance of breeding values or, more commonly, known as the additive genetic variance), variance due to the dominance interactions between the alleles at a locus (V_D, the dominance variance) and variance arising from the interactions between different loci (V_I, the interaction variance). Together V_D and V_I are commonly known as non-additive genetic variance.

V_A/V_P denotes the proportion of the phenotypic variance that is due to the genes passed on by the parents. Termed 'heritability in the narrow sense', or h^2, this is the measure of degree of resemblance between relatives that is of most value in breeding and genetic conservation programmes (Falconer & Mackay, 1996, Chapters 7 and 8).

As selection proceeds, genetic variability will decrease and, as a direct result, so will heritability (Falconer & Mackay, 1996, Chapter 11). It is commonly found that heritability of a character may differ at various stages during the lifetime of an animal (Simm, 1998, Chapter 8). For example, laying hens show their highest rate of egg production at the outset of their laying career when they approach the biological limit

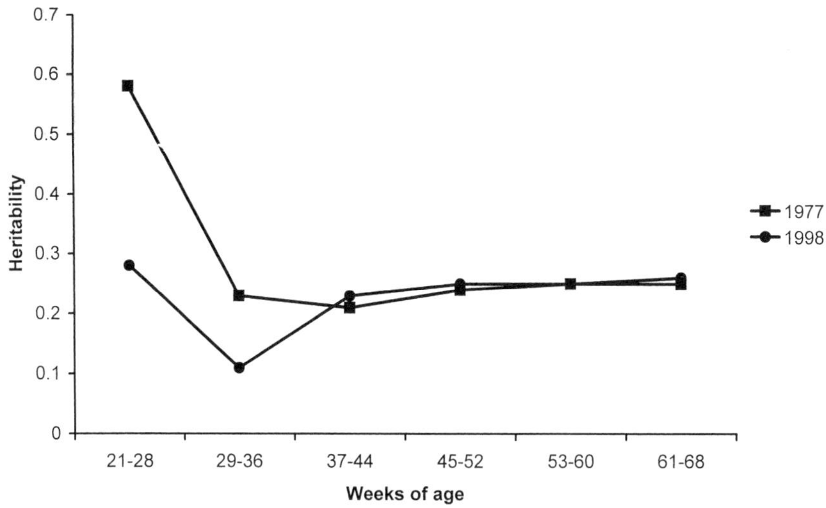

Figure 1.7 Dependence on age of the heritability of rate of egg production of laying chickens has been reduced by breeding programmes. Drawn from Preisinger (2002).

of one egg per day. As a result of selection for egg production the heritability of their rate of lay in these opening weeks has been much reduced (Figure 1.7) and their responsiveness to management inputs has increased. Genetic variation has been depleted as homozygosity of favourable genes has increased, but genes controlling egg production in the later weeks seem to have been less accessible to selection.

Heritability can also depend on the season, i.e. the proportion of the phenotypic variance explained at the time of measurement by additive genetic variance will vary from time to time. In N'Dama (taurine) cattle in west Africa, heritability of gastro-intestinal nematode faecal egg count increases with age, and is also greater at the beginning and end of the rainy season than when parasitism is at its maximum at the peak of the rainy season (Zinsstag *et al.*, 2000). In goats and sheep this increase in heritability may be associated with a phase of acquisition of immunity (Vagenas *et al.*, 2002). If heritability for a trait varies over the life cycle, in wild populations one would expect it to be lowest at the points in the life cycle when selection operates, as selection will have eroded the genetic component of phenotypic variance by tending to fix the most favourable genes that are expressed at that time. In Canadian wild sheep (Réale *et al.*, 2000b) a 25-year field study has shown that heritability of body mass is very low at two years of age, being higher for lambs and yearlings and for sheep over two years old. Presumably body mass at two years old is strongly related to fitness and selection has removed genetic variance in this character, driving favourable genes to fixation. Lamb and yearling weights still have a large genetic component perhaps implying inconsistent selection from year to year as a result of environmental variation.

The general finding (Falconer & Mackay, 1996, p.162) has been that 'life history' traits (including, for example, survival and fecundity) have lower heritabilities than other traits relating less directly to 'fitness', i.e. those which might be classified as morphological, behavioural or physiological. Presumably, natural selection, acting over a long period, will deplete additive genetic variation in fitness traits faster than in other traits, so, assuming heritabilities of all traits are the same at the beginning, at any given time heritabilities of fitness traits will be lower (Roff, 1997, p.63).

In young domesticated sheep, heritability also changes with age (Figure 1.8a,b). Except at one month of age, forelimb length shows a strong genetic component, suggesting it would be relatively easy to breed taller or shorter sheep, and also that height at the shoulder is comparatively independent of environment. However, at one month of age, before the lamb is weaned, the environmental component is strong

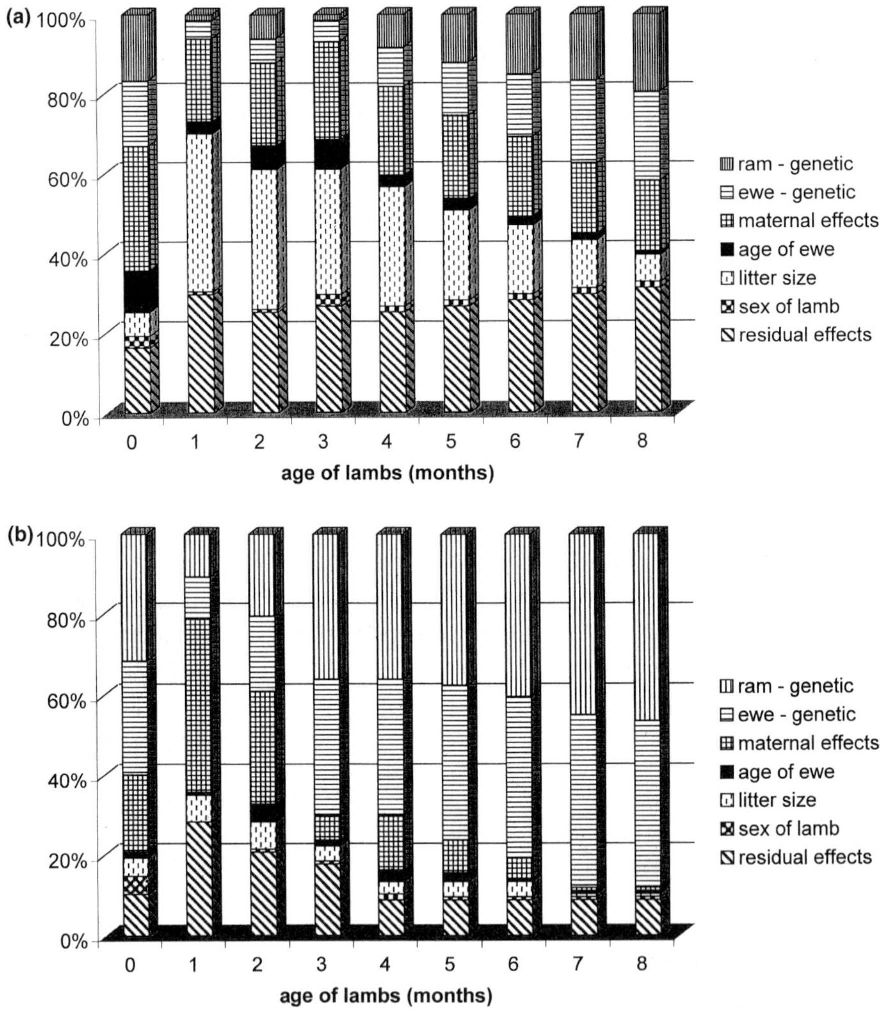

Figure 1.8 Factors affecting total variation in body size of lambs. Body weight (a) is much less influenced by genetic factors than is forelimb (b) cannon bone length (elbow – flexed wrist). Redrawn from Hunter (1957) and Hammond (1963). (Reproduced from Hunter (1957) with permission from Cambridge University Press.)

– clearly because forelimb growth up to this age has depended on the mother's milk. Variation in live weight is much less dependent on genes.

Genetic processes under domestication: interactions between genes

Recombination

During meiosis, the homologous chromosomes of maternal and paternal origin form duplicate chromatids, which may join together and exchange sections of their length (Nicholas, 1987). This process of recombination enables organisms to respond to novel disease and parasite challenges. Chiasmata (the structures whose presence indicates that recombination has occurred) are more numerous and rate of recombination consequently higher in animals with a greater age at maturity. An evolutionary reason could be because longer-lived animals face more challenges and recombinant genotypes are more likely to have the genetic variation enabling them to meet these challenges. Domesticated animals have more chiasmata than wild mammals of the same life span (Burt & Bell, 1987; Ollivier, 1995). This higher rate may have been indirectly selected because it will help to remove negative correlations between characters that are being selected in novel combinations. It could also be due to selection favouring high recombination rates, which may be accentuated in changing environments and where genetic drift is important (Notter, 1999; Otto & Barton, 2001).

Linkage effects

When an allele is heavily selected for, closely linked alleles can become fixed, a process known as a *selective sweep* (Chapter 7). It is not yet known if these are common in livestock (Goddard, 2003). Quite frequently confirmed in livestock, in contrast, is the phenomenon of two alleles at different loci occurring together within an individual more frequently than would be predicted by chance. This is known as *population allelic association* or *linkage disequilibrium* (Frankham *et al.*, 2002).

If there is a locus A where there is heterozygosity *Aa*, and a mutation takes place at a nearby locus B (*B* mutates to *b*) the new mutant will be associated with one of the alleles at A (say, *A*) and not with the other (*a*). Recombination is unlikely between the two loci so there will be an excess of gametes of type *Ab*. However, recombination will occur so eventually the deficit will disappear and then the proportions of gametes *AB*, *aB*, *Ab* and *ab* will be predictable simply from the allelic frequencies of *A*,*a*,*B* and *b*.

Linkage disequilibrium can arise from various causes including mutation or (quite commonly in livestock) as a result of small effective population size; fusion among genetically divergent populations is another possible cause, but this has not been studied in livestock. In cattle, where artificial insemination (AI) is very important, gametes bearing the particular combination of alleles of a popular bull will be detectable for many generations. In Holstein-Friesian cattle linkage disequilibrium is evident in all the autosomes (Georges, 1999), which is consistent with the known heavy use of particular individual sires in this breed.

Artificial selection

Domestication is intimately linked with artificial selection. Under artificial selection, fitness is defined as compliance with the goals of the breeder and selection is directional towards a rapidly changing extreme; selection can take the mean for a phenotype far beyond the original range (Falconer & Mackay, 1996, see Chapter 12).

Under selection, animals with a high breeding value are favoured. These possess genes which, when passed to progeny, cause the latter to deviate most favourably from the population mean. The greater the variation in breeding values, or the shorter the generation interval, the greater the selection intensity, or the greater the accuracy of selection, the greater the response to selection. These four factors determine the predicted rate of response, which consists of the increase in the frequency of favourable genes until the population reaches its selection limit. In any population, inbreeding will increase homozygosity and genetic drift will change gene frequencies, so in a smaller population selection response decreases and rate of progress towards the selection limit increases.

Genetic drift affects variation in response to selection, and in a small population (to be precise, if N_e is small – see Chapter 9) then it is a matter of chance whether the response is large or small. Nicholas (1987, p. 424) shows how, if a given response to selection is to be achieved, then N_e has to be relatively large. Increase in inbreeding must be kept low, otherwise there is a risk of harmful alleles becoming homozygous and this is another reason for a high N_e.

Artificial selection is the primary technique for the genetic improvement of livestock and it has been practised for centuries. It consists of deciding which of the available animals should be used for breeding. In livestock with small litter sizes this usually means selection of males but the high reproductive rate of pigs and chickens means selection of females is also possible in these species.

In traditional societies there is a wide range of breeding practices. Kohler-Rollefson (1997) gives examples. Some societies, including most camel pastoralists, restrict or prohibit sales of female animals. Others have special patterns of exchange: in Lesotho and west Zambia, reportedly, a cow camel may be placed in a herd with a locally renowned bull and after some years she returns home with her improved progeny. Rajasthani pastoralists do not allow inbreeding and this mirrors their pattern of marriage which tends to involve marrying a woman from outside the community, while in the endogamous Arabian Bedouin, inbreeding of livestock is permitted. An important component of livestock breeding practice is castration of surplus males. In some societies it is said to be the better animals that are castrated for work or fattening while in others it is the less vigorous males. However, there is no comparative account of castration of livestock.

Modern animal breeding guides the choice of breeding stock by defining the selection goal, deciding what to measure and what to select, designing the breeding programme, implementing it and monitoring progress. Many texts deal with this in detail; evidence of genetic improvement and its economic benefits is best established for pigs and poultry, then dairy cattle, but also for sheep and beef cattle (Simm, 1998). The most important practical technique is AI, which permits better testing of candidate sires as well as contributing to the dissemination of improved genotypes. Computers and appropriate software enable data to be combined so selection can be on the basis of performance records from large numbers of animals.

Selection within breeds may also be influenced by a sexual bias in which animals are used in trade. The Kuri cattle of Lake Chad, West Africa, provide a striking example (Figure 1.9). It was noted that the Buduma people who keep this distinctive and locally adapted taurine breed supply bulls to neighbouring pastoralists for use in their herds of zebu cattle because the Kuri has good conformation. However, zebu bulls were not used in Kuri herds (Quéval *et al.*, 1971). Recent DNA work (Hanotte *et al.*, 2000; Meghen *et al.*, 2000) has shown that while the Kuri cattle have today a sizeable genetic endowment from zebu cattle, the bringing in (introgression) of genes known to be carried only on the zebu Y chromosome has not taken place. Certainly since the 1950s, the normal exchange for one Kuri bull was one or more zebu heifers. The resulting crossbreds had a long lactation and high fertility. Kuri bulls were preferred as herd sires (Quéval *et al.*, 1971). Thus the Buduma were able to select within the purebred segment of the breed for conformation, and rely on crossbreeding to maintain herd fertility. In contrast, trade in Asian (mainly Chinese) pigs in the late eighteenth and early nineteenth centuries was largely in females, as shown by historical records and the prevalence of Asian mtDNA types in some European

Figure 1.9 Kuri cattle. This cow is of traditional Kuri type but her calf was probably sired by a Red Bororo bull.

breeds (Giuffra *et al.*, 2000). Perhaps females were easier to handle on the six month long voyage from China to Europe.

Marker-assisted selection and quantitative trait loci

When electrophoresis of proteins (separation of variants by exploiting differences in their electrostatic charge) was first developed, it became possible to observe gene products directly. It was soon appreciated that linkage meant that electrophoretically detectable, biochemical variation could in principle act as markers for genes with advantageous quantitative effects. Developments in immunology revealed blood groups in livestock which also held this promise.

In cattle the B blood group system was linked with milk yield but the proportion of the phenotypic variance accounted for was too small to be of practical use (Haley, 1995). Genome scans using microsatellites are now being made in order to find linkages with loci which have significant effects on quantitative traits (QTLs) and which are difficult to measure on the farm, such as disease resistance, carcass composition or survival (Merks *et al.*, 2000). The loci themselves can now be detected directly (Goddard, 2003). In the first study of this kind (reviewed by Haley, 1995) a porcine QTL was located on chromosome 4, which influenced carcass characteristics. Other QTL affecting intestine length and growth rates were found on other chromosomes. Van Laere *et al.* (2003) showed how a single nucleotide substitution in an allele at a QTL increases meat production, at the expense of fat, by 3–4% in pigs. This allele is at high frequency in breeds of pig that have been heavily selected for lean growth.

Correlated traits

Selection for one trait can often lead to change in others. The extent of the change depends on how much the characters share the same genes, and when the characters are quantitative traits this degree of sharing is called the genetic correlation (Nicholas, 1987). Some genetic correlations are economically favourable, such as that between milk yield and protein yield. In cattle selection for lifetime cow fertility appears to have accelerated sexual maturity in both sexes (Mackinnon *et al.*, 1990). In sheep, pigs and mice a female trait, ovulation rate, can be increased by selecting males for testicular size (Land, 1973) while in cattle scrotal circumference is correlated with rapid attainment of puberty and with likelihood of conception at induced oestrus (Coulter & Foote, 1979). Others are not favourable, for example between many production traits and metabolic, reproductive and health traits in meat poultry, pigs and dairy cattle (Rauw *et al.*, 1998; Royal *et al.*, 2000). In chickens, egg production and egg size are negatively correlated in most breeds (McMillan, 1990). High body weight (desirable for meat birds) does not mean increased egg production (Clayton, 1972; Hartmann, 1989). Although larger birds have a higher rate of ovulation, this exceeds the capacity of the oviduct and uterus to complete formation of the egg.

In dairy cows, there are strong genetic correlations between milk yield and production efficiency (production/feed intake ratio: van Arendonk *et al.*, 1991).

Selection for production automatically improves gross feed efficiency, which is why the latter is not included in selection programmes. Provided nutritional requirements are met the efficiency of the individual animal increases with the level of production (Figure 1.10). In all studies of this kind it is important to be clear what measure of efficiency is being used. For example, Friesian cows are more efficient than Jerseys at producing milk volume but there is no difference in efficiency of producing milk energy (Gibson, 1986).

Another term, *association*, is often used when there is a single-gene trait associated with a quantitative trait, so that selection for one leads to change in the other. There are three possible causes – linkage between the single gene and a QTL, multiple effects of a single gene or population structure. In Brown Swiss cattle, weaver disease, which leads to postural and locomotory failure, is caused by a gene which is tightly linked to a QTL for increased milk production (Georges *et al.*, 1993). In Quarter Horses hyperkalaemic periodic paralysis is associated with some desirable aspects of conformation. A sodium channel mutation is involved which 'may redirect ion traffic in such a way as to induce [muscular] hypertrophy and thus create an inseparable linkage of the disease and a valued phenotype' (Womack, 1992).

Sometimes genetic defects have spread in livestock and companion animals because of the extensive use of popular sires. Such animals would have been perceived as having exceptional merit for some valued characteristic but unfortunately in many cases they also carried a harmful gene. Bovine leukocyte adhesion deficiency is a genetic defect of Holstein cattle; a single autosomal recessive gene causes adhesion molecules on the surface of leukocytes to be poorly expressed and responses of leukocytes to infection are handicapped. Cattle that are homozygous for the gene are especially susceptible to bacterial infection and are likely to die before one year of age. The gene traces back to a single ancestor (Osborndale Ivanhoe, born 1952) and has been widely spread in north American Holsteins (Shuster *et al.*, 1992; Wanner *et al.*, 1998).

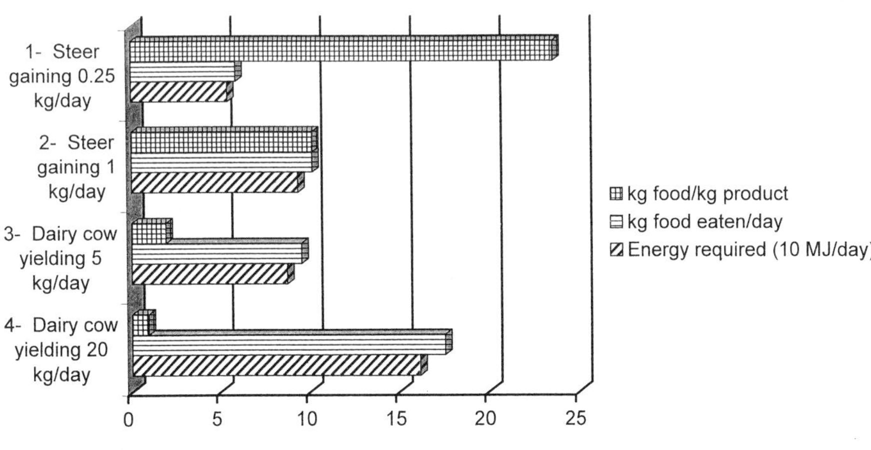

Figure 1.10 Increase in efficiency as production level increases. As milk yield per day increases, the consumption of food and energy increase, but production efficiency increases as well. The same is true of body mass gain. Data from 590 kg Friesian milk cow, and 400 kg beef steer, respectively (MAFF, 1975).

Natural selection

Under domestication many animals die, in spite of the farmer's care, of disease or some other infirmity before reaching reproductive age. Together with failure to conceive, this is often seen as representing the action of natural selection. Artificial selection is the process that is applied to the survivors to decide which of them are allowed to breed.

Darwin (1868) clearly believed both natural and artificial selection had shaped breeds: 'it can hardly be doubted that in different countries, varieties differing in constitution and in various characters would succeed best, and so be naturally selected' (vol. II, p.259), while 'the slight differences, however, which characterize each individual animal and plant would in most, probably in all, cases suffice for the production of distinct races through careful and prolonged selection' (vol. II, p.271). The two processes only differ in the kinds of characteristics to which they are applied, and in the intensity of selection and they are not really separable, for example animals may be selected for breeding because they are predicted to produce offspring that are less likely to die young.

Prompted by new findings on the large amount of biochemical variation within and between breeds, Mason (1973) re-examined the question and concluded that biochemical and physiological differences among breeds were subject to natural selection – in other words, that there is not purposive selection for these traits. The morphological differences among breeds were the result of artificial selection and the geographical distribution of breeds was more due to human migrations than to ecological factors.

Physiological differences among breeds have been taken as evidence of natural selection (Clutton-Brock, 1999; Price, 2002). For example, ecological adaptations of livestock breeds have been shown at high altitudes. Chickens whose haemoglobin has particularly high oxygen affinity have been identified in the Peruvian Andes (Velarde *et al.*, 1991). In Chinese grazing livestock, respiration and heart rates, haemoglobin concentration and packed cell volume (also called haematocrit: the proportion of the volume of the blood that comprises red cells, which is depressed in anaemia) increase as altitude rises (Cheng, 1984).

Under domestication, disease challenge would be increased because crowding and malnutrition, aggravated by stress (Lord *et al.*, 1998), are likely. Animals may be exposed to the diseases of other species, adding to the selective pressure. This would predict that juvenile mortality is higher in domesticated animals than in wild ancestors. Data are lacking, but in Peruvian camelids from sites dated at 5500–6000 BP, a high proportion (73% in one site) of dead animals were juveniles. In nature about 35–40% of dead animals of these species would be young (Browman, 1989). This over-representation could be due to high neonatal and juvenile mortality presumably caused by disease.

Reasons for the diversity of genes in the major histocompatibility complex (see Chapter 5) are complex, but natural selection is thought to be a major cause (Apanius *et al.*, 1997). Comparative quantitative genetics can provide independent evidence that natural selection can be important under domestication. When alleles are at high or low frequency, such as results from selection, their additive contribution V_A to total genotypic variance V_G decreases and variance due to dominance V_D can increase (see Chapter 7, and Falconer & Mackay, 1996, pp. 127–129). This implies that traits which are found to have high dominance variance have been subject to selection. Roff (1997, pp. 65–68) found that in wild species dominance variance was more important in 'fitness traits' (essentially those relating to survival and reproduction: 54% of phenotypic variance due to dominance effects) than in behaviour (24%), physiological traits (27%) or morphology (17%). In domesticated species, dominance variance contributes 39% of the phenotypic variance for fitness traits, and 42% for morphological traits.

This wide difference in wild species implies that fitness traits are under consistently strong selection, but selection for other traits may be inconsistent in strength or direction. In contrast, in domesticated species, morphological traits (which include the traits of particular interest to breeders) and fitness traits have been under equally strong selection.

Patterns of geographic variation in genetic polymorphisms can imply that selection is acting; and if there is no obvious artificial selection, the inference is that natural selection is operating. Barker *et al.* (1997a) illustrate this reasoning. If there is no selection and genetic drift and migration are the only factors affecting allelic frequencies then the extent to which different polymorphisms vary in their allelic frequencies among the different populations (measured as variance of F_{ST}: Chapter 7) should be similar. In eleven populations of water buffalo, variance of F_{ST} for protein-coding polymorphisms was much greater than that for microsatellites. If it is assumed that the microsatellites are selectively neutral, the inference is that the protein-coding allele frequencies reflect the action of natural selection. Similar results were found with Asian goats (Barker *et al.*, 2001).

Finally, there are many examples of increased reproductive rate and of increased sexual precocity in domesticated animals. Egg size and number are greater in domesticated sea trout and salmon, which also have larger body size (Petersson *et al.*, 1996) and domestic goats have the ability to become pregnant while lactating and can thus breed twice a year, unlike any wild *Capra* (Parkes, 1993). These attributes will have been selected in their own right, but will also be selected indirectly as a side effect of selection for other attributes. When there is selection for any characteristic, generation time is likely to shorten (Fisher & Holt, 1944). This is because if candidate animals are chosen on the basis of their superiority to the rest of the population, the population mean for the characteristic will evolve during the lifetime of a given animal and older animals will therefore show less superiority. This leads to selection – which might be described as 'natural' or 'unconscious' – for early maturity.

Sexual selection under domestication

Sexual selection is said to be operating when the animal that passes on its genes is the one which is more competent at mate choice, combat, or sperm competition. The selective pressure is applied not by the environment but by the other animals in the population. In domestication, sexual selection has been abolished and all breeding decisions are made by the farmer. Male-male competition is sometimes encouraged in breeding meat chickens and is quite common in sheep breeding (Price, 2002) but otherwise it is unusual in farmed animals. In feral populations sexual selection may resume its importance (Chapter 6).

Some aspects of reproductive biology may have been fixed genetically before domestication; this could be the case with semen quantity and quality for which there is not a strong heritable component in livestock (Land *et al.*, 1983). Some other aspects are now responding to selection, implying that in the wild state they were not under strong selection, and as a result genetic variation has persisted. For instance, male libido has a strong genetic component, at least in cattle (Chenoweth, 1981) but this is probably only expressed under the very artificial conditions of an AI unit. In the wild, mating success would probably have been a function of age and social status rather than of libido.

A phenotype highly relevant to success in combat or in attracting mates is body size. The wild ancestors of livestock were strongly sexually dimorphic, one probable cause being sexual selection and another, maybe, competition for forage. This dimorphism persists, but is less marked, in livestock breeds. In wild ungulates, adult males are usually larger than females and the two sexes often occupy different habitats at certain times of the year, especially when food is scarce. Smaller-bodied grazing animals are more able to meet their daily needs from a short sward, while larger animals are better adapted to a lower-quality diet (Mysterud, 2000). Sexual body size dimorphism may have evolved as a response to competition for forage. In many wild species the degree of dimorphism is reduced under poorer growth conditions (Badyaev, 2002) but this seems unlikely to apply under domestication, when there would be less competition for food between the two sexes, possibly because fewer males need to be kept, or because supplementary food is given.

Degree of sexual size dimorphism can be considered as proportionate sex difference in body mass, (M–F)/F, and the overall mean for large wild herbivores is 0.548 (range: 1.44 for kangaroos to 0.04 for the roe deer: Mysterud, 2000). Domesticated species are compared with their wild relatives in Figure 1.11.

Clearly the ratios for the domesticated animals are lower than those of their wild counterparts, indicating reduced sexual dimorphism. This is especially obvious for the horse and water buffalo, reflecting the use of both sexes for work, which will discourage divergent selection.

Weight data are lacking for some species but further evidence for sexual size dimorphism being less in domesticated animals comes from measurements of skull length. Bohlken (1964) concluded that the sexual size dimorphisms of domesticated cattle, Bali cattle, gayal and water buffalo were less than those of the wild precursors (aurochs, banteng, gaur *Bos gaurus* and wild water buffalo, respectively). For unknown reasons the aurochs showed great variability in body size, and the range of withers (shoulder) heights for aurochs was 168–177 cm for bulls, 155–164 cm for cows (Muzzolini, 1985). Modern cattle breeds seem to show a plateau for

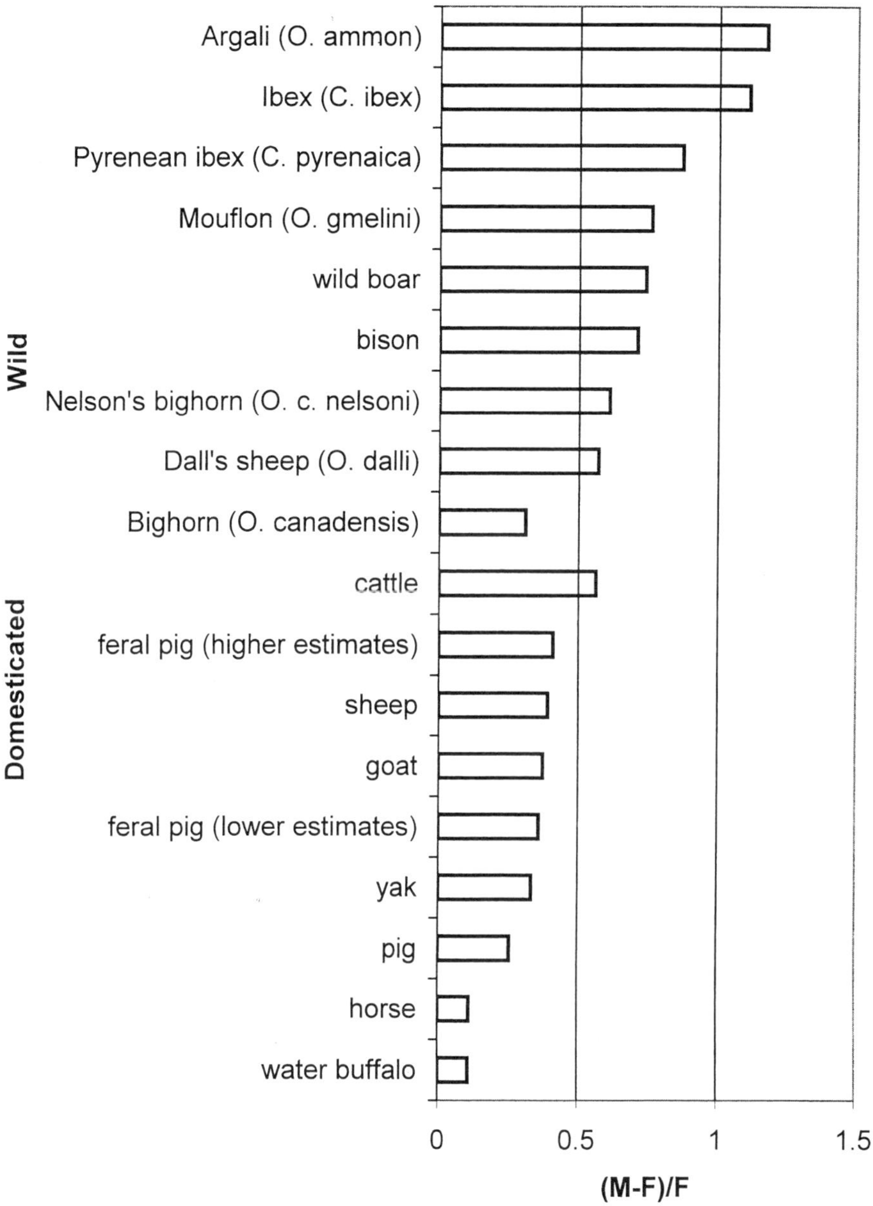

Figure 1.11 Sexual size dimorphism in domesticated and related wild species. Sexual size dimorphism has, in general, been much reduced in domesticated mammals. Data are from Mysterud (2000), for wild species, and Simon & Buchenauer (1993) for domesticated species (yak and water buffalo: six and eight Chinese breeds respectively; Cheng, 1984). Feral pig weights are from Clarke & Dzieciolowski (1991b).

female withers height and there are very few where bulls are over 160 cm (Simon & Buchenauer, 1993).

The apparent failure of any modern bovine breed to have exceeded the size of the aurochs does need to be explained. Possibly under domestication genes for gigantism were eradicated from the species, or maybe genes for some degree of dwarfism were selected for, thus removing the capacity for evolution in this respect during breed differentiation. There are several kinds of dwarfism – for example the limbs can be disproportionately shortened (achondroplasia), body dimensions can be reduced in proportion to each other (proportionate dwarfism) or body proportions of the juvenile can be retained into adulthood (a form of neoteny). All are seen in livestock (Hall, 1991). Perhaps different dwarfing genes were fixed in different domestications.

Degree of sexual dimorphism may vary among groups of breeds. In European sheep (calculated from Simon & Buchenauer, 1993), (M–F)/F is greatest in the Merino group (mean 0.536, 27 breeds) and the lowest in heath (0.390, 28 breeds), mountain (0.340, 94 breeds) and 'local coarsewool' (0.364, 34 breeds). Continental and British longwools (38 breeds), shortwools and Down breeds (20 breeds) and milk sheep (17 breeds) had intermediate values. Perhaps this reflects the very long history of selection of Merinos for wool yield (Hall & Clutton-Brock, 1988), which would favour the keeping of large, heavy-fleeced, castrated sheep, necessitating selection for large male body size. At the other end of the scale maybe the harsh environment of the mountain breeds has imposed selection for an ecologically optimum body size which is similar in both sexes.

Conclusions

Molecular genetics has enabled many hypotheses about domestication to be tested, and has given new insights into when and where domestication took place. It is clear that under domestication new evolutionary regimes apply and evolution can be much more rapid than in the wild. Possibly, because of the novel stressors that operate in the domesticated state, there are higher rates of molecular and chromosomal evolution. Quantitative genetics, and scientific animal breeding, have given insights into the rates and consequences of evolution under domestication. Artificial selection, the natural environment being held at bay, leads to the evolution of useful traits, but the ways in which it is practised vary considerably and can lead to unexpected correlated responses. Natural selection probably acts on domesticated animals, but sexual selection, which is very important in many wild species, does not apply. The cultural factors influencing domestication, and the nature of the human-animal interactions involved in the process, remain largely unknown.

Chapter 2

Origin and differentiation of breeds

Introduction

What does the term 'breed' mean, and does its definition matter? How did breeds arise, and what is their cultural significance? Does this vary from one human society to another? What are the interactions between breed differentiation and human history and culture? What are the prospects for interdisciplinary approaches?

Definitions

Many definitions of the term 'breed' have been proposed. According to Lauvergne (1993) the term or its counterpart in translation first began to appear in print in the seventeenth century (in relation to cattle in France in 1600, and to horses in Germany in 1672). Some authors have emphasized the cultural background to breeds, others their biological attributes. Exemplifying the former, Kohler-Rollefson (1997) says that in developing countries breeds should be viewed as the products of complex social and cultural processes. She adopts an anthropological approach:

'[Domestic animal populations] may be regarded as "breeds" if they fulfil the following criteria:

- They are subjected to a common utilization pattern
- They share a common habitat/distribution area
- They represent largely a closed gene pool, and
- They are regarded as "distinct" by their breeders.'

A widely used definition is that of Clutton-Brock (1992a):

'A breed is a group of animals that has been selected by man to possess a uniform appearance that is inheritable and distinguishes it from other groups of animals within the same species. It is a product of artificial choice of characters that are not necessarily strategies for survival but are favoured by humans for economic, aesthetic, or ritual reasons, or because they increase the social status of the owner of the animals'.

This leads to some confusion because many breeds are not of uniform appearance and the definition of Simm (1998) is perhaps more versatile:

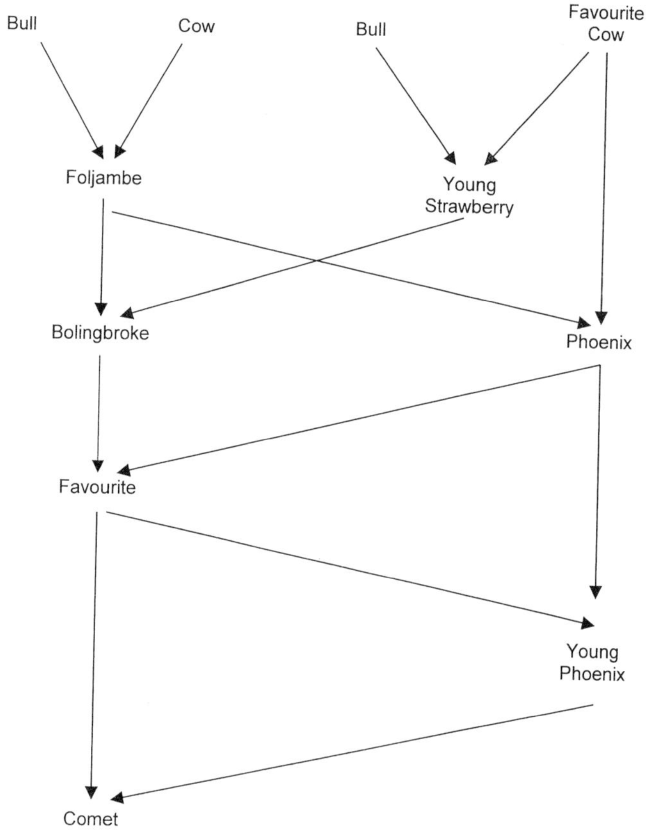

Figure 2.1 Pedigree of Favourite and Comet. The Shorthorn bulls Comet (born 1804) and his sire Favourite were two of the most famous animals in livestock history. Favourite was the son of a brother-sister mating. Also, one of his granddams was the dam of the other. He was therefore highly inbred (19.2%). He was mated with his own dam to produce Comet (47.1% inbred), regarded as the best Shorthorn bull ever. Favourite is considered the 'father' of the Shorthorn breed. (Reproduced from Wright (1977) with permission from the University of Chicago Press.)

Change in function of breeds

As farming conditions change in a specific area, the demands on the local livestock change and the breeds will either have to adapt to the new requirements or, if they cannot, they must find a new commercial niche elsewhere if they are not to become replaced. Only if some people have a vested interest in the continuance of the breed will attempts be made to keep it going – this is one of the major functions of breed societies (see Chapter 8). In some cases, stratified crossbreeding systems have become established (see Chapter 4) and breeds have become specialized as either sire or dam breeds. If a new husbandry system is introduced breeding programmes may be redirected towards meeting its requirements. If such programmes are going to be slower than breed replacement, then the breed is likely to be graded up (see Chapter 4) to a more commercially adapted type.

Genetics of breed formation

Just as in speciation, breed formation requires a group of animals to be reproductively isolated and to become distinct. In recorded histories of breed formation, reproductive isolation has been established, followed by development of a distinctive phenotype. Reproductive isolation is vital because only one migrant per generation is sufficient to prevent divergence (Frankham *et al.*, 2002, Chapter 13).

In chickens, breeding for fighting performance also led to greater body size and improved meat conformation – this was particularly marked in Asiatic breeds. One of these, the Indian Game (Figure 2.2), was imported into south-west England in the

INDIAN GAME.

"AGITATOR" AND "LADY WHITFIELD."

THE PROPERTY OF MR. G. T. WHITFIELD, COLEBRIDGE, GLOUCESTER (Hon. Sec. of the Indian Game Club).

Cock, winner of 2nd prize Birmingham; 3rd prize Crystal Palace. Hen, winner of 1st prizes at Crystal Palace, Birmingham, &c., &c.

Figure 2.2 Indian Game chickens, bred for cockfighting, became the foundation for modern meat breeds. Engraving from Brown (1892).

nineteenth century and later exported to the USA where it became known as the Cornish Game. It is 'the principal origin of the few male lines used as the sires of the thousands of millions of meat chickens that are produced today' (Clayton, 1972). Most of the main egg-producing breeds came from the Mediterranean and the one that has contributed most to egg-producing genotypes in the world today, the White Leghorn, was introduced to the USA from Italy in 1853.

As breeds differentiate, divergence of phenotype can come about through natural or artificial selection, genetic drift and non-genetic processes, all of which can act at the same time. The herd or flock, and its human guardians and commensal species, clearly constitute a biological community, which experience suggests can be a unit of selection. Group selection is an evolutionary mechanism which is not clearly demonstrable in nature. However, Roth (1996) and Homewood and Lewis (1987) describe how, in drought conditions in Kenya, the environment may appear equally harsh for all herds, but some will suffer little loss while others may be wiped out. This is presumably due to some herders being more proficient, experienced or privileged than others. Here, the unit of selection appears to be the herd as a whole – and perhaps the associated human households as well.

Group selection is used in some commercial breeding schemes, for example the use as the unit of selection for caged laying poultry, the cage of four birds (Muir, 1996). Traits that are being selected are not solely the genes for egg production of a single individual.

Genetic variance due to mutations V_M will also accumulate – though it has negligible effects on phenotypic variance within inbred lines, it will contribute to between-line variance (Falconer & Mackay, 1996, p.270). Once reproductive isolation is established, natural selection can only be a cause of breed differentiation if it is stronger than the net effect of these other factors. Evidence of natural selection on single-gene loci mainly comes from the existence of polymorphism, and classical models of single-locus genetic systems suggest that polymorphism may be maintained by spatial variation in selection in a panmictic (freely interbreeding) population. However, for quantitative characters it now seems clear that environmental heterogeneity will generally not maintain genetic variation (Via & Lande, 1987). This has an important practical consequence relevant to the conservation of livestock biodiversity – if reproductive barriers are removed between previously distinct populations and a new overall selection regime is imposed, one should not expect local adaptations to persist.

Reproductive isolation is fundamental to the establishment and maintenance of breed identity. It can be inferred from historical records and confirmed by molecular genetics. For example, the genetic divergence of Icelandic cattle from Scandinavian breeds, predictable from their history of 1000 years of separation, is confirmed by microsatellite studies (Kantanen *et al.*, 2000) which have shown reduced heterozygosity and allelic diversity in the former. Hall and Ruane (1993) contended that it was the difficulty of maintaining reproductive isolation in goats and water buffalo, compared with, for example, sheep and cattle, that may have led to there being fewer breeds of the former two species (relative to their world populations).

Relative antiquity of breeds

Breeds can become differentiated and achieve their own identity very quickly. For example the formation of the Cambridge sheep breed began in Britain in 1964 and the first breed society sale was held in 1985 (Owen, 1996).

Some breeds are long established, others relatively new, and while it is difficult to distinguish them in inventories, their relative values especially for conservation are bound to be influenced by the length of their histories. For example, in the UK the Rare Breeds Survival Trust (RBST, see Chapter 8) will only recognize breeds that have existed for '40 years plus 6 generations' (the RBST priority list is updated every spring in its journal, *The Ark*).

Ashton and Ashton (2001) used contemporary vocabulary to distinguish long-established breeds of duck ('classic ducks') from recently-formed breeds ('designer ducks'). Even within a breed some lineages may be seen as 'classical' and others as non-classical, as is the case with Lipitsa horses (Kavar *et al.*, 1999). In the context of the developing world, Blench (2001) compares 'research station constructs' – breeds which 'have no natural habitat but only persist in fields outside research stations' – unfavourably with breeds that have 'co-evolved' with particular farming systems. Documented history has a vital part to play in the establishment of the relative interest and importance of breeds.

Cultural background to breed differentiation

All forms of subdivision of human communities (by kinship, ethnicity, language or economic group) could lead to distinctive livestock by imposing restrictions on matings. The phenotypes of breeds have been claimed to indicate cultural attitudes towards animals – exaggerated characteristics of breeds can enhance their owners' prestige (Clutton-Brock, 1999).

Military conquest and colonization has often led to changes in livestock breeds which can be inferred from historical records and often discerned in patterns of genetic variation. Totally new species may be introduced, as in the post-Columbian New World, or new breeds may be introduced. For example, Portugal and Spain have been heavily influenced by north African conquerors, and Portuguese cattle exhibit mtDNA of African type (Cymbron *et al.*, 1999).

In a study of 277 African societies, Guglielmino *et al.* (1995) found that 47 cultural traits (aspects of social organization relating to family and kinship, economy, social stratification, labour division by sex, house design, livestock husbandry, among others) showed various degrees of predictability from three sets of factors. These were ecological setting (determined from a vegetation atlas), language group, and whether the societies tended to be geographically clustered or not. Some traits were almost entirely predictable from one factor, for example traits of family and kinship were mostly predictable from language grouping, as would be expected from their transmission by parents or group pressure. The traits relating to animal husbandry were dependent on all three factors implying that cultural heritage, ecological factors and the practices of neighbouring communities all had an influence.

In the Old World, human cultural diversity and the capacity of an area to support a large human population are both correlated with livestock biodiversity. Areas that are peripheral or remote also tend to have distinctive breeds (Hall & Ruane, 1993).

There are many breeds of livestock that are associated with particular ethnic groups. In west Africa, the Fulani people have a strong pastoralist tradition and are associated with certain breeds of sheep and cattle (Blench, 1994). In north America, the Navajo people developed a husbandry system and wool industry based on the Churro sheep from Spain that became widespread in New Mexico about 200 years ago (Christman *et al.*, 1997). In northern Italy the Orobica goat is very different genetically from neighbouring breeds and this could be due to it having originated in the south and having been introduced with migrating people in the nineteenth century (Ajmone-Marsan *et al.*, 2001; Crepaldi *et al.*, 2001).

Social structures clearly define the operation of livestock systems. For example, among the Maasai of Kenya, cattle sometimes change hands through purchase but more often through customary transactions such as gifts and in exchange for a bride. Herds are divided into matrilines whose names reflect the circumstances of the acquisition of the founder of each of the lines, and are also assigned to particular households within the polygamous Maasai family. Cattle have names reflecting their lineages, these are the names that Maasai men usually use, while women and children who are closely involved with herding and milking tend to use individual phenotypically descriptive names. Ryan *et al.* (1991) claim that the lineage system of classification 'reflects salient aspects of Maasai social relations: it is through the cattle genealogies and the remembered transactions in cattle that past and present social relations of humans are preserved'. Coat colours of Maasai cattle have ecological relevance (see Chapter 3) and are also associated with human social structures (Finch & Western, 1977). Indeed, classification of the natural, human and spiritual worlds by colour (red, white, black) is very widespread in traditional societies worldwide and relates to livestock in various ways. For example, in northern Ghana among the Anufo people (Kirby, 1997) different coloured animals are appropriate for sacrifice on different occasions and at different shrines.

In France the beef breeds originated in areas where traditionally métayage, a form of sharecropping, was practised (Spindler, 1996). This was a partnership between landlord and tenant in which the former owned the land and buildings and sometimes the livestock too, while the latter supplied the labour and the produce was divided. The difficulties of controlling and dividing the production of a dairy herd encouraged landlords to promote beef production and this is claimed to be the reason for the development of the famous French beef breeds – the Charolais, Limousin, Blonde d'Aquitaine and Maine-Anjou. Another example of a system of tenure promoting breed differentiation is given by the North Ronaldsay sheep (of the Orkney Islands, Scotland; see Figure 2.3) where the removal of the sheep to the foreshore following land tenure reorganization around 1840 forced them to develop an ability to subsist on seaweed (Hall & Clutton-Brock, 1988). The main physiological challenges are the

Figure 2.3 North Ronaldsay sheep. Confined to the foreshore of their small island, the northernmost of the Orkney archipelago off mainland Scotland, North Ronaldsay sheep exhibit physiological and behavioural adaptations to their diet of seaweed.

low copper content and high arsenic content of the diet (Wiener *et al.*, 1978; Hansen *et al.*, 2003).

There would seem to be much potential for further studies of this kind, and on the human background to breed differentiation. Kohler-Rollefson (1997) provides a justification of this as an area of potential interest, stating 'Based on anthropological data, the notion is developed that the sociocultural corollaries of animal breeding, that are at work in traditional societies, result in genetically well circumscribed livestock populations . . . '. In the case of camels, under normal circumstances in the herds studied by Kohler-Rollefson (1995) female breeding stock are never sold. This keeps female composition stable over generations, leading to breed formation. A combined molecular genetics and anthropological approach could well be effective in elucidating the links between cultural environments and biological consequences.

Peoples strongly associated with particular breeds may nevertheless switch to different breeds if necessary. Blench (1994) contends that in Nigeria, Fulani pastoralists have tended to switch from zebu breeds associated with semi-arid west Africa, such as the Red Bororo, to another zebu, the White Fulani. This breed is seen as more tolerant of humid conditions, which coincides with a tendency for Fulani herds to move into the more humid southern areas in response to pressures on land in the more arid north.

Comparative linguistics can be very informative about livestock history generally, though it has not yet contributed to understanding breed differentiation. Human genetic diversity and linguistic patterns support the idea that neolithic farmers spread out from a centre of origin of domestication (Cavalli-Sforza *et al.*, 1994). Refinements have been added to this model, for example some variation in language distribution can be explained by proposing a spread of horse-owning pastoralists from the southern Ukraine (Piazza *et al.*, 1995).

The words given by people to animal species may, when analysed linguistically, give clues as to the history of those animals which can be checked against other evidence from archaeology, history or genetics. For example in northeast Nigeria (Blench, 1995), an area with a complex livestock history, three of Africa's four language phyla come together. Words for cattle are very diverse, suggesting multiple introductions possibly beginning 5000 BC. In contrast, there is more uniformity in words for chicken, and the principal one appears to have come from Arabic implying the species was introduced with the earliest trans-Saharan caravans about AD 500. Many languages here have words for camel, but the speakers 'only have indistinct ideas about the appearance of the camel itself', implying that camels were more numerous until recently. In New Guinea and in the islands of the Torres Strait that separate it from Australia the different ethno-linguistic groups have very different words for pig, suggesting a great antiquity of establishment (Baldwin, 1983). This is discussed further in Chapter 6.

The interaction of human history and livestock genetic diversity

In north-west England, the people of the Lake District carry distinct blood groups which link them with Norwegian people (Figure 2.4). This area has a distinctive breed of sheep, the Herdwick, which is thought to have originated from Scandinavia (Hall, 1996). There are other examples of parallels between human genetic distinctiveness and that of their livestock. For example, Astolfi *et al.* (1983) describe the two distinct breeds of cattle in Sicily, the Cinisara of the northwest and the Modicana of the eastern part. The human populations of these two areas are also genetically distinct (Cavalli-Sforza *et al.*, 1994). The two distinct types of yak in Sichuan, China, are the valley type and the plateau type and these coincide in their distributions with those of the Kham Tibetan and Anido Tibetan people respectively (Wu N., 2000). Crossbreeding with other yaks and hybridization with cattle are rare with the valley type, because of the difficult terrain. Plateau type yaks are often crossbred because their area is geographically accessible. Similarly, Andalusia in Spain has greater human diversity than elsewhere in the Iberian peninsula (Côrte-Real *et al.*, 1996). With 8 of the 28 Spanish breeds of sheep and goats being native to this area (Muñoz & Tejon, 1986), it is also stated to have 'a special wealth of autochthonous populations (native breeds)' (Rodero *et al.*, 1992).

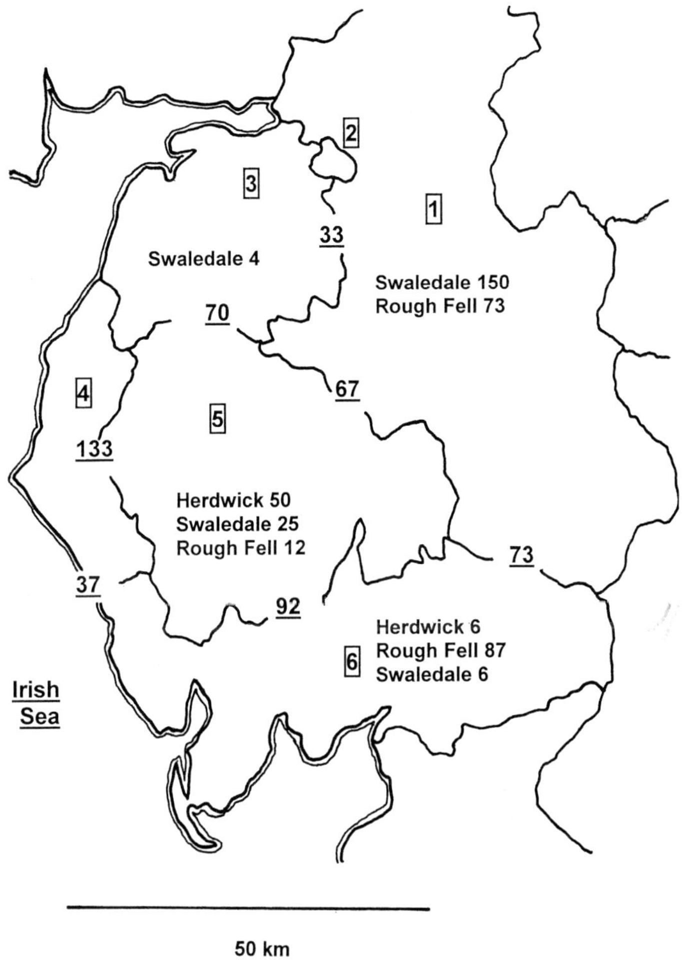

Figure 2.4 Cumbria, UK. Genetic distances (the underlined figures) between the human populations of the six regions show that the central region is especially distinctive and it also has a distinctive sheep breed, the Herdwick. Numbers of pedigree flocks of each breed in each area are given.

The obvious test is whether there is any parallel between the genetic distances between ethnic groups and between their livestock. The only studies where this has been explicitly considered appear to be by Tanabe (1991) where genetic similarities between Japanese dog breeds are paralleled by genetic affinities between corresponding human groups.

China would be a good place to test the hypotheses as it has massive livestock biodiversity (Cheng, 1984) and 56 officially recognized human ethnic groups. There are at least three independent centres in China where agriculture and associated Neolithic culture arose, and people of the main ethnic group, the Han, can also be classified on the basis of their surnames which can be used to trace male lineages back 4000 years (Piazza, 1998). There could be a fertile field of genetic anthrozoology linking these genetic histories.

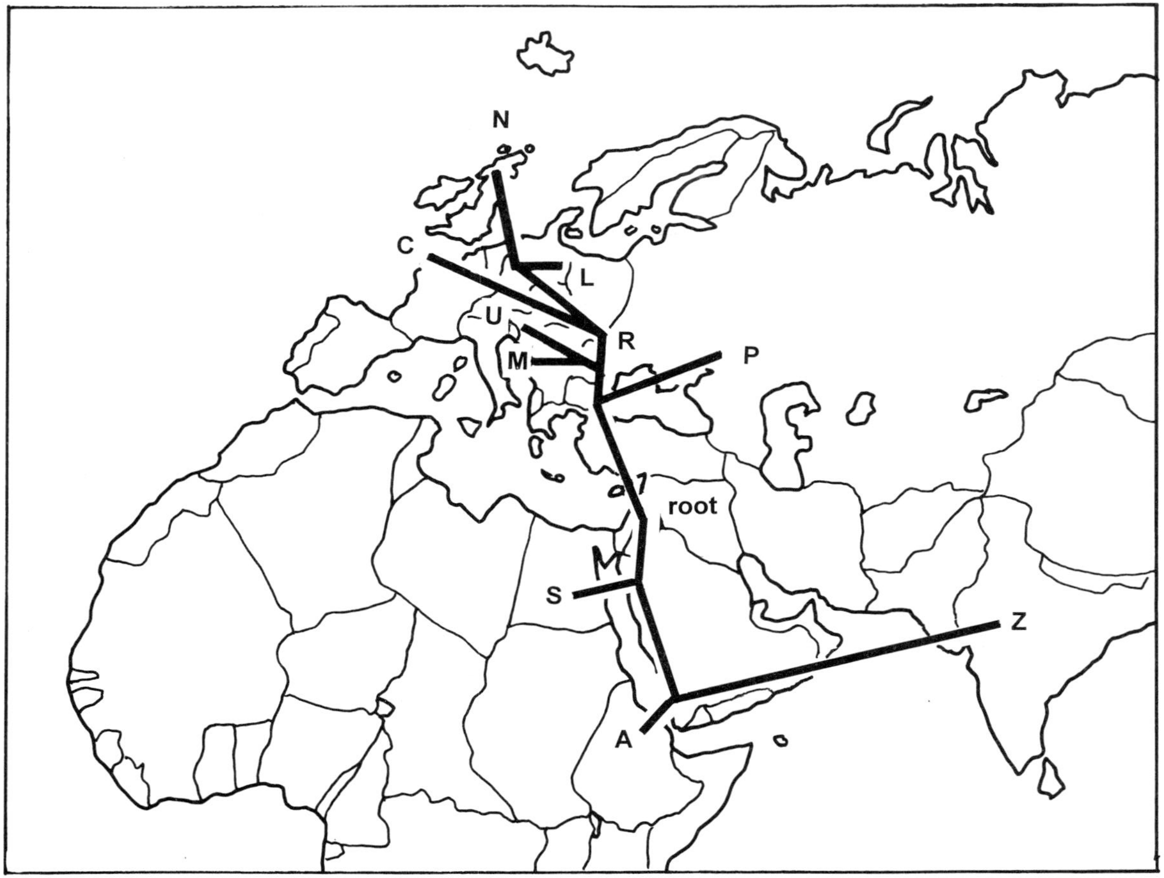

Figure 2.5 The geography of cattle phylogeny. The phylogenetic tree for ten major cattle breed groups can be superimposed on a map in a way that is consistent with geographic distribution of the breeds. From Manwell and Baker (1980).

In a very influential paper Manwell and Baker (1980) produced a phylogenetic tree of 216 cattle breeds from biochemical polymorphism data which they superimposed on a map to yield a correspondence between the genetic distances among breed groups and their geographical locations (Figure 2.5). This correspondence is consistent with the notion of livestock husbandry entering Europe from southwest Asia. Medjugorac *et al.* (1994) considered ten Balkan and Alpine cattle breeds in a similar way and also found that genetic distances were consistent with probable colonization history. Similar conclusions have been drawn from studies made on a more local scale, Kantanen *et al.* (2000) showed how the genetic distances of Scandinavian cattle from Icelandic cattle are in proportion to their geographical distances, suggesting distinctiveness depends on time of isolation. However, the genetic distances among 15 cattle breeds of Spain and Portugal show no correlation ($r = 0.04$, d.f. $= 104$) with the geographical distances between their traditional centres of distribution (calculated from data of Cañon *et al.*, 2001 and Porter, 2002). This suggests migration between these breeds has been mediated not by spatial proximity but by other factors, presumably decisions by breeders as to what were appropriate introgressions to make.

Conclusions

There are many definitions of the term ‘breed’. Now that under the Convention on Biological Diversity genetic resources are seen as the property of individual countries, the one that will be used in this context will probably be the proprietorial definition – ‘a breed is whatever a government says it is’.

The process of breed differentiation has been a cultural one in which reproductive isolation has been followed by the establishment of a breed identity that has been in response to cultural and commercial pressures. Breeds are human artefacts and in the Old World, especially, their histories and those of the people who keep them have been intertwined for many centuries.

Part 2
The Utilization of Livestock Biodiversity

Genetic resources are necessary for farming to respond to present and future needs. Some new challenges will be met by increasing inputs or by improving management but accessible genetic variation is necessary for sustainable development. Conservation costs money and economic justification is needed for animal genetic resources to be conserved. The first step towards this justification is an appreciation of breed adaptation. At least some of this adaptation could be due to gene interactions that may not conform to the model of gene action that has been generally used for the selection of productive traits.

Why livestock biodiversity is threatened needs to be understood. Farmers behave in economically rational ways, so there will be good economic reasons for the erosion of livestock biodiversity, which usually happens as a result of crossbreeding. At the same time, crossbreeding programmes can be designed that exploit breed attributes in a sustainable manner, thus providing an incentive for the maintenance of pure breeds.

Livestock biodiversity is more likely to be conserved if ways are found to reap economic benefit from its present value. In the developed world, where food is a declining proportion of household budgets, special features of individual breeds may increase their market appeal. In the developing world, the value of livestock biodiversity may reside in its local adaptations that will contribute to the sustainability of farming systems.

Chapter 3

Ecological adaptations of breeds

Introduction

Husbanded animals will be sheltered from as many rigours of the environment as their owners can contend with, but are there key environmental features that determine what livestock breeds can be productive in a given area? Weather, vegetation, seasonal changes, disease and parasites are all important to livestock but how clear-cut are differences between breeds in their abilities to cope with environmental challenges?

Genotype–environment interaction

Breeds usually perform better in some environments than in others, so the appropriate choice of a breed is important for profitable farming. Many of the reports of ecological adaptations of breeds are essentially descriptive and empirical. In some, the underlying physiology has been elucidated, and in a few the molecular genetic basis has been determined.

Quantitative genetics approaches the issue by acknowledging that some phenotypic variance can be due to genotypic value and the environment being correlated, or a specific change in the environment having a greater effect on some genotypes than others. Then $V_P = V_G + V_E + 2\ cov_{GE} + V_{GE}$ (Falconer & Mackay, 1996, p.131). Here, cov_{GE} is the covariance (a measure of correlation) of genotypic values and environmental deviations and V_{GE} is a variance term attributable to genotype–environment interaction.

Genotype–environment interaction is also called phenotypic plasticity or environmental sensitivity. Ideally a breed would show a greater V_{GE} when the environment is good, and a lower V_{GE} when the environment is bad, i.e. it would have the capacity to respond to improvements in husbandry yet would be relatively unaffected by poor conditions.

Selection for sensitivity can be achieved when artifical selection and the natural selection imposed by the environment are acting in the same direction (Falconer & Mackay, 1996, p.324; Kolmodin *et al.*, 2003). This is called *synergistic selection*. Practical examples are scarce but it may well have been important in domestication, breed differentiation and livestock breeding. Even simple husbandry practices such as rearing young males separately, with good feeding, from the main group and selecting them at a young age for breeding would amount to selection for increased performance in a good environment. At the same time, females might be selected in a poorer environment for traits relating to survival (Figures 3.1, 3.2).

Body size and physiological adaptation

In nature, differences in body size explain many of the ecological and other differences among animal species. In livestock, does body size influence economic efficiency and environmental adaptation?

Animal production efficiency may be expressed in economic terms (total input costs/total output receipts, gross margin per hectare, etc.) or in biological terms (as an input/output ratio of, usually, energy or protein). Biological efficiency of poultry, rabbit and pork production is much greater than that of lamb or beef production (MJ feed energy to produce 1 kg of meat protein is 336 for broiler chicken, 438 for rabbit, 633 for pork, 1787 and 1849 for lamb and beef: Dickerson, 1978). However, the economic efficiency of poultry, rabbit and pork is much lower than these figures suggest because their feed, capital and labour requirements are higher. Economic differences between species are due not to differences in body size, but to differences in reproduction, digestive processes, growth rate and body composition.

Figure 3.1 Survival in a harsh environment. Northern Nigeria: a thin Red Bororo cow is herded south to market. Strategies of the pastoralists who keep cattle like these are generally aimed at maximizing or maintaining herd size.

Figure 3.2 Response to a benign environment. Southern Nigeria: a fat Red Bororo bull is offered for sale. The breed evidently has the capacity to survive poor conditions and yet to respond when they improve, suggesting variation in environmental sensitivity.

When livestock breeds have been viewed from this perspective it has become clear that efficiency comparisons should be made over the whole life cycle of the animal. For example the rapid growth to slaughter condition of large-bodied cattle or sheep breeds may be offset by their having more difficulties giving birth, and longer gestation and lactation lengths. Based on this understanding, Dickerson (1978) advised that the breeder should first decide what body size of mature breeding female is best suited to the environment, the husbandry system and the market; this should determine breed choice and breeding plans can then be designed to optimize performance.

Adaptations to extreme temperatures

Many studies have been made on breed adaptations to the thermal environment, especially on livestock in hot conditions. There are many adaptive strategies, both behavioural and physiological.

Avoiding the sun

Time spent in the shade correlates reliably with measures of sensitivity to the sun in several species and breeds. In three genotypes of cattle, longer time spent in the sun by individual animals was correlated with a higher sweating response (Finch *et al.*, 1982). When grazing, Angus cattle had a higher harvest rate (g organic matter/kg body weight per minute of grazing) than Brahman x Angus and Tuli x Angus, but the Angus carried out less daytime grazing than the latter two genotypes which are both more drought adapted (Sprinkle *et al.*, 2000). The Aberdeen Angus developed in the UK, while the Brahman is a zebu breed developed in the south-west USA and the Tuli is an east African breed. Mamber goats are not as heat-adapted as Bedouin goats, losing body weight and failing to control rectal temperature unless they are given shade, which Bedouin goats do not need (Mualem *et al.*, 1995). In hot environments, ruminants have to contend with water restriction and heat load. These reduce appetite and although the digestibility of the feed is increased this is not enough to compensate for the reduced intake. This means that production and growth decline unless the animals are sheltered, or a breed is used that is adapted to the conditions.

Silanikove (1992) reviewed the physiological mechanisms linking these environmental factors with effects on appetite and digestion. Essentially, appetite is reduced through the action of elevated body temperature and it may also be reduced through the increase in gut fill, which results from reduced gut motility. It is reduced thyroid secretion that causes this reduced gut motility. The digestive tract is itself a site of metabolism and a source of heat (Sykes, 1994), and zebu cattle are generally thought to have a relatively smaller digestive tract than taurine (Sprinkle *et al.*, 2000). Some cattle breeds increase night-time grazing in response to elevation of afternoon rectal temperature. Sprinkle *et al.* (2000) present data suggesting a simple equivalence, perhaps a 1°C rise of this temperature leading to an extra 100 min of night-time grazing.

Deflecting or storing heat

Coat structure is highly relevant to protection against solar radiation in wild and domestic species (Grenot, 1992). The fleece of sheep shelters them from radiant heat stress, with respiratory rate and skin and rectal temperatures all being higher when wool length is shorter (Parer, 1963) and no wild species has such extreme development as the long-woolled sheep.

Coat colour can also be an important factor. In Kenya, pastoral herds of zebu cattle in hotter environments have higher proportions of light-coloured cattle (Figure 3.3). Dark-coloured cattle seemed to perform better in cooler Kenyan climates apparently because they expend less energy shivering in the morning (Figure 3.4). Under drought conditions, mortality of light-coloured cattle was higher in cooler environments and mortality of dark-coloured cattle was higher in hotter environments. Black coats of cattle absorb 80%, and white 43% of short-wave radiation. Black goats absorb heat 1.8 times more rapidly than white, which is believed to give them a metabolic advantage in cold weather (Finch *et al.*, 1980).

In the subtropics, Holstein cows with large areas of white in their coats tend to produce more milk than darker-coloured cows, and in Florida the mean area of white coloration was 26% of the body surface (Becerril *et al.*, 1993). This percentage is

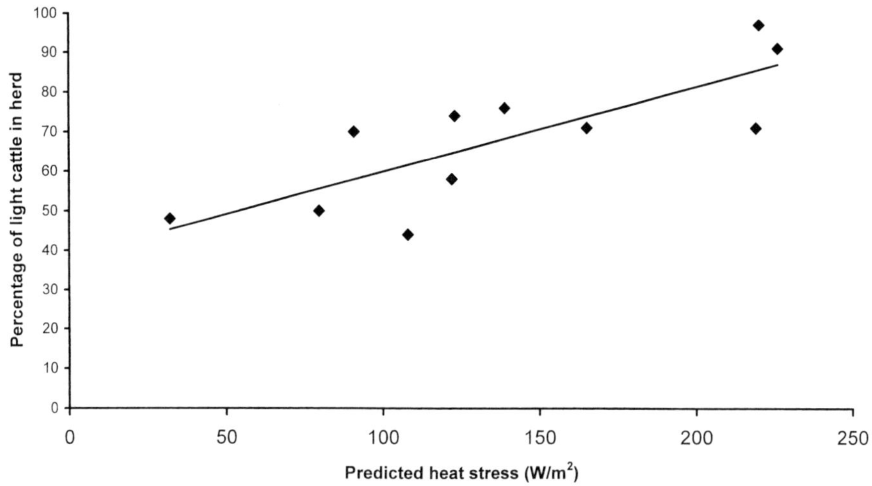

Figure 3.3 Cattle in hot climates: coat colour as an adaptation in Kenyan cattle. Regression of percentage of cattle that are light in colour, on heat stress (W/m^2). Each point represents one herd. The regression line is significant. (Reproduced from Finch & Western (1977) with permission from the Ecological Society of America.)

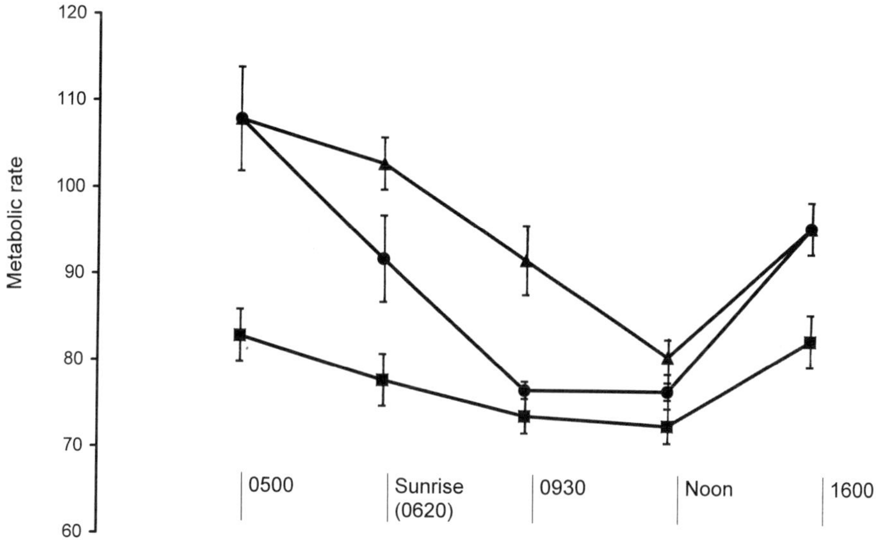

Figure 3.4 Cattle in hot climates: ability of Kenyan cattle of different body colour to warm up in early morning sunshine. In the cool climate, shivering kept metabolic rate of white cattle (triangles) high for longer than that of black cattle (circles). In the warm climate there was little effect of coat colour (squares). (Reproduced from Finch (1981) with permission from Cambridge University Press.)

highly heritable ($h^2 = 0.72$). One might surmise that white colour means less heat stress and thus more milk, but the situation may be more complex. In Germany (which has a lower average temperature than Florida) mean proportion of white coat colour in German Holstein and German Simmental cattle was higher (45%) and also highly heritable ($h^2 = 0.88$; Reinsch *et al.*, 1999). It was found to be influenced by a

QTL, located on chromosome 6, where QTL for milk production traits in Holsteins and other breeds have also been mapped.

Dissipation of heat

A thin coat is advantageous in large mammals presumably because of their relatively small surface area. Heat gain during the day is accepted, stored and then radiated at night (Grenot, 1992). Gaughan *et al.* (1999) suggested that the Brahman may have a lower respiratory rate and rectal temperature than cattle of pure or part Hereford genotype, at least partly because the former has a higher surface area per unit body weight and other skin characteristics permitting the rapid loss of stored heat.

Most research into routes of heat dissipation in livestock has considered panting and sweating. Cattle normally take 20–60 breaths per minute, 80–120 indicates moderate heat stress and > 120 implies excessive heat load (Gaughan *et al.*, 1999). To explain why Tuli crossbred cattle could keep a lower rectal temperature than Brahman crossbreds apparently by increasing respiratory rate, these workers suggested that Tuli cattle may have larger lungs, relative to body size.

Sheep can take over 300 breaths per minute under heat stress (Grenot, 1992), however, there are breed differences. In Lebanon, native fat-tailed Awassi sheep were found to have lower respiratory rate, heart rate, and body temperature than progeny of their crossbreeding with Finnish Landrace and Texel (Netherlands) sheep (Sleiman & Saab, 1995).

Kasa *et al.* (1999) ran goats of three breeds (Anglo-Nubian, Saanen and Toggenburg) on a treadmill (3 km/h for 1 h, on six alternate days). Sweating rate rose to the same degree in all, but there were differences in respiratory rate with the Anglo-Nubian showing the smallest increase and the highest rectal temperature. This apparent tolerance of heat load is consistent with its history as a breed of hot countries (Hall & Clutton-Brock, 1988). The Black Bedouin breed is unusual among goats in that it relies on sweating rather than panting to dissipate heat (Dmi'el *et al.*, 1979). Its maximal sweating rate of $188\,cm^3/m^2h$ is at least seven times that of other breeds and is due more to a high secretion rate of individual glands than to a high density of sweat glands. This high sweating rate is related to the thermal load the breed incurs through its dark coat. Sweating responses are higher in Brahman and Brahman x Hereford-Shorthorn cattle than for Shorthorn (Finch *et al.*, 1982). As rectal temperature increased, sweating rate for the Shorthorn reached a plateau, but it kept going up for the other genotypes (Figure 3.5).

In Nigerian cattle, the highest rate of evaporation was recorded in the White Fulani (Amakiri & Mordi, 1975), which is associated with the hotter and less humid areas. The West African Shorthorn, a taurine of the humid coastal forests, had the lowest rate while the N'Dama and the imported Friesian (also taurines) had intermediate rates. The West African Shorthorn also had the highest density of sweat glands in the skin (2208 per cm^2, Amakiri, 1974). Why the breed with the most sweat glands should have the lowest sweating rate, and whether this is connected with its humid habitat, is an interesting point. Hall *et al.* (1996) questioned whether sweating would

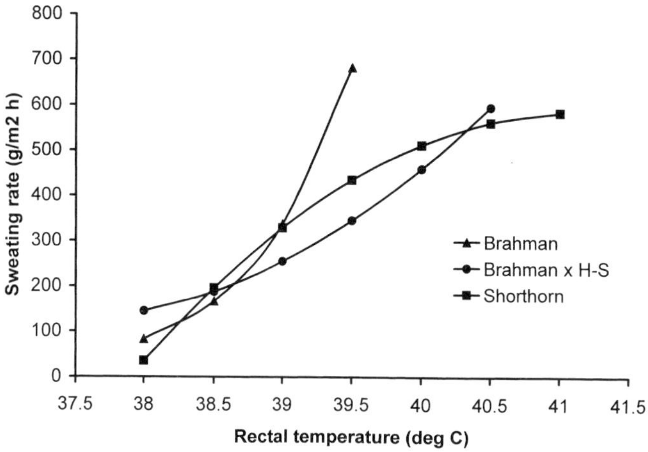

Figure 3.5 Sweating responses of cattle. Relationship of sweating rate and rectal temperature for Brahman (zebu breed), Shorthorn (taurine) and zebu x taurine. The Brahman restrains temperature rise by increased sweating while in the Shorthorn sweating ceases to defend body temperature. (Reproduced from Finch, Bennett & Holmes (1982) with permission from Cambridge University Press.)

be useful for dissipation of heat in very humid environments, and suggested that the coat characteristics of the West African Dwarf goat, which lives in the same humid habitat, may reflect an emphasis on sebaceous gland secretions rather than on sweating. Sebaceous secretions probably protect the skin against bacteria, help to keep the skin and coat in good condition and may be very relevant in this habitat.

Rehydration

Some breeds have the capacity to rehydrate rapidly. Camels can drink water amounting to a third of their body mass in ten minutes and Bedouin goats can take on 40% of their body mass (Grenot, 1992). In contrast, Mamber goats when kept in the sun and watered once daily only increased their water intake by an insignificant amount above normal, but Bedouin goats drink 2.26 times more water under these conditions than when kept in the shade. Tolerance of infrequent watering and capacity to rehydrate rapidly are exhibited by several breeds of ruminant (Nicholson, 1985).

Cold weather

Most work on cold adaptation in livestock has been on neonatal mammals. In Britain alone, between three and four million lambs die each year because of starvation and/or hypothermia. Cold resistance can be measured by part-immersing a lamb in a cool water bath and recording the time taken to reduce the rectal temperature by a fixed amount. Generally cold resistance appears proportional to birth weight (Figure 3.6) but breeds with similar birth weights can differ markedly in incidence of hypothermia in the field. It is possible that some breeds such as the Scottish Blackface, Welsh Mountain and Cheviot, which are all hill breeds, are more cold resistant than would be expected while the Southdown and Merino (lowland breeds) are less so. There is

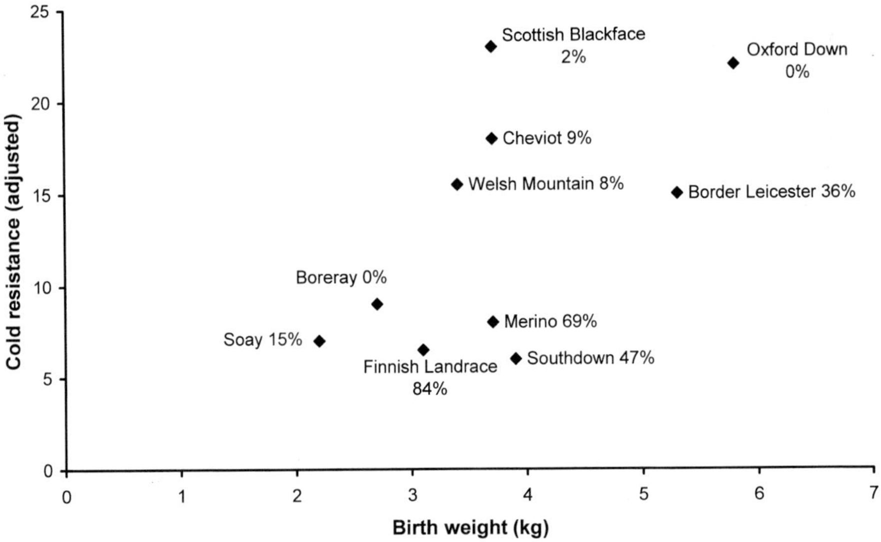

Figure 3.6 Cold resistance of lambs. Generally cold resistance of lambs increases with body weight, but there are clear breed differences. Incidence of hypothermia (as percentage) is given for each breed. (Slee, 1984; Samson & Slee, 1981).

within-breed variation, with heritability of cold resistance being 0.3 in Scottish Blackface (Slee & Stott, 1986). The young lamb or calf generates metabolic heat by shivering, and also by non-shivering thermogenesis. The latter takes place within deposits of brown fat and is evoked by noradrenaline, which is itself secreted in response to cold exposure. In Scottish Blackface, about 66% of the metabolic heat originates from the brown fat and in Merino, about 31% (Slee *et al.*, 1987); there are breed differences in these processes in cattle too (Martin *et al.*, 1999).

There are breed differences in the times lambs take to stand following birth, and the time to reach the udder (Slee & Springbett, 1986) – these are paralleled in cattle. Godfrey *et al.* (1991) compared Brahman and Simmental x Brahman-Hereford calves. When placed in a cold (4°C) environment within 20 minutes of birth and prior to suckling, the former showed lower rectal temperature and behavioural vigour. They had higher blood concentrations of energy-carrying constituents, which they were apparently not utilizing effectively, perhaps because of a relatively high concentration of hormones including cortisol, which is implicated in the stress response.

Although newborn Meishan piglets (from China) were 16% lighter than Large White piglets (Europe), they had similar resistance to cold, remaining comfortable down to 6–7°C. This tolerance was apparently due to a higher plasma free fatty acid concentration in the former, which could have been linked to a higher lipid content in the colostrum (Ledividich *et al.*, 1991). Piglets do not have non-shivering thermogenesis (Herpin *et al.*, 2002).

In temperate regions livestock are housed in winter for management reasons. While poultry, goats and pigs need some shelter, horses, sheep and cattle usually do not. There is very little evidence of breed differences in tolerance by adults of cold conditions. In upland Germany Wassmuth *et al.* (1999) compared Limousin x Friesian cows with Galloway cows in outdoor wintering, where the environmental temperature went to minus 20°C, although the latter is usually considered an especially hardy breed (from upland Scotland) there was no obvious difference in cold tolerance.

However, when the annual cycle is considered it becomes clear that some genotypes show seasonal patterns of change of body condition that adapt them to harsh winters. The Blue-Grey cow is a crossbred (Whitebred Shorthorn sire, Galloway dam) mainly kept in the hills of southern Scotland and mated to beef bulls such as the Charolais for spring calving. Studies in the 1980s showed that, compared with another important crossbred beef cow, the Hereford-Friesian, the Blue-Grey tends to be fatter at calving with higher reproductive performance, losing less weight during lactation and weaning a lighter calf. Clearly the Blue-Grey partitions less of her energy towards milk production and more towards maintaining her body reserves. She has a much longer reproductive life – in one trial ten year old Blue-Greys all became pregnant while only 60% of Hereford-Friesians of this age did so (Wright *et al.*, 2002). Now that the British Friesian has become, essentially, a north American Holstein (Chapter 4), with accentuation of dairy qualities and loss of beef conformation, one would expect the advantage of the Blue-Grey in overall durability to have become even more pronounced.

Angora goats in South Africa are kept to produce fine fibre (Cronje, 1992). Compared with the Boer goat, which is a meat-producer, Angora goats show poor adaptation to cold, with abortions and slow growth. This used to be blamed on lack of adrenal function, but it now seems that as a result of selection for hair-protein synthesis they are unable to mobilize body protein quickly enough in cold conditions.

Adaptations to the vegetation environment

Grazing animals modify their foraging strategies according to their internal state and the strategies available to them are determined by their morphology and behavioural repertoire (Hodgson & Illius, 1996). Information is steadily accumulating on breed differences, for example Wright *et al.* (1994) compared the performance of three genotypes of beef cow at different herbage allowances. Most previous work had measured outputs but not inputs (so efficiency could not be calculated). By manipulating the sward height of pastures, the effect of nutritional environment on output of weaned Charolais cross calf per hectare per year (land-use efficiency) was deduced. Aberdeen Angus x Friesian, and Hereford x Friesian, showed higher land-use efficiency on the short sward while that of the Welsh Black was lower. This was because the Welsh Black cows had lower milk yield and their calves were weaned at a lower weight and, with their high body mass, (535 kg at weaning; Aberdeen Angus x Friesian 487 kg, and Hereford x Friesian 508 kg) they were the least energetically efficient. In an extension to this work (cited by Wright *et al.*, 2002) performances of purebred Welsh Black and Charolais-cross steers were compared when they grazed either improved permanent pasture or 'semi-natural' pasture dominated by purple moor grass, *Molinia caerulea*. Performance of the genotypes did not differ on the improved pasture, but live weight gain of the Welsh Black steers was significantly higher on the other, implying better adaptation to this environment.

Locating forage

A very important feature of extensive systems is the necessity for animals to travel long distances in search of food and water, especially as the grazing season progresses. As the available biomass of browse falls below 40 kg/ha, goats of the semi-arid Sahel in Mali trek over 10 km/day (Sangaré & Pandey, 2000). Sometimes the presence of crops increases the distance animals must trek. West African studies (Schlecht *et al.*, 1999) showed that in the wet season cattle travelled 26 km/day to get to grazing lands beyond the cropped area, while in the dry season when crop residues were available, the daily 'grazing orbit' was less than 15 km.

In many pastoral systems, different resources are exploited at different seasons, another reason why grazers must be able to move long distances. Anderson and Urquhart (1986) found cattle with zebu genes to range further than taurine genotypes in semi-arid rangeland in New Mexico. Nicholson (1987) describes how Boran cattle may be particularly efficient at walking (energetic costs of locomotion were calculated as 0.6 compared with 2.0 kJ/kg km for taurine cattle in Britain). Whether this is due to metabolic adaptation or perhaps to morphology and biomechanics (Sahelian and savanna ruminants have long legs, Hall, 1991) does not seem to have been studied.

Selection at the level of the plant

Winder *et al.* (1995) recorded incidence of different forage plant species in faeces of Brangus cows and calves in New Mexico. Sires of these cattle were known and the heritabilities of diet composition were calculated. Percentages of different species, and numbers of different species in the faeces were generally significantly heritable in autumn, and less so in winter and summer. The authors suggest there are genes for aspects of diet selection, but the results could also be a consequence of heritability of birth weight. If certain sires tend to produce heavier calves, these might have different dietary patterns as a result of precocial development of grazing patterns.

Prehension of food

In mammals, food requirements are related to body weight to the power 0.75 ($W^{0.75}$) while gut capacity is related to $W^{1.0}$. This is an example of allometry – doubling the body weight increases food requirements by 68% but increases gut capacity by 100%. This implies that large herbivores would be able to tolerate more bulky foods than smaller ones (Illius & Gordon, 1987). At the same time, incisor breadth is found to be related to $W^{0.36}$ – halving the body weight reduces incisor breadth by 71%. When swards are long, bite weight is linearly related to live weight, increasing by 0.7 mg/kg live weight in sheep (Penning *et al.*, 1991) but when the sward is short, bite weight is closely related to width of mouthparts (Illius & Gordon, 1987). These considerations are highly relevant to competition between grazing herbivore species and many breed differences in foraging are becoming explicable from differences of body size and mouth morphology.

In the case of sheep and probably other relatively small herbivores, when forage is scarce, a broad incisor arcade would be favoured. When selective grazing is necessary (for example if there is an accumulation of plant material only some of which is of adequate nutritive value) a narrow arcade would be favoured. This indeed seems to be the case with the Soay sheep of St. Kilda, Scotland (Illius *et al.*, 1995) where winter survival is influenced by arcade breadth. Narrow-mouthed animals are more likely to die when grazing is bad, but they may be at an advantage during the population recovery phase when vegetation, both living and dead, is abundant and selectivity is advantageous. This is an example of countervailing selection, where different genotypes are favoured according to which fitness component is being considered.

Similar results are reported by Osoro *et al.* (1999) who kept ewes of two Spanish breeds – the Latxa (42.4 kg body weight) and the Gallega (33.1 kg) on two hill pastures with different proportions of heather *Calluna vulgaris* and grasses. Latxa ewes selected a more digestible diet and had higher rates of liveweight gain in the early part of the season when grass was more available, but later in the season the Gallega showed more intensive selection and a greater degree of adaptation, confirming that smaller-bodied breeds are better suited to poorer grazing.

Ruminal processes

Some dietary adaptations may depend on gut flora. For example *Leucaena leucocephala* is a Mexican leguminous shrub that is now widespread in tropical countries where it has been promoted as a valuable forage plant, but with the disadvantage that it contains a high concentration of the toxic amino acid mimosine. Livestock eating it sometimes exhibit mimosine poisoning, resulting from the rumen microbes metabolizing it only partially, to another toxic compound, 3,4-DHP. When rumen liquor from fully mimosine-metabolizing Indonesian goats was transfused into Australian goats that were suffering mimosine poisoning, the latter recovered. Similar results were obtained with cattle. It was subsequently found that 3,4-DHP degrading microbes could spread between cattle via droppings so herd protection from mimosine poisoning could build up. In this case, the dietary adaptation was due to the rumen symbionts (Jones & Lowry, 1984; Pratchett & Triglone, 1989; Hammond, 1995).

The juniper shrub (*Juniperus* spp.) contains high concentrations of essential oils, which make its foliage unpalatable to livestock, though goats will eat it when other forage is scarce. Working in Texas, Pritz *et al.* (1997) found that on average Spanish goats ate more juniper than Angora goats. The former digested it less completely (digestibility being high at low intake, and reduced at high intake, is commonly observed with forages generally, (Iason *et al.*, 1995)). On balance, both breeds obtained energy from juniper but they expended more nitrogen in digesting it than they gained from it.

Ruminal activities in two Spanish sheep breeds, the Churro and the Merino, have been compared (Ranilla *et al.*, 1997, 1998). The Churro more readily digested foods with higher cell wall content, possibly because the rumen was rather larger and they were better at keeping the rumen pH relatively high, which is a more favourable condition for digestion of fibre. There were no strong differences between the breeds in rumen dynamics, mainly due to high within-breed variability. Similarly in the UK, Cheviot sheep, which are associated with relatively poor environments, have higher digestive efficiency than Suffolk-cross sheep (Givens & Moss, 1994) especially when the animals receive a low digestibility diet.

It has long been thought that zebu cattle are better able than taurine cattle to utilize poor forage (Kennedy, 1982), and most studies have considered this question from the viewpoint of rumen processes. Early *in vitro* experiments compared the abilities of the respective ruminal fluids to ferment different diets. The general experience has been that these abilities do not differ between zebu and taurine (Hunter & Siebert, 1986). More recently, with the development of theories on how roughage intake is regulated by interplay of rumen dynamics and energy metabolism, experiments have been conducted on the different components of the system. Hunter and Siebert (1986) described how taurine (Hereford) cattle were more able than zebu (Brahman) to increase intake when offered good quality roughage. Zebu cattle have a smaller bite size than taurine cattle (McDowell *et al.*, 1996). For intake to increase in this way, rate of removal of material from the rumen (by fermentation or by exit of undigested particles) would be expected to increase. However, this was not obviously the case, leading to the suggestion that taurine cattle might allow a greater gut fill (with a longer retention time for particles in the rumen), and thereby higher intake, through having less sensitive stretch receptors in the rumen wall.

Kennedy (1982) in contrast concluded that retention time was longer in Brahman than Hereford cattle, but in this study the animals were much smaller. Perhaps this hypothetical loss of sensitivity of the taurine stretch receptors only happens as the animals grow. It is possible that zebu cattle increase their forestomach volume in response to decreasing digestibility of the diet but this needs further study (Schlecht *et al.*, 1999).

Some breed differences in feed intake may have a physiological explanation. Fiems *et al.* (1999) described how dual purpose (traditional) Belgian Blue and White cattle have higher voluntary feed intake than double-muscled cattle of the same breed. They suggested the latter may have a smaller gastrointestinal tract, or incur greater heat increment during digestion.

Seasonal and spatial patterns of resource utilization

In most pasture based livestock systems there is an annual cycle of food type, quality and quantity. Different strategies such as pastoralism, transhumance or switching of forage type are adapted to deal with this seasonality and one of the factors that could lead to the success of such strategies is the degree of adaptation of the livestock used.

Animals possess reserves which enable them to cope with periodic underfeeding and part of the art of husbandry is to ensure that these reserves prove adequate for survival. The general strategy is to feed for animal growth when good quality forage is available and to use lower quality feeds at other times, when weight loss can be accepted (Bayer & Waters-Bayer, 1998). The seasonal cycle of forage abundance should mesh with the cycle of animal production. These two cycles are well integrated in, for example, temperate zone sheep production. A seasonal cycle of appetite is particularly clear in northern breeds of sheep (Iason *et al.*, 1994) and horses (Fuller *et al.*, 2001).

A highly visible consequence of seasonality of food supply is loss of body weight often leading to emaciation. Many domesticated animals have to cope with periodic underfeeding and they do this by mobilizing body reserves of fat and protein. Loss of body solids also reduces total energy requirement. Price (1977) found Hereford bulls and steers survived loss of 36% of live weight. The steers (castrated males) were less lethargic and emaciated due to their higher fat reserves. Wild boar can survive the loss of 30% of their body weight (Spitz, 1986). Livestock are at least as capable as wild ungulates of tolerating weight loss. In Nigeria, Alaku & Moruppa (1983) recorded a 20% loss in body weight of goats during the dry season. Pregnant Scottish Blackface ewes, kept on hill pasture without supplementary feeding may lose over 20% of their maternal body mass during winter (Russel *et al.*, 1968) – there is evidence of breed differences in the ability to recover lost weight. Iason *et al.* (1992) found that Welsh Mountain sheep have less body fat than the Beulah Speckled Face, at the same percentage of mature live weight, especially after periods of feed restriction. This implies that during the recovery from adversity, the latter are more likely to store fat. In Zimbabwe, Afrikander cows cease reproductive activity (no oestrous cyclicity, risk of abortion etc.) once they have lost 19% of body weight while for Mashona cows this threshold is 32.5% (cited by Chicoteau, 1991).

In sheep, during underfeeding body fat loss is higher in individuals that were initially fatter, indicating that efficiency of energy utilization is lower in these animals. Fat-tailed breeds carry more fat at a given body weight than thin-tailed breeds, in the tail and under the skin, and the tail fat is mobilized more slowly than other stores. This may be why some fat-tailed breeds survive well during underfeeding (Chilliard *et al.*, 2000).

Breeding seasons

The wild ancestors of mammalian livestock are either known to have been seasonal breeders (wild boar, mouflon) or are highly likely to have been (aurochs, horse etc.). Under domestication, most if not all species have become much less seasonal in their reproduction. Considering sheep, in the mouflon, mating activity peaks in October but many modern breeds breed much earlier (Figure 3.7) while equatorial sheep breed all year round. With a five month gestation for sheep, late breeding is an advantage in colder climates as it ensures lambs are born when spring grazing has become available but there is a risk that puberty will not be reached until autumn of the next year. Thus, under domestication it appears that factors inhibiting the start of the breeding season

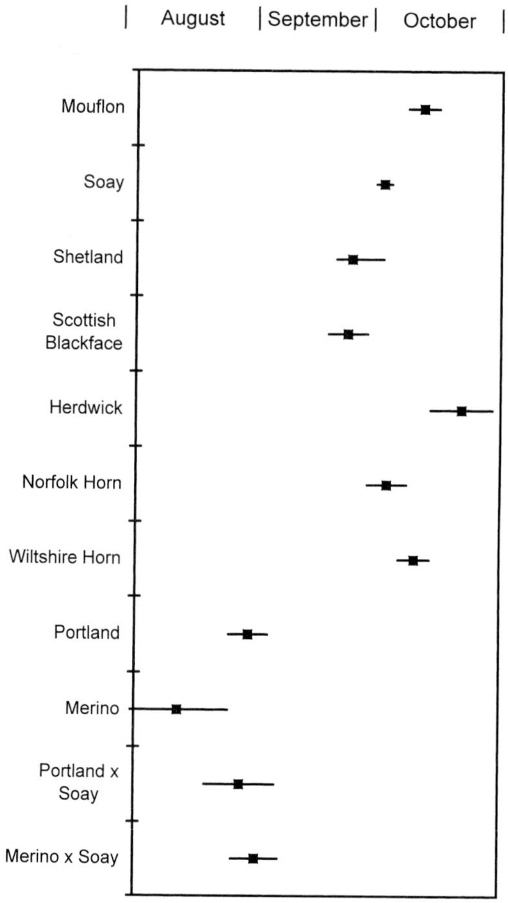

Figure 3.7 Breeding season of sheep. Timing of maximum testis diameter in rams all kept together near Edinburgh (mean and SEM indicated). The ancestral breeding season was presumably that of the mouflon or Soay. The lateness of the Herdwick is consistent with a Scandinavian origin. Sheep of more southerly origin (Merino, from Spain, and Portland, from Dorset) breed earlier. The lateness of the Norfolk Horn and Wiltshire Horn may be because in their relatively benign native areas there was no strong selective pressure for early lambing. Data from Lincoln (1989).

have been selected against, to different degrees in different breeds. Responsiveness to nutritional cues also varies – in Merino sheep, a 'Mediterranean breed', testicular growth responds to photoperiod, but nutrition dominates this response, while in the Suffolk, a 'temperate breed', the response to photoperiod is far more important, with only slight changes in response to nutrition (Martin *et al.*, 2002). In the temperate zones cattle do not show reproductive seasonality but poor nutrition or heat stress will delay puberty and decrease cow fertility (Tucker, 1982). There may be some effect of photoperiod (Hansen, 1985). In the tropics imported taurines tend to show sexual precocity and a relatively short calving interval. Lactational anoestrus is longer in zebu cattle (Figure 3.8a,b). If zebus are fed well all year round, they cease to show seasonality but otherwise their conceptions tend to be in the wet seasons when food is

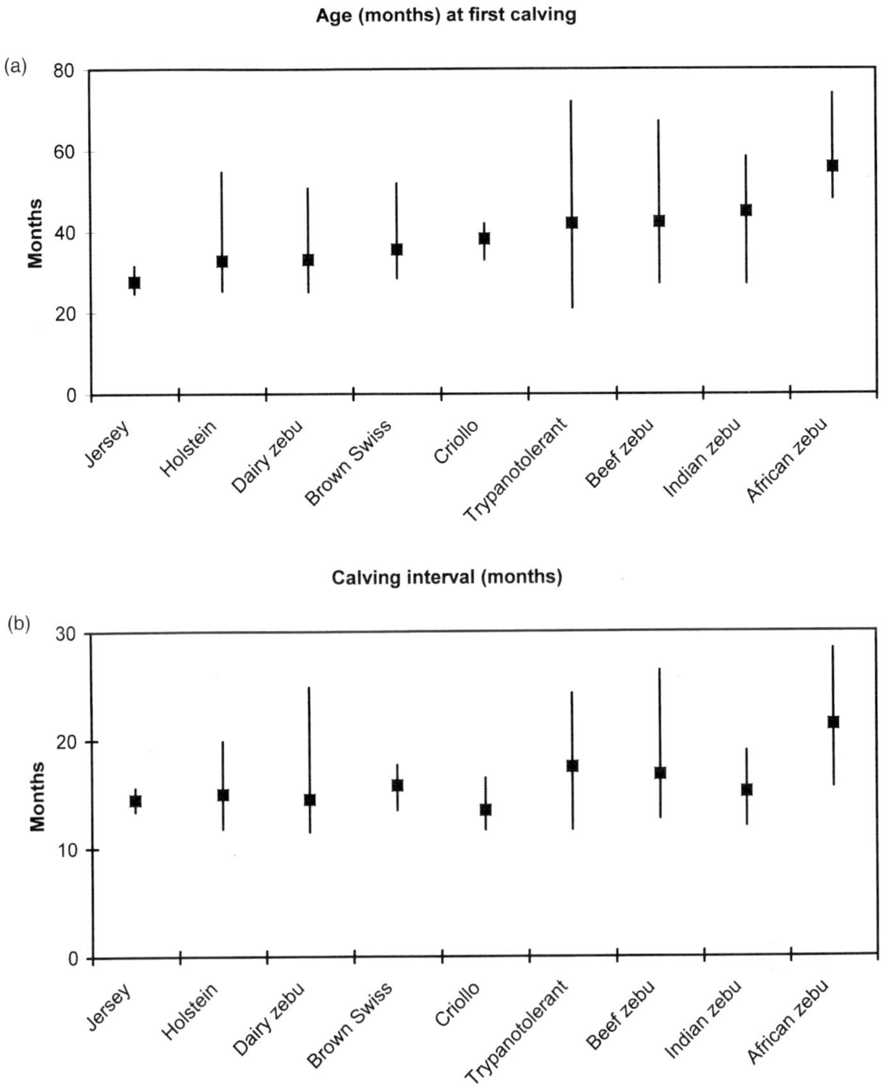

Figure 3.8 Reproduction of cattle in tropical environments. Summary of studies on cattle in the tropics (from Chicoteau, 1991). Jersey, Holstein and Brown Swiss are taurines of temperate zone origin, Criollo and trypanotolerant breeds are tropical taurines. Age at first calving (a): summary of 186 studies; calving interval (b): 132 studies. Mean and range are given.

plentiful. In contrast, taurine cattle imported from temperate zones will tend to conceive during the cool season (Chicoteau, 1991).

Ovulation rate and litter size

Heritabilities for numbers of ewes lambing and for litter size are low (around 0.07 and 0.1 respectively). Heritability of ovulation rate can be much higher that that of litter size – at least in some breeds (0.45 in the Finnish Landrace and 0.57 in the Galway; Land *et al.*, 1983). Some breeds of sheep are associated with high ovulation rate; in the Romanov, Finnish Landrace and D'Man, this reflects the action of a relatively large number of genes each with a small effect. In contrast, in the Booroola Merino (Australia) and several other breeds there is a single gene or closely linked group of genes (Webb *et al.*, 1999; Goddard, 2003). In the Booroola Merino, and perhaps in others, the high ovulation rate may be due to a loss of control within the ovary. This gene is thought to have entered Australian sheep in the nineteenth century from the Garole sheep of Bengal.

Coping with parasites and disease

A host animal may be completely impervious to a parasite or pathogen, or the host may permit survival and reproduction but with the pathological effect of the infection being kept within limits so the host's fitness is not diminished. Resistance refers to 'traits that prevent infection or limit its extent', and tolerance to 'traits that do not reduce or eliminate infection but instead reduce or offset its fitness consequences' (Roy & Kirchner, 2000).

Resistance and tolerance will evolve in different ways; it is predicted that resistance traits would tend to evolve to be polymorphic, and tolerance traits to be fixed. This is because the incidence of infection declines as a resistance gene spreads through the population, reducing the advantage of possessing it. In contrast the incidence of infection rises as a tolerance gene spreads, increasing its fitness advantage. The genetic basis of disease tolerance and resistance is usually discussed with reference to the major histocompatibility complex (MHC). This comprises tightly linked genes coding for glycoproteins that bind to foreign proteins and present them to circulating T cells, setting off an immune response (Edwards & Hedrick, 1998). It is the best characterized aspect of the genetics of disease resistance, but there are other genetic mechanisms.

MHC Class I genes deal mainly with proteins produced inside cells, for example by viruses, and Class II genes deal with those from extracellular pathogens, for example bacteria (Schild & Rammensee, 2000). Birds have Class III and Class IV genes as well which have additional functions. The bird MHC genome is much smaller than that of mammals (Hughes & Hughes, 1995). The MHC is highly polymorphic and the consensus is that this reflects evolution to cope with pathogens which themselves have the capacity to evolve (Apanius *et al.*, 1997). The process in mammals has differed from that in birds. In mammals, new gene combinations arise frequently; this is not so in birds implying that the MHC might not have evolved as a set of defences against pathogens in general but perhaps in response to a single selective pressure (Parham, 1999). Links between genes at the MHC and particular diseases first became evident in chickens where the heterozygosity in MHC genes, which was observed in small inbred populations, was one of the early indications that this polymorphism could be associated with selective advantage (Kaufman & Wallny, 1998; Rothschild *et al.*, 2000). Particular MHC alleles confer resistance to Marek's disease (Nicholas, 1987, Chapter 8). The tightness of the link between disease challenge and possession of specific genes testifies to the lower plasticity of the avian MHC (Parham, 1999).

In cattle the MHC (also known as the bovine lymphocyte antigen – BoLA – complex) is implicated in the response to many diseases. Different MHC haplotypes are associated with resistance and susceptibility to clinical mastitis in various dairy breeds (Nordic Red and White, and Holstein: Rothschild *et al.*, 2000). Polymorphism is widespread and Ellis and Ballingall (1999) discuss it in relation to certain diseases. As studies in experimentally infected mice show MHC involvement in the response to these diseases, polymorphism of the MHC in cattle may be at least partly pathogen-driven and there are many breed specificities (examples in Maillard *et al.*, 1993). There is evidence of direct selection on specific MHC alleles in Soay sheep (Paterson *et al.*, 1998). In this study, microsatellite markers within the MHC itself were characterized.

In pigs, the MHC (or swine lymphocyte antigen – SLA – complex) is the best characterized part of the genome, and it is involved to some extent in response to pathogens and to the parasitic nematode *Trichinella spiralis*. Some haplotypes (combinations of alleles) are characteristic of particular breeds or groups of breeds of pig but the functional significance of this is not known. It also affects some production traits including carcase composition, growth rate, fat distribution patterns and sexual development (Chardon *et al.*, 2000).

The diversity of genes in the MHC (some loci have over 100 alleles: Edwards & Potts, 1996) is a key component in the mammalian immune response and it is thought that polymorphism in it confers resilience towards disease challenge and that polymorphisms in MHC genes are older than species themselves (Holmes & Ellis, 1999). It is also possible that it is variation, rather than the haplotype itself, that is the important determinant of disease resistance (Coltman *et al.*, 1999). The conclusion is that MHC diversity is only one of the lines of defence against infection. Diversity of MHC in domesticated mammals may well indicate increased disease challenge or it could reflect the relatively large numbers of lineages in the original domesticated populations (see Chapter 1).

Differences within and between breeds

There are seven classes of disease in livestock – epidemic infectious diseases, vector borne diseases, parasitic diseases, production/management diseases, genetic diseases, genome-associated diseases and cancers. Most research has concentrated on within-breed studies. This is because many sectors of the livestock industries are committed to particular breeds and it is more practicable for them to develop disease resistance by within-breed selection rather than by bringing in genes from other breeds. For all seven classes of livestock disease there is at least some evidence of between-breed differences in susceptibility.

A full understanding of the dynamics and evolution of resistance to diseases and parasites is not yet available – genetic epidemiology is in its early days (Schork *et al.*, 1998; Woolhouse *et al.*, 2002). Host genotype may, at least in mathematical models, interact with epidemiological processes, suggesting that selection response would be higher than is predicted by quantitative genetic theory alone (Bishop & Gettinby, 2000).

Epidemic infectious diseases

Examples include foot and mouth disease (FMD), rinderpest and contagious bovine pleuropneumonia, all of which are viral. Control usually involves mass slaughter or vaccination and is often made difficult by the persistence of disease reservoirs. For example, in 1994 a rinderpest virus last reported 30 years previously reappeared in buffalo and kudu in Kenya; it had survived as a mild disease in local cattle (Mariner & Roeder, 2003). However, zebu cattle are more resistant than taurines to rinderpest and it is thought the rinderpest epizootics were introduced to Africa from Asia with zebu cattle during the 1300 years since the first large-scale influx of zebu cattle, leading to eradication of local taurines in east Africa (MacHugh *et al.*, 1997).

Whether resistant breeds will always be welcome in modern agriculture is not clear. They could be accused of being a reservoir of disease, which could then infect intensively farmed populations. Susceptible genotypes could be of value for sentinel populations; chicken flocks are being established in California which will be monitored for the arrival of the West Nile virus (Hoag, 2003).

As livestock production becomes more intensive the risks of epizootics increase. These may have been made worse by the decrease of breed diversity in the more industrialized husbandry systems. Emergence of bovine spongiform encephalopathy (BSE), and spectacular outbreaks of classical swine fever and FMD, together with the Nipah virus (viral encephalitis) which led to one million pigs being destroyed in Malaysia in 1999, can all be attributed to increased intensification (de Haan *et al.*, 2001). Avian flu in the Netherlands might have entered indoor flocks from free-range hens that have had contact with wildfowl (Abbott, 2003). The response of the industry might be to increase biosecurity rather than to breed for disease resistance. Even numerous breeds can find themselves threatened by disease eradication programmes if they are geographically localized. The threat to the Herdwick and Rough Fell sheep of Cumbria and Lancashire during the British FMD outbreak of 2001 is described by Bowles *et al.* (2003). According to Wright *et al.* (2002) '...prior to the outbreak of foot and mouth disease the South Country Cheviot Sheep Society figures showed a population of breeding ewes of approximately 65 000. Up to 17 April 2001 it was estimated that some 16 000 ewes had been slaughtered...The annual ram sale in Lockerbie averages around 440 rams. Almost 25% of the breeders selling at this sale have had their flocks destroyed...'.

Vector-borne diseases

Maillard *et al.* (1993) described how Brahman cattle were imported into the West Indies from the USA around 1952. They are susceptible to dermatophilosis and cowdriosis, both transmitted by the tick *Amblyomma variegatum*, a native of west Africa. In contrast the native Creole (taurine) cattle at least some of which have genetic affinity to west African cattle (Magee *et al.*, 2002) are highly resistant to these diseases.

The African trypanosomiases (becoming renamed collectively as trypanosomosis) is a complex set of diseases caused by the protozoan *Trypanosoma* that is transmitted in the saliva of tsetse flies *Glossina* spp. There are at least 36 species of tsetse fly and four species, with many strains and clones of *Trypanosoma*. Wild animals can act as a reservoir of infection, tsetse flies remain infective for long periods, multiple infections are possible and a single parasite can exhibit many different antigenic types (d'Ieteren *et al.*, 2000). There are trypanocidal drugs and many more or less effective techniques for controlling tsetse but little prospect so far of a vaccine. Some livestock breeds can 'survive, reproduce and remain productive under trypanosomosis risk without the use of trypanocidal drugs' (d'Ieteren *et al.*, 2000) especially if they are properly fed and not overworked. These trypanotolerant breeds (the best known are the N'Dama and West African Shorthorn cattle, and West African Dwarf sheep and goats, of humid and subhumid west Africa) have been exported to many areas nearby and in central Africa where livestock farming had not previously been possible due to trypanosomosis (Shaw & Hoste, 1987).

Trypanotolerance involves two apparently distinct processes (d'Ieteren *et al.*, 2000), the control of parasite development and the mitigation of the main pathological effect which is anaemia (Figures 3.9a,b). These two processes could correspond respectively to the resistance and tolerance aspects defined by Roy and Kirchner (2000). It might be suggested that the former corresponds to the action of the immune system and the latter to other defence mechanisms but the distinction is unlikely to be clear-cut.

Breed comparisons, when resistance and tolerance are considered separately, may be feasible, in view of experiments like those of Katunguka-Rwakishaya *et al.* (1997). They compared the responses of Scottish Blackface and Finn Dorset sheep to experimental infestation with *Trypanosoma congolense*. Both sheep genotypes exhibited the same degree of parasitaemia, but the former suffered a higher degree of anaemia. Under controlled conditions, when cattle of a breed considered to be trypanotolerant and also trypanosusceptible cattle were infected with cloned *T. congolense* both types showed similar parasitaemia and anaemia in the early stages. However, the trypanotolerant cattle began after 30–50 days of infection to show limitation of parasitaemia and reduction of anaemia; these capacities increased upon renewed challenge (d'Ieteren *et al.*, 2000).

Breeding and screening programmes are under way with a view to mapping bovine trypanotolerance genes (Kemp & Teale, 1998). Meanwhile, the genetics of trypanotolerance in the mouse, as a model species, are becoming clearer. Experimental populations derived from resistant and susceptible grandparental strains (assumed to be homozygous for resistance and susceptibility alleles respectively) have been tested for trypanotolerance in conjunction with linkage mapping. This has enabled genes conferring resistance to be located in the genome. In mice, three loci seem to be involved (Kemp *et al.*, 1997), on different chromosomes, while genetic background may also influence the expression and dominance of the traits. Studies of this fundamental kind should improve understanding of the host-parasite interaction and

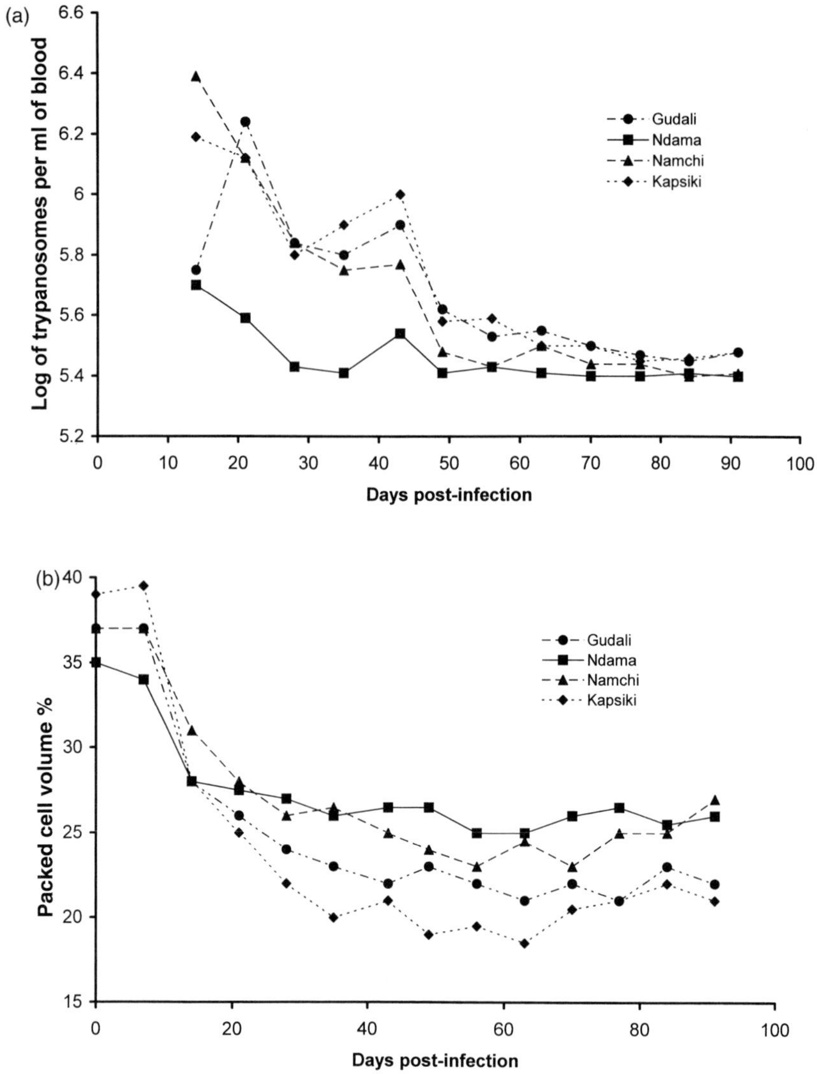

Figure 3.9 Course of experimental infection with *Trypanosoma congolense* in trypanotolerant (N'Dama and Namshi) and non-trypanotolerant (Kapsiki and Gudali) west African cattle. N'Dama cattle show the most effective control of parasitaemia (a), followed by the Namshi. They are also similar in their ability to resist anaemia (b). The Gudali is a zebu breed, the others are taurines (the Kapsiki shows the greatest degree of zebu introgression). (Reproduced from Achukwi, Tanya, Hill *et al.* (1997) with kind permission of Kluwer Academic Publishers.)

may be exploitable through marker-assisted selection, and possibly even through biotechnology (Kemp & Teale, 1998).

Work on cattle is 'beginning to reveal' resistance QTLs (Teale *et al.*, 1999). Trypanotolerance in west African taurine cattle seems to diminish in proportion to zebu introgression (Achukwi *et al.*, 1997). Perhaps therefore these resistant QTLs have an additive effect. It also suggests that it is important to keep trypanotolerant breeds free of zebu introgression if this genetic resource is to be conserved and utilized sustainably (see Chapter 4).

Parasitic diseases

Macroparasites are essentially helminths or arthropods and they reproduce outside the body of the host. Infection can last a long time and reinfection of the host is possible. Some hosts carry heavy macroparasite burdens, others light, and for the purposes of epidemiological modelling the interaction with the host can be expressed quantitatively (for example egg output, effects on host, death rate of parasite, etc.). In contrast, microparasites (bacteria, viruses, protozoa, fungi) reproduce within the host, infection is brief and in survivors reinfection can be prevented by an immune response. The interaction with the host is usually not readily describable in quantitative terms although hosts in a population can be classified as 'susceptible', 'infected' or 'recovered and immune' (Bishop & Gettinby, 2000).

Many studies have claimed to show breed differences in ecto- and endoparasite resistance. The evidence is only reliable if animals from many sires have been used (Baker, 1991); between-sire differences can be as large as reported between-breed differences, and there can be different resistance mechanisms in different selection lines. Large experiments are needed – Baker *et al.* (1999) used 1691 ewes of three genotypes and 54 rams of two breeds to demonstrate breed differences in parasite-induced pre- and postweaning mortality of lambs and other effects (Red Maasai ewes were more resistant than Dorper). Inbreeding can reduce parasite resistance too (Coltman *et al.*, 1999; Meagher, 2000). Coltman *et al.* (2001) found the genetic correlation of growth with parasite resistance to be positive in some populations of sheep and negative in others. Studies on sheep in Scotland, Australia and New Zealand have not produced a clear relationship between resistance and productivity and this hampers inclusion of resistance in breeding programmes (Vagenas *et al.*, 2002).

Genetic approaches to disease and parasite resistance are beneficial in many ways. Many veterinary drugs have ecotoxicological effects. For example, avermectins are injected into mammalian livestock to give protection against nematode and arthropod parasites; their residues can affect some dung-eating insects and can inhibit nutrient cycling in pastures (King, 1993; Hutton & Giller, 2003). Nematodes develop resistance to anthelmintics and this has been most important commercially in sheep and goats, especially in Australia and South Africa (van Wyk *et al.*, 1999). There is a real likelihood of nematode parasites of small ruminants becoming uncontrollable in many areas in the near future. Anthelmintic resistance has been much less important in cattle and pigs.

Native breeds are not always more resistant to parasites than exotic breeds. In Kenya, Saanen goats (from Switzerland) were more resistant to experimental infection with *Haemonchus contortus* larvae than local breeds. This could be explained by the Saanen having been developed in a grassland environment, where infection by ingestion of faecal material is relatively likely. African goat breeds are more likely to browse than to graze, and as browsers are relatively unlikely to ingest each others' faeces there would be expected to be a lack of evolutionary pressure towards resistance in their case (Gall, 1991).

Resistance mechanisms are becoming better understood. Sheep can control fecundity and body length of *Ostertagia* worms by parasite-specific immunoglobulin (breed differences have not been documented) (Stear *et al.*, 1997a,b). In the case of ticks, resistance involves reducing the tick's ability to feed and is associated with inflammatory responses at the feeding site (Wakelin, 2000). The bite of the tick evokes an immune response; this immunity can be transferred by vaccination and some vaccines have been developed (Brossard, 1998). Resistance can operate at many different stages in the relationship that develops between host and parasite. The host can have avoidance mechanisms, for example grooming behaviour is not merely reactive. While the bite of the tick can induce release of histamine, causing irritation which evokes a grooming response, animals also groom themselves according to an internal timing mechanism that is modulated by body size and tick challenge. Smaller-bodied animals groom more frequently and body mass accounts for about two-thirds of the variation in grooming rate across species (Mooring *et al.*, 2000). Neural and endocrine aspects of grooming have been reviewed by Spruijt *et al.* (1992); breed differences in tick resistance might be profitably investigated from a behavioural viewpoint.

In many cases disease resistance could in principle be bred for. However, there may be an overriding commitment to a particular genotype, and crossbreeding with a resistant breed would so depress yields that the economic benefit would be non-existent. A good example comes from the cattle industries of Queensland, Australia. Here, tick infestation reduces milk production by dairy cattle (each female tick feeding on a Holstein-Friesian cow reduced daily milk yield by 8.9 cm^3 and over 15 weeks reduced liveweight gain by 1.0 g; Jonsson *et al.*, 1998). Zebu introgression has been used in the beef industry (see Chapter 5) to confer resistance to ticks, but this has not happened in the dairy industry. Although the Jersey and the Australian Milking Zebu (derived from Brahman x Jersey) both compete with the Holstein-Friesian in net income, most farmers feel themselves to be committed to the latter. Thus, most interest has been concentrated on within-breed selection.

Production diseases

Production diseases are defined as those caused by management systems. These include nutritional diseases, trauma and infections arising from the conditions under which animals are kept. Many production diseases influence productive life (Rogers *et al.*, 1999). There is some evidence of breed differences, for example young Duroc and Yorkshire (Large White) pigs are more resistant to clinical and subclinical disease than Landrace and Hampshire, though this depends on management (Henryon *et al.*, 2001).

The economic and welfare consequences of the prevalence of mastitis in dairy cows are now very apparent (Heringstad *et al.*, 2000). White blood cells and udder epithelial cells are present in high concentration in the milk of cows suffering udder damage. These can be counted and recorded along with milk yield of individual cows. There is some evidence (Miglior *et al.*, 1995) that in Canadian Holstein dairy cattle the more inbred animals have a greater susceptibility to mastitis but this does not seem to be the case in US dairy cattle (Thompson *et al.*, 2000a,b). Holstein bulls that transmit more milk yield also have daughters with a higher incidence of mastitis (Rogers *et al.*, 1999). Bulls that are candidates for use in AI can be characterized for mastitis resistance and this is done in a coordinated way in Denmark, Finland, Norway and Sweden (Lundén *et al.*, 1993; Heringstad *et al.*, 2000). As a result the Nordic Red and White breed group can be described as a high yielding dairy type, which has benefited from having been selected partly on the basis of disease resistance. A lower incidence of mastitis in dairy cows (currently it affects 40% of British dairy cows each year; Coghlan, 1996) is clearly desirable. Selection criteria that would achieve this have been outlined (Phillips, 1997) and the point is made that rapid milk ejection, which has been advantageous in modern dairying, may become of less importance if robotic milking systems allow cows to be milked unattended. Modern milking machines could, by measuring conductivity of milk, enable early detection of mastitis (Coghlan, 1996). Thus there are at least two non-therapeutic routes to reduction of mastitis; technological advance and the application of genetics.

Dystocia, or difficulty giving birth, can also be seen as a production disease. Breeds have been claimed to differ in susceptibility to it (for example, Silva & Noakes, 1984; Schwabe & Hall, 1989). High calf birth weight is a major cause, and one of the advantages of the introduction of Tuli cattle to Australian rangeland beef production (see Chapter 5) could be the low heterosis that crosses with this breed exhibit for birth weight (Frisch & O'Neill, 1998a).

Genetic diseases, genome-associated diseases and cancers

Retroviruses have RNA as their genetic material; when they infect a cell this is transcribed onto host DNA and replication follows. If the retroviral gene is actually incorporated into host DNA, a heritable insertion mutation can result (Nicholas, 1996). If organs are transplanted from pigs into human patients, in principle the latter could be challenged by novel virus diseases (Platt, 2001). Le Tissier *et al.* (1997) confirmed that genes of retroviral origin are present in all pig breeds so far tested.

Maedi-visna (sheep), caprine arthritis encephalitis virus (goats), immunodeficiency viruses (bovine and equine), and the Jembrana virus of Bali cattle (Wareing *et al.*, 1999) are all retroviruses of the genus *Lentivirus* (DeMartini *et al.*, 2000). There are breed differences in susceptibility to maedi-visna but historical or environmental reasons for this do not yet exist.

Scrapie is the transmissible spongiform encephalopathy (TSE) of sheep and goats, induced by an infectious protein particle known as a prion (Nicholas, 1996a) – it is fatal, with a long incubation period, and is found throughout the world but not in Australia and New Zealand. The prion protein controls development of the condition. It is encoded by the prion protein gene PrP which is a normal part of the mammalian genome, and is expressed in most organs of both healthy and diseased animals

(Goldmann, 1998). There is polymorphism in sheep and goat PrP, and amino-acid substitutions at three positions in the molecule influence disease onset, but how this influence is exerted is unknown (Hunter *et al.*, 1997). The positions and the amino-acids that occur are as follows: 136 (valine, coded as V, or alanine, A); 154 (arginine, R or histidine, H) and 171 (glutamine, Q, or R, or H). Thus, the allele coding for the prion variant with alanine at position 136 and arginine at 154 and 171 is known as allele ARR. Not all combinations of the different alleles are of wide enough occurrence to have been recorded. Scrapie risk (assessed by infection challenge) has been assessed in relation to prion variant. Perhaps surprisingly, whether or not a given variant is associated with risk of scrapie can depend on which breed is being considered. The alleles that are important in Britain and their characteristics are given in Tables 3.1a,b. The results of PrP genotyping of British rare sheep breeds are reported by Townsend *et al.* (in press).

Australian and New Zealand sheep are free of scrapie but highly susceptible PrP genotypes do exist in sheep there (Hunter *et al.*, 1997). This suggests that it is not purely a genetic disease but that an infectious agent, present in every other sheep-farming nation, may be necessary for scrapie to be prevalent. In contrast, there is no obvious association between PrP genotype in cattle and incidence of BSE (Hunter *et al.*, 1994).

Table 3.1 (a) Alleles at the *PrP* locus coding for variants of the prion protein in sheep and the associated risks of scrapie (from Dawson *et al.*, 1998).

ARR	Low risk of scrapie. Dominant in all breeds, so ARR heterozygotes are at low risk.
AHQ	Low risk of scrapie in breeds or flocks where the VRQ allele is found. Dominant.
VRQ	Only found in certain breeds, where it confers high risk of scrapie. VRQ/ARQ heterozygotes also at high risk.
ARQ	In breeds which do not carry VRQ allele, linked with increased risk of scrapie.
ARH	Like ARQ allele.

(b) The PrP alleles found in various sheep breeds in Britain are as follows (also from Dawson *et al.*, 1998):

ARQ and ARR	Berrichon du Cher, Cotswold, Hampshire Down, Romney, Soay, Suffolk, Vendeen
ARQ, ARR and VRQ	Bleu du Maine, Border Leicester, Charollais, Lincoln Longwool, Poll Dorset, Rouge de l'Ouest, Wensleydale
ARQ, ARR and AHQ	Bluefaced Leicester
ARQ, ARR, AHQ and VRQ	Beulah Speckled Face, Cheviot, Clun Forest, Dalesbred, Devon Closewool, Herdwick, Scottish Blackface, Shetland, Swaledale, Welsh Mountain
ARQ, ARR, AHQ, VRQ and ARH	Texel, Lleyn

Transmissible spongiform encephalopathies are of obvious veterinary and public health interest. In the UK, the government is committed to reducing scrapie (motivated by fear of the consequences of BSE getting into the national sheep flock; Butler, 2001) and this is proceeding by genotyping of individual rams, rather than by purging breeds of the susceptible alleles. The overall aim of the National Scrapie Plan, launched in July 2001, is to increase the frequency of the ARR allele and to reduce that of VRQ which is the allele most associated with susceptibility. Ferguson *et al.* (2002) estimate that the infection prevalence of scrapie in the UK is between 0.1 and 1% and that 'roughly one third' of sheep have a genotype defined as BSE-susceptible. The least susceptible sheep, ARR/ARR, can still be infected by BSE if inoculated intracerebrally, i.e. the ARR prion can be converted to a pathological form *in vivo*, adding extra complexity to hypothetical mechanisms of genetic resistance to TSEs (Houston *et al.*, 2003). Implementation of the National Scrapie Plan could have serious consequences for rare sheep breeds by reducing the number of rams available for breeding. In July 2003 the RBST negotiated an option for rare breed owners to delay this implementation, and obtained funding for research on the impact of the Plan on rare sheep breeds (Townsend, 2003a).

Breed differences in susceptibility to cancers are well documented in laboratory rodents (Festing, 1997) – for example diethylstilbestrol reduces number of tumours in outbred Sprague-Dawley rats but caused tumours in 70% of inbred ACI rats. Considering livestock, nearly 3 000 Hereford cattle were examined in north America; 27% had lesions indicative of eye cancer. Pigmentation in the corneoscleral area of the eye was found to be heritable and to have a protective effect (Anderson, 1991).

Conclusions

Livestock breeds show clear physiological adaptations to different physical environments, which underlie their roles in traditional production systems. Extremes of temperature, humidity, and different vegetation environments, often reveal these adaptations. While body size is an important factor, there are very real differences between breeds.

Although there are differences between breeds in resistance to parasites and diseases, the reasons for these are not generally understood partly due to methodological difficulties and the low priority attached to such studies. Genetic diversity *per se* is often important, and generally within-breed variation is better understood. Indeed this is usually the level at which the adaptation is exploited. Breed differences are well documented in relation to susceptibility to trypanosomosis, scrapie and ticks, though in all of these the mechanisms of resistance or tolerance require further study.

Chapter 4
Crossbreeding and coadaptation

Introduction

There are two ways of exploiting genetic variation to increase performance or fitness in livestock – selection and crossbreeding. Crossbreeding is much quicker, for many traits an improvement in performance is virtually guaranteed for the first crossbred generation through heterosis. If uncontrolled, crossbreeding leads to loss of breed identity. How can it be managed sustainably, so heterosis can continue to be exploited? Evidence is accumulating that genes at different loci can interact in relatively complicated ways to determine phenotype. Crossbreeding could disrupt these interactions. How does this relate to the sustainable development and conservation of breeds?

The value of crossbreeding

The advantages of crossing different breeds to restore fitness and viability or to produce animals that combine the advantages of both breeds were well known to Darwin (1868, vol. II, chapter 17). Systems of crossbreeding to produce superior cattle have been operated by practical farmers for many years (Figure 4.1; Hall & Clutton-Brock, 1988; Simm, 1998). In these systems farmers keep pure breeds and sell crossbred progeny either to other farmers for further crossbreeding or for slaughter. This, the first kind of crossbreeding, is known as regular crossing and it has the aim of producing a particular type of progeny by mating specific genotypes. In naming a crossbred, conventionally the sire is given first, e.g. a Hereford-Friesian or Hereford x Friesian is the progeny of a Hereford bull and a Friesian cow.

Crossbreds often form the bulk of the livestock population. For example, in Scotland there are 1.9 million breeding cows. Of these 0.34 million (18%) are purebred dairy cows, 0.27 million (14%) are purebred beef cows and the remainder are crossbred beef cows (Wright *et al.*, 2002). The second general version of crossbreeding is the development of selected inbred lines and their methodical crossing to produce breeding stock for sale to farmers. This began in the 1930s, in the USA, when poultry and pig breeders began to apply the approach pioneered by maize breeders. The third kind of crossbreeding is the mating of breeds to produce a new synthetic breed. A milder form of this is the bringing in of genes from one breed into

Figure 4.1 The Newbus Ox, engraved in 1812 by William Ward after a painting by Thomas Weaver, was sired by a Shorthorn bull out of a Highland cow. (© Christie's Images Ltd., 1978.)

another, termed introgression. Finally, transgenesis is the process of bringing in genes, usually from another species, by direct, artificial transfer of genetic material.

Essentially, crossbreeding can be used to improve efficiency through the use of specialized sire or dam breeds or to combine in one animal the merits of several lines or breeds (Simm, 1998). Historians have always been more interested in pure breeds than in crossbreds, probably because the former are much better documented, and the contributions of individual breeders are so much clearer, and a general account of the development of livestock crossbreeding has yet to be written.

History of heterosis

Also known as hybrid vigour or heterozygote advantage, heterosis is the phenomenon of the performance or fitness of offspring being greater than the average performance of the two parents (Nicholas, 1987, Chapter 18). That the different merits of different breeds can be combined in an additive way by crossing them has certainly been known for a long time. In heterosis, the combination of the parental genes is not simply additive. The notion that heterosis, as the converse of inbreeding depression, can be of benefit to animal production is almost certainly recent. Perhaps enthusiasm for pure pedigree breeding, and for upgrading, delayed a full appreciation of heterosis.

In the 1930s the general opinion was that under relatively mild inbreeding, within-breed selection could overcome the depressing effect of inbreeding and that, if anything, slight inbreeding would be beneficial and tend to fix genes with desirable dominant effects in each generation (Cundiff, 1977). However, experimental evidence of between-breed heterosis began to accumulate. In the USA, national policy for beef cattle breeding from 1908 onwards was to upgrade to purebreds, with the 'British breeds' (mainly Shorthorn, Hereford and Aberdeen Angus). While this was happening much of the improved performance must have been due to heterosis but was instead credited to the breed that was being used for upgrading.

In the southern USA, Brahman cattle began to be used widely and experiments from the 1950s showed the importance of heterosis when these were crossed with British breed cattle. The major benefits were improved fertility, longevity and lifetime production of calves (Cundiff, 1977).

The 'purebred epoch' of American beef cattle was brought to an end, in the 1960s and 1970s, by the spread of dwarf genes as an accidental result of selection for compact animals (Willham, 1982). In the 1970s, following British trials on the Charolais which began in 1961 (Lerner & Donald, 1966), large scale experiments on beef cattle were started in the USA, Australia, New Zealand, France and Canada to study the potential of continental European breeds for siring good beef crossbreds (Morris *et al.*, 1993; Roughsedge *et al.*, 2001). This was the background to the appreciation of the importance of heterosis in practical breeding.

The amount of heterosis in a given breed cross often depends on the environment and, generally, knowledge of the phenomenon is empirical rather than theory-based (Barlow, 1981). Heterosis tends to be greater in life history traits (such as survival and fertility) than in production traits (such as meat milk or egg yield) (Clarke, 1995; Hartmann, 1989; Emmerson *et al.*, 1991). When parasite resistance has been studied in crossbreeding trials most studies have shown the F1 generation to have greater resistance than the more susceptible parental stock, though it is not always so (Baker *et al.*, 1999). Frisch and O'Neill (1998b), studying F1 crosses of African, European and Indian cattle, found consistent heterosis for tick resistance, but not for worm egg counts.

Heterosis has also been expected to be greater when breeds are genetically distant (Falconer & Mackay, 1996). Genetic distance can predict the increment in heterozygosity of a crossbred population, over the mean heterozygosity of the parent populations, and for practical purposes heterosis shows some proportionality with this increment (Roughsedge *et al.*, 2001), though not universally (Cepica *et al.*, 1995). Breeds do not have to be very distantly related for heterosis to exist. Ageeb (1991) described how there was heterosis for milk yield (3.3%) and for weaning weight (10.1%) when local Sudanese cattle breeds (the Kenana, Butana and Western) were crossed. When poultry breeds are crossed, heterosis may be observed for production traits but not consistently, and generally only for specific crosses (Emmerson *et al.*, 1991).

Sometimes, heterosis is only apparent when the system is considered as a whole. Frisch (1987) describes how in zebu–taurine crossbreds, the zebu parent is superior for environmental resilience while the taurine parent has better growth characteristics. For both attributes, the offspring is no better than the midparent value, but the performance of the crossbred as a whole is better than the midparent value. This can be a reflection of crossbreds performing better than the midparental value in some aspects and not in others; in Peru when local Criollo ewes were crossed with rams of various imported and improved breeds, or with Criollo rams, the purebred Criollo lambs had a poorer growth rate but survived better. Consequently the crossbred lambs performed no better overall (Burfening & Carpio, 1995).

A number of genetic interpretations of heterosis are possible. 'Overdominance' is the best known – the heterozygote out-performs both homozygotes. In other words the two alleles do not interact additively (if they did, performance of the heterozygote would be the arithmetic mean of the parental homozygotes). Also suggested (Falconer & Mackay, 1996, Chapter 2) are pleiotropy (a gene may have multiple effects, some increasing fitness and some reducing it, so the best gene combination is the heterozygote) and linkage disequilibrium (loci may be closely linked so a favourable allele at one segregates with an unfavourable allele at another, so again the heterozygote is favoured). 'Partial dominance' is another possible cause – performance of the heterozygote is higher than the mean performance of the parents; this could happen when it is advantageous for an organism to possess a mixture of gene products.

Essentially, '...selection theory is based on the accumulation of favourable additive effects from generation to generation...crossbreeding theory is largely (though not exclusively) based on the arrangement of favourable dominance effects in each generation.' (Cunningham & Syrstad, 1987). The relationships between parental, F1 and F2 performances that might be predicted, are illustrated in Figure 4.2.

Crossbreeding to a regular system

If heterosis is to be capitalized upon, pure breeds must be maintained and a systematic crossbreeding programme operated. This can be expensive, so it is often the case that production of a synthetic breed is more economical.

There are very many crossbreeding systems. One that is widespread is based on the females of the F1 generation being mated with a third breed. Being heterozygous at a number of loci these females would be expected to show heterosis for mothering ability (Simm, 1998, Chapter 3). Their offspring would benefit from this and would also exhibit their own individual heterosis because they too have heterozygosity. For the system to function, a regular supply of the crossbred F1 females must be maintained, which provides an incentive for keeping pure-breeding populations of the two parental breeds.

Heterosis can be maintained with more elaborate crossbreeding systems such as three-way crosses and rotational crossing (Wiener, 1994; Simm, 1998). The design of the system depends on the structure of the livestock industry. In British sheep farming the stratified crossbreeding system (Figures 4.3a,b) exploits maternal and individual heterosis (Simm, 1998, p.92).

The British stratified sheep crossbreeding system is so frequently given as a standard example that it is important to remember that it is not necessarily fixed

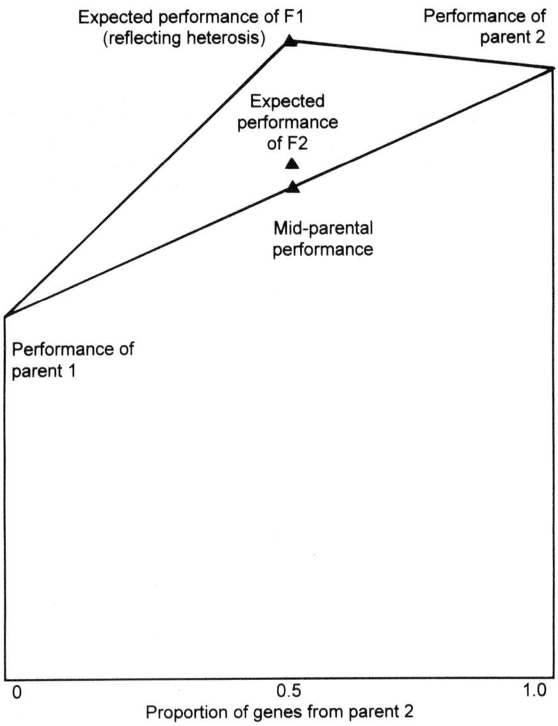

Figure 4.2 The 'Greek Temple' model popularized by Professor Paddy Cunningham illustrates the expected performance of the progeny (F1 generation), of the matings within the progeny (F2), and of the parental (P1 and P2) generations. This expectation is on the assumption that gene effects are additive, and that the F1 shows heterosis.

forever. Numbers of ewes of terminal sire breeds are so much greater than those of crossing sire breeds that it must be suspected that many hill sheep farmers are using terminal sires like the Suffolk and Texel directly on their ewes, rather than using crossing sires.

In UK sheep farming low fecundity is the major constraint so heterosis for female fertility is a major benefit. Often, in dairying, the heterosis in milk production that comes from crossbreeding is not enough to provide greater profit than the best purebred herds. In much of the world dairying is based on the Holstein-Friesian, which has a higher milk yield (measured by volume) than any other breed. When, as under the current system the yield of liquid milk is at a premium, crossbreeding is not viable as the milk yield of crossbreeds is less than that of the purebred Holstein-Friesian. In New Zealand, however, the national herd of 3.1 million cows is 57% Holstein-Friesian, 16% Jersey, 18% crossbred Holstein-Friesian x Jersey, 2% Ayrshire and 7% other breeds. A modelling study (Lopez-Villalobos *et al.*, 2000a,b,c) shows how, with slight changes in relative profitabilities of meat and milk and in economic weighting of fertility, crossbreeding systems of various designs could be profitable. Each breed has particular features of value – high milk yield and large body size (Holstein-Friesian), high butterfat and milk protein

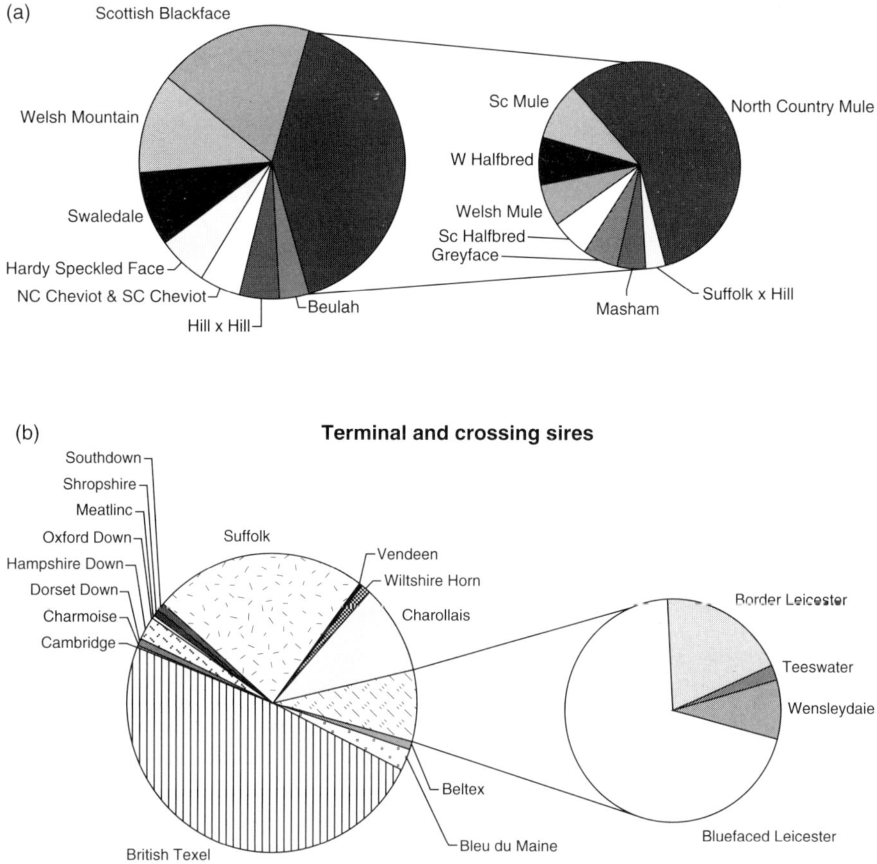

Figure 4.3 Stratified system. The UK breeding ewe population is 16.43 million ewes (Meat and Livestock Commission, 21 March 2003). About half of these are hill purebreds and half (male group) are lowland crossbreds (Simm, 1998). The proportions of the major genotypes are depicted (a). Typically a hill ewe is kept in her flock of birth for her first four matings, which are to rams of her own breed. Then she is mated with a 'crossing sire' (b) (Border Leicester group) to produce crossbreds. Male lambs are slaughtered; female progeny are usually kept by lowland farmers and mated with 'terminal sires' to produce meat lambs.

content (Jersey) while the availability of other breeds enables more efficient crossbreeding systems to be devised. However, for such systematic crossbreeding to operate the dairy industry would need to become organized on a different basis from that which applies at the moment. Most New Zealand dairy farmers have mixed herds and tend to use semen from the best bulls regardless of breed, so there is clearly crossbreeding in progress. Certainly, the retention of viable pure breeding units of the minority dairy breeds has given New Zealand farmers, who are responsible for 25% of the international trade in dairy products, flexibility to respond rapidly to changing conditions. These minority breeds are clearly valuable genetic resources.

Production of synthetic breeds

As a regular crossbreeding system is expensive to maintain, an attractive option may be to develop a synthetic breed. The starting point is selection from within an F1 progeny group (Cunningham & Syrstad, 1987). Genetic considerations and practical aspects combine to indicate what proportion of the genes of a new synthetic breed should come from each founder breed.

Introgression, upgrading and the formation of synthetic breeds are similar. They involve an introduced breed being mated with a resident breed. Introgression relates to crossbreeding conducted with backcrosses to the resident breed, with a view to a small proportion of the genome of the introduced breed being incorporated into the resident breed which otherwise remains much the same as before. Upgrading relates to backcrosses being made repeatedly to the introduced breed, so the resident breed becomes progressively more and more like the introduced breed. A synthetic breed is formed when the introduced and the resident breed are crossed, and selection is then applied to the crossbred progeny.

The Gascon pig was used to create a synthetic breed for use in Haiti (Molénat *et al.*, 1993). From 1978 the entire pig population of Haiti was exterminated in order to eradicate African swine fever. Initially, replacements were of American breeds but these were not successful and subsequently under a French aid programme (MacKenzie, 1993), pigs of a new synthetic line were provided. This line was genotypically 25% Gascon, 25% Chinese and 50% Creole (from Guadeloupe). Between 1987 and 1992 around 20 000 animals were distributed. Zepeda Sein (1989) stated that the project was too expensive and had to be abandoned, but in 1993 it was estimated that 200–250 000 pigs carried some genes from this new breed.

Grading-up to an exotic breed can mean the loss of genes for local adaptation. In the tropics many attempts have been made to create adapted, productive dairy cattle by crossing temperate zone breeds with local cattle (mainly zebus) and then selecting from the crossbred population to develop a new synthetic breed. Rutledge (2001) analysed 80 such attempts, the enhanced performance of the F1 generation is not maintained in the progeny of matings within that generation. It was suggested that zebu and taurine cattle each had tightly coadapted genomes which had diverged during their separate development, and the gap is too great to be bridged. In contrast, in spite of their separate domestication histories (see Chapter 1), Asian and European pig breeds have been successfully combined into synthetic breeds. In 1992 the Manor Meishan hybrid pig was introduced (Porter, 1993, p.244). It is ¼ Meishan, ¼ Large White and ½ Landrace and it is intended to combine the high prolificacy (30 piglets per year) of the Chinese breed with the growth and meat qualities of the European breeds.

Many new synthetic breeds of sheep have been formed, with an eye to specific markets, especially since about 1950 (Maijala & Terrill, 1991). The Cambridge sheep is an example (see Chapter 2).

The beef industry of northern Australia gives good examples of the use of cross-breeding in exploiting resources, in this case 156 million hectares of tropical pasture

(Vercoe, 1996). It began, using British breeds, in the mid nineteenth century but when the cattle tick *Boophilus microplus* was introduced accidentally from Java in 1872 the diseases it carried almost wiped out the herds (Frisch, 1990). Cattle rearing became totally dependent on acaricidal dips and as the British breeds were poorly adapted to the environment in other ways as well, experimental crosses were made with zebu breeds. In the 1950s zebu (Brahman) cattle were imported from north America. Although their growth rate and reproduction under benign conditions were 20% lower than those of the British breeds their resistance to ticks, other parasites, heat stress, eye diseases and poor nutrition meant they grew faster in harsh conditions. By 1986 it was clear that systematic crossbreeding of British breeds and Brahman would yield an F1 generation that was both productive and hardy, and to maintain this advantage a third breed was needed in the crossing scheme.

On the basis of published data the Boran and Tuli of east Africa were chosen for further investigation. The former is considered a pure zebu breed and the latter a sanga breed. With elaborate biosecurity precautions (Frisch, 1990; Frisch *et al.*, 1997) the two breeds (43 Boran from Zambia, and 39 Tuli from Zimbabwe) were introduced to Australia in 1990. Other approaches have been used including the development of a tick-resistant taurine breed, the Belmont Adapteur (Frisch *et al.*, 2000), by selection from the current breeds.

Many of today's African cattle breeds seem to have arisen as the result of introgression. Hanotte *et al.* (2002) expressed microsatellite variation as principal components which they superimposed on a map of Africa. The most important PC correlates with genes of zebu provenance and implies introduction (mainly after AD 700; Hanotte *et al.*, 2000) via the Horn of Africa. Humpless cattle were already present. The second PC, which relates to taurine affinity, shows a geographical distribution which is consistent with an earlier spread of taurine cattle from an apparent centre of radiation in the eastern Sahara. Present day breeds in Africa show less zebu introgression in tsetse fly areas (Figure 4.4) as would be predicted from the susceptibility to trypanosomosis that accompanies zebu affinities (see Chapter 3). The zebu introduction must have been by bulls as no zebu mtDNA haplotypes have been found in Africa (Hanotte *et al.*, 2000). Zebu microsatellites have also entered Near Eastern breeds (Loftus *et al.*, 1999) and even the Hungarian Grey shows some zebu admixture. One of the world's major beef industries, in Brazil, is based on zebu cattle, mainly of the Nelore breed (Ferraz & Eler, 2002). From about 1823 until 1962, zebu cattle (totalling about 6300) were imported from India and today about 128 million cattle in Brazil have zebu affinities. The pre-existing taurine Criollo breeds are nevertheless appreciated as genetic resources and are extensively used for crossbreeding.

Breed development programmes

One of the most spectacular recent examples of breed development has been the use of north American Holstein semen to increase milk yields of European black-and-white dairy cattle (Friesians). This process started in the early 1970s because of pricing policies which made the beef producing abilities of black-and-white cattle less

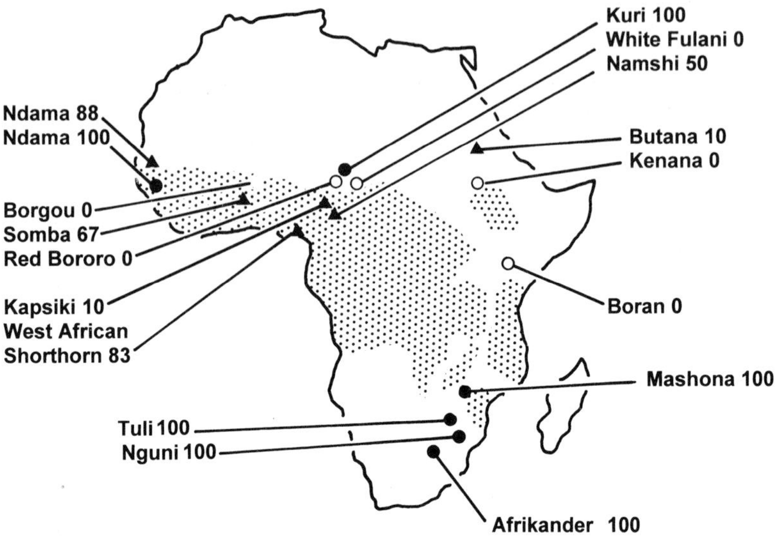

Figure 4.4 Zebu influence on African cattle breeds. Frequency of taurine and zebu Y chromosome alleles at the INRA locus in sub-Saharan cattle. For each breed the centre of distribution and proportion of taurine (grey) or zebu (black) allele is indicated by the symbol. The light grey shading indicates the distribution of tsetse flies. From Hanotte *et al.* (2000).

important, and which put a premium on milk-producing qualities (about 20% higher in the Holstein) and better udder conformation (Cunningham, 1983). Though Friesians and Holsteins still show genetic distinctiveness (Hanslik *et al.*, 2000) this is being eroded and most European populations of black-and-white dairy cattle will soon have been upgraded to the north American Holstein. While direct comparisons of reproductive performance of Holstein and Friesian genotypes have not been made, there is evidence linking Holstein introgression with measures of reduced fertility (Hoekstra *et al.*, 1994; Royal *et al.*, 2000), notably the interval from calving to conception (Lidauer & Mäntysaari, 1996). But it must be remembered that genetic change is not the only possible cause of reduced fertility in dairy cattle – increased stress through new management approaches may also be implicated (López-Gatius, 2003).

Holstein semen is widely exported, so genetic progress in this instance is benefiting from continuing evaluations and genetic improvement in north America. This kind of approach is often called a 'breed improvement programme' and is usually regulated by the breed society, which will attach importance to maintaining the traditional breed type as well as incorporating desirable new genes. For example, in some breeds animals produced in the course of such a scheme can only be registered subject to certain conditions. The Hereford breed of beef cattle is one of several that have undergone this process (Blott *et al.*, 1998) and in certain segments of the breed Canadian Herefords are strongly represented. In Britain the 'Traditional Hereford', which is free of this introgression, is listed by the RBST.

British Southdown sheep have received genes from French and New Zealand animals and the breeding careers of the ten rams imported between 1966 and 1980 were analysed by Hall (1989b). Although the different populations were different in certain respects (otherwise there would have been no point in the matings, which were done to match the conformation of the Southdown more closely to modern requirements) the progeny were 'within the standards'. Many sons of these rams were registered and there was quite strong selection among them (Figure 4.5). Only a few of these sons sired registered progeny themselves.

It was the same in Thoroughbred horses, where the famous stallions (see Chapter 7) who founded paternal lineages passed on this influence to future generations through only very small numbers of male descendants (Cunningham *et al.*, 2001). The Byerley Turk's influence was all through his great grandson Herod (foaled 1758), the Darley Arabian's, through Eclipse (1764) and the Godolphin Arabian's, through Matchem (1748).

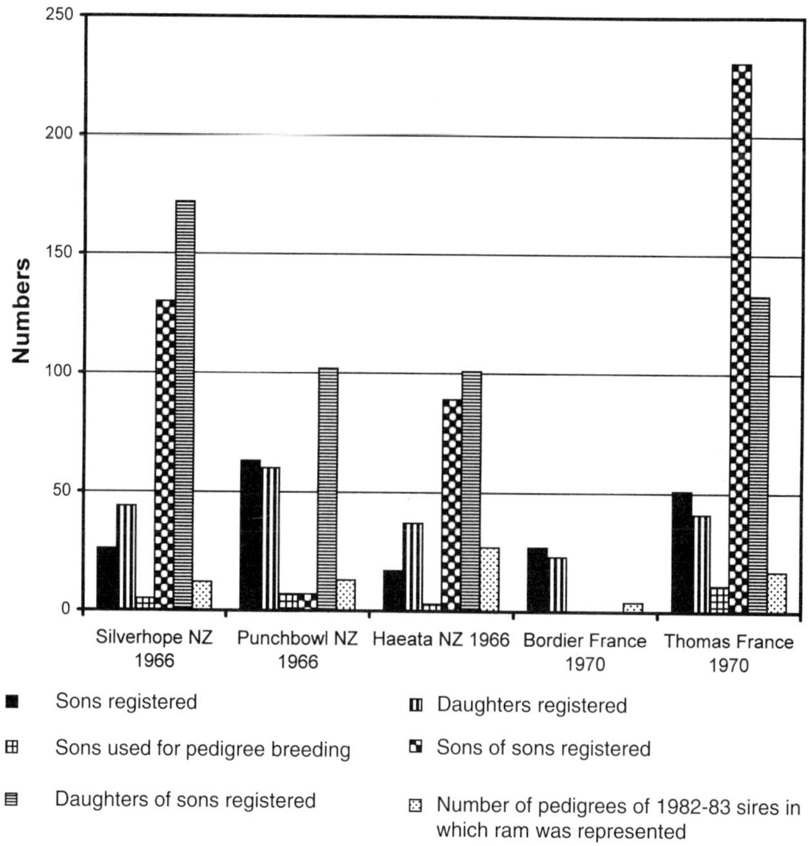

Figure 4.5 Breeding careers of Southdown rams imported into the UK. Between 1966 and 1970 seven Southdown rams were imported from New Zealand and France. (a) Most rams sired many lambs which were then registered. (b) Only a very few of these sons were used for pedigree breeding. (c) Their sons (i.e. the grandsons of the original imports) were registered in large numbers in the case of a few rams. (d) Finally, of the 74 rams which sired 1982 and 1983 lambs, many (83%) showed descent from one or other of the imports.

The different ways in which genes have been brought in from other breeds are well illustrated by pedigree analysis of the eight main French dairy breeds since about 1970 (Boichard *et al.*, 1996). The French Friesian, first imported from the Netherlands in the eighteenth century, is by far the dominant breed with a national distribution and 2.3 million females born between 1988 and 1991. Importations from abroad, especially the USA, meant that by 1992 the breed was genetically 90% Holstein (78% American Holstein). The second most numerous, the Normande (also a national breed with 358 900 females born 1988–91) and the least numerous, the Tarentaise (a local breed with 5817 females born 1988–91) have remained closed. The French Brown (local, 13 856 females born 1988–91) is now effectively the same as the American Brown Swiss while the French Simmental (local, 13 947 females born 1988–91) has had a clear policy of avoiding 'holsteinization' and is now roughly equally of French Simmental, Swiss Simmental and German Simmental affinity. The Abondance and Montbéliard (local, 12 345 and national, 331 600 females born 1988–91, respectively) both owe about 6% of their genes to the American Red Holstein, a proportion which is likely to remain stable. The French Red Pied Lowland has traditionally been a relatively open breed and now includes a high proportion of red Holstein and Meuse-Rhine-Yssel genes, and finally the Salers, Aubrac and Maine-Anjou have ceased to function as dairy breeds (Figure 4.6).

In the tropics, Holstein-Friesian cattle produce an average of 3500 kg of milk per lactation (milk-recorded herds in Kenya and Barbados) with some individuals managing 5000–6000 kg. Tropical zebu dairy breeds produce much less (Sahiwal in Pakistan 1500–2000 kg; Kenana in Sudan 1500 kg) while Holstein-Friesian x Sahiwal crosses produce 2500–3000 kg (Philipsson, 2000). The appeal of using the Holstein-Friesian for crossbreeding or upgrading to increase milk yields is obvious. In many native African cattle breeds, for example the Mashona of

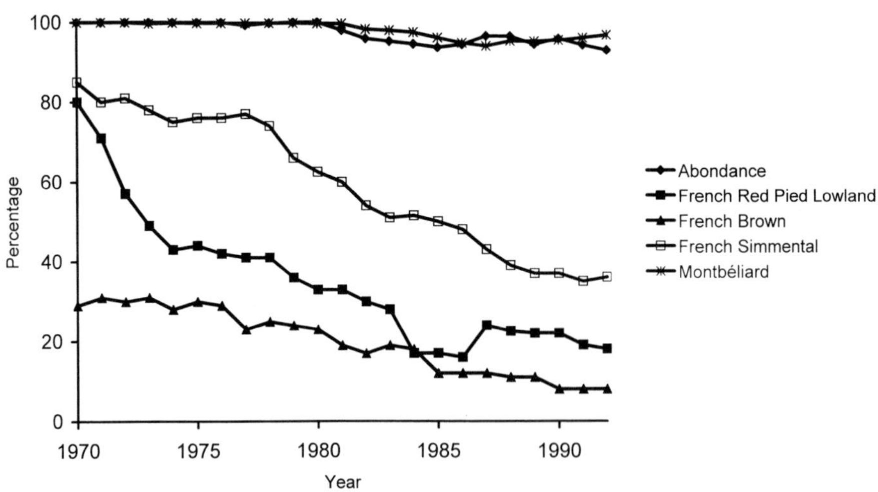

Figure 4.6 Introgression into French cattle breeds. Introduction of genes from outside into French cattle has shown considerable differences between breeds (redrawn from Boichard *et al.*, 1996).

Zimbabwe (Perry *et al.*, 1987), milk yields are very low (in this case 1 litre per day, which is probably a result of nutrition and management), the teats are small (only 3–4 cm long) and the milk ejection reflex is poor. Selection for dairy performance would be very difficult in this breed and crossbreeding would be the only option.

Molecular indicators of introgression

Breed development programmes are thoroughly documented when detailed pedigree records are kept, and the influence of particular introductions can be traced. When there are no records, molecular methods can provide insights into introgression. There have been some surprising results. Polziehn *et al.* (1995) found bovine mtDNA in north American bison from Custer State Park, South Dakota. These animals were descended from a group caught in the wild in 1888 which, like most of the bison that survived the near-complete extermination of the species in north America in the late 1800s, were subsequently kept on cattle ranches that also crossed bison with cattle. Female progeny of a bison bull mated with a domestic cow are usually fertile. Although it had been believed that all hybrids had subsequently been removed, this was evidently not so.

There also seems to have been introgression of genes (possibly including the halothane gene, which is implicated in a stress response; Nicholas, 1987) from domestic pigs into European wild boar (Giuffra *et al.*, 2000; Lundstrom, 1995). The highly endangered wild water buffalo has also been hybridized, to an extent, with domesticated and feral water buffalo (Flamand *et al.*, 2003). Introgression of genes from farmed crustaceans and fish, especially salmon, into wild stocks is a major conservation issue (Levin & Williams, 2003).

Breed history can be illuminated by molecular study of introgression. MacHugh *et al.* (1998) attributed the high allelic diversity of European populations of Charolais and Holstein-Friesian cattle to rapid increase of their numbers through grading-up of other breeds. The allele for black colour in Chinese Meishan and British Large Black pigs differs by four nucleotide substitutions from the consensus of alleles at this locus in European wild and domesticated pigs. Thus implying introgression of this allele from Chinese pigs into the Large Black (Giuffra *et al.*, 2000). If alleles are detectable in a population that are characteristic of another population, then introgression might be implied and this means authentic populations of breeds of interest can be identified. In west Africa genes from zebu bulls from the drier, trypanosome-free areas are often mated with the native breeds of the forests and wooded savanna, increasing body size but reducing trypanotolerance (see Chapter 3). Purebred populations of these native breeds must be identified if genes for trypanotolerance are to be utilized (Camus *et al.*, 1981). This cannot be done, reliably, by inspection alone. Bradley *et al.* (1994) and Achukwi *et al.* (1997) showed how a Y-chromosome specific probe can be used and MacHugh *et al.* (1997) extended this work using zebu-specific microsatellite alleles.

Introgression between breeds will usually be accompanied by selection for some phenotype and clearly the breeds will converge in that respect, but so far as variation that is not under selection is concerned, the breeds will retain their distinctiveness so long as the number of migrants is low. Many breed societies acknowledge this by providing grading registers whereby migration into the breed from outside is controlled.

Sometimes a desirable gene may be detected in an individual animal or in a relatively minor breed, and it is proposed to introgress the gene into a commercial ('recipient') breed. Some constraints might be envisaged (Hillel *et al.*, 1993). The process must be rapid and only a minimal amount of the genome of the donor should be introgressed. Apart from the desirable gene the new stock should be homozygous for the desirable gene and should not be inbred – time and expenses must be minimal. Theoretical models, devised for poultry, show that even if the donor is heterozygous, the process can be completed within two to four years. It depends on there being a reliable marker for the desirable gene.

The polled gene in Canadian Charolais cattle is associated with microsatellite markers (Schmutz *et al.*, 1995). Polled is dominant to horned and it is obviously useful to be able to distinguish homozygous from heterozygous polled animals. However, for traits with a more complex genetic background the process takes much longer and is very sensitive to the number of genes involved in the introgression and to female reproductive capacity. Indeed, for quantitative traits it could be cheaper to set up a breeding scheme to get the donor population to the same standard as the recipient population (van der Waaij & van Arendonk, 2000).

Biotechnology

Cunningham (1999) reviews the potential and actual uses, in livestock, of new biotechnologies. These are reproductive and genetic technologies, recombinant hormones, and the modification of the immune system to remove constraints on production. Transgenic technology involves alien genes being inserted into the genome of recipient animals to enable the latter to produce novel proteins – a hormone, whose action leads to a change in control of growth or reproductive processes, or an end-product. For example Devlin (1994) microinjected a gene construct of sockeye salmon (*Oncorhynchus nerka*) origin into eggs of coho salmon (*O. kisutch*). The transgenic salmon were on average 11 times heavier than non-transgenic controls and their production of growth hormone was year-round rather than concentrated in the spring. The alpha-1-antitrypsin gene was introduced into a sheep around 1988, to provide (in the sheep's milk) a cheap supply of this enzyme. By 1998, there were 800 grand-daughters descended from this sheep (Moffat, 1998), and in 2003, 3000 (Ward, 2003) producing variable amounts of the enzyme, which could be used to treat cystic fibrosis if a commercial-scale plant to purify the protein could be financed (Hogan, 2003). It is now possible to make specific modifications to the pig genome which will help in the development of xenotransplantation (Prather *et al.*, 2003).

Non-therapeutic compounds (known as nutraceuticals) can also be produced in this way. For example, the human lactoferrin gene has been inserted into cattle leading to production of milk claimed to be especially good for babies. It has extra lactoferrin, which is reported to increase iron absorption and act as an intestinal bacteriostat. Other human proteins can be secreted in cow's or ewe's milk including calcitonin and lysozyme. As these are not therapeutic products they are not subject to clinical trials. The global market for nutraceuticals is predicted to be $500 billion by 2010 (Pintado & Gutierrez-Adan, 1999).

Transgenesis is expensive (Wall, 1997), with each transgenic animal currently costing $25 000 (pig), $60 000 (sheep) or $500 000 (cow). Chickens are attractive candidates for transgenesis, partly because there is already effective technology for separating and purifying egg yolk proteins, but much further research and development is required (Zajchowski & Etches, 2000). World requirements for pharmaceutical products would be satisfied by only a small number of transgenic animals so the benefit to agriculture as a whole would be limited. It is a different matter with nutraceuticals or with qualitative improvements to food end-products, where the quantity of product is much greater even though the value is lower.

These issues are of particular relevance to dairying where there are concerns over possible dietary health risks (Maijala, 2000). Quantitative changes in milk composition that could be achieved by transgenics relate to protein content, fat content and lactose concentration (Pintado & Gutierrez-Adan, 1999). Considering proteins, an increase in kappa-casein would be desirable because this protein traps fat and water during cheese production thus increasing cheese yield. As the mammary gland has only a finite ability to synthesize protein, kappa-casein production might be increased if production of other less desirable proteins is suppressed. Milk content of beta-lactoglobulin, which is the main allergen in milk so far as humans are concerned, could be reduced, perhaps by a transgenic procedure (Boland *et al.*, 2001). This protein has no known function and is absent from human milk, but is ubiquitous in cow's milk.

Reducing fat content of milk is difficult in conventional breeding because it is correlated with protein content. It is tempting to suggest that fat might simply be removed during processing but there are also transgenic approaches such as decreasing acetyl coA or stearoyl coA activity (reducing overall fat synthesis, and presence of saturated fatty acids, respectively). Finally, lactose levels could be reduced by ribozymes inactivating synthesis of alpha-lactalbumin (one of the proteins in the lactose synthetase complex) or by inducing lactase to be expressed in the mammary gland (Pintado & Gutierrez-Adan, 1999).

In food production, consumer acceptance of biotechnology is a prerequisite (http://europa.eu.int/comm/dg12/biot.html). Conventional breeding, as well as competing economically with biotechnological approaches (Van Vleck, 1999), may therefore be favoured by the market.

Genetic architectures

In livestock breeding, production traits like milk yield and growth rates have usually been selected on the basis of a genetic concept of many genes having small, additive effects. Studies of inbreeding and crossbreeding have given support to genetic concepts that attach importance to interactions among genes. These two genetic architectures – the additive-dominance model and the coadaptation model – trace back to R.A. Fisher and Sewall Wright respectively. In the 1920s and 1930s Fisher argued that natural selection acted primarily on single genes, to favour alleles that increased fitness across all genetic backgrounds. In contrast Wright proposed that it could act to retain favourably interacting gene complexes.

As pointed out by Cunningham and Syrstad (1987), selection for quantitative traits within breeds is based on a simple additive model of gene action and response to selection is usually in accordance with prediction. This is not so with crossbreeding, where heterosis depends on the gene frequencies in the different breeds, and on dominance interactions, and often – or perhaps usually – effects arising from interactions between genes at different loci (epistatic interactions).

If coadaptation is important, genetic conservation becomes a question of conserving favourably interacting gene complexes as well as of conserving alleles and, unless pure breeds are protected, the coadaptation will be lost through recombination. At the same time, it is very clear that when breeds are crossed there is often heterosis. A major challenge is how to take advantage of this, while not losing coadaptation, and crossbreeding systems need to be designed for this.

Interactions between single loci and the genetic background

The original meaning of the term epistasis was the phenomenon of gene action at one locus affecting gene expression at another. Evolution of maize, with its very strong apical dominance, from teosinte (the much more bush-like wild progenitor), depends on five major genes which interact epistatically (Doebley *et al.*, 1995; Goodnight, 2000b). The growth hormone construct, mentioned above as inserted into the genome of salmonids, had a much greater effect in wild genotypes than in ones which had already undergone intensive selection for growth rate (Devlin *et al.*, 2001).

A chicken line that was segregating for B blood groups was crossed with various distinct homozygous chicken lines and the progeny exposed to Marek's disease (to which MHC alleles, linked to B blood groups, confer resistance). In one line, birds with the B21 blood group were resistant, but in another line they were not (Rothschild *et al.*, 2000). This implies the resistance genes interact with the genetic background in this case, but there are other chicken diseases where the genetics of resistance do not involve such interactions (Karaca *et al.*, 1999).

There have also been experimental demonstrations of genes with undetectable main effects, which interact epistatically to produce a clear phenotypic effect. A donor strain of mouse was used to breed strains for subsequent crossing, each of which carried different combinations of the donor strain genome. In some instances, while neither of the two parental strains was susceptible to cancer, the progeny were, indicating that the susceptibility arose from epistatic interaction among sets of genes (Frankel & Schork, 1996).

Interactions inferred from variance of quantitative traits

Quantitative genetics as applied to animal breeding has not needed to invoke epistasis in order to develop theoretical explanations and predictions (Falconer & Mackay, 1996). Coadaptation has not been seen as relevant in selection for production characteristics, and its possible role in other fitness traits in livestock has not been extensively investigated. While productive traits tend to conform to the additive model, there is growing evidence (Barker, 2000) that traits relating to fitness including environmental adaptation can be better explained by models which include non-additive (dominance and epistasis) effects.

The existence of epistasis can be inferred when the performance of progeny is different from that predicted by the additive-dominance model. One of the most obvious instances where this model does not work is in the crossing of zebu and taurine cattle with a view to producing tropically adapted dairy breeds. The F1 generation shows heterosis, but the F2 and subsequent generations do not, possibly because meiosis in the F1 generation breaks up coadapted gene complexes (Sheridan, 1981; Rutledge, 2001). This has also been observed in wild populations where it has been inferred as a cause of the outbreeding depression, which is sometimes observed when genetically distinct populations are mated (Frankham *et al.*, 2002, Chapter 15). Mice in a hybrid zone inhabited by two distinct species *Mus musculus* and *M. domesticus*, had far higher worm burdens than those outside the zone (Sage *et al.*, 1986, 1993). F1 crosses between these species showed heterosis (Moulia *et al.*, 1995) suggesting that in the hybrid zone it was recombination subsequent to the F1 generation that had caused the loss of advantageous combinations of genes. However, there are no theoretical grounds for saying that heterosis in the F1 generation will automatically be followed by recombination losses in the F2 generation (Roff, 1997, p.336).

Under some circumstances, breaking up these gene complexes could lead to enhanced performance. Some of these genes may have regulatory functions which restrict the expression of other genes (Notter, 1999). The mutation in a quantitative trait allele of pigs, which increases rate of growth of lean muscle (see Chapters 1 and 7) is in a part of the gene that functions in regulation, rather than coding for an active protein (Van Laere *et al.*, 2003). Indeed, genetic improvement, by which is meant enhanced performance, may actually represent progressive loss of this regulatory control. The extension of the sheep breeding season (see Chapter 3) may be another example.

Turning to within-breed experimental studies, the additive-dominance model predicts the change in performance as the inbreeding coefficient increases. When loci combine additively and dominance is the only source of non-additive variation the change of mean on inbreeding should be directly proportional to the inbreeding coefficient F (Falconer & Mackay, 1996). Change of mean should therefore be a straight line when plotted against F. If the relationship is not linear, then there must be epistatic interactions, whose magnitude is indicated by the degree of deviation from linearity. Working on three breeds of sheep (and crosses among them) Wiener *et al.* (1992a,b) were able to remove the contribution of dominance variance revealing epistasis as influencing conception rate, litter size, litter weight at birth, live weight, certain body traits and some fleece characteristics. Arthur *et al.* (1999) found that inclusion of epistatic effects improved their prediction of performance of crossbred cows. Mackinnon *et al.* (1996), studying a dairy herd composed of a three-breed cross, found epistatic effects on milk production were implied but were not statistically significant. They considered recombination losses were most likely to be important where the breeds had each been highly selected for the trait of interest.

Roughsedge *et al.* (2001) and Wolf *et al.* (2000) also suggested there may be a relationship between genetic divergence of populations, and the degree of epistasis in crosses between them. The importance of epistasis has not actually been quantified in livestock, but R.L. Wu (2000) achieved this in tree populations. Epistasis accounted for 6–9% of the genetic variance within a given poplar species, and for 44% of that of the interspecific F1 generation.

Epistasis may be more important under some conditions than others (Blows & Hoffmann, 1996), for example when there is strong directional selection, a propensity for epistatic interactions driving the phenotype in that direction could be selected. Under stabilizing selection, epistasis of this kind might be selected against. These predictions, clearly relevant to livestock, have been tested in *Drosophila serrata* in relation to fitness traits in crosses among lines with and without a history of selection for resistance to desiccation. It was indeed found that epistasis became more important in determining fitness phenotypes as their expression deviated further from the mean. Perhaps stabilizing selection reduces the importance of epistasis, and directional selection increases it.

Consequences for conservation

Is a breed a repository of specific alleles of potential or actual value, or does its phenotype reflect valuable epistatic interactions? The answer to this question bears directly on how livestock biodiversity is conserved and utilized. If epistasis is important, then conservation must be based on maintaining the integrity of breeds, as the value of conserved genes would reside to a great extent in their harmonious interaction with other genes.

Discussing beef cattle, Gregory & Cundiff (1980) acknowledged that studies had not been done on the importance of epistasis. It had been thought it was probably unimportant for the kinds of traits considered in the post-1945 breeding programmes, mainly because selection goals had changed during recent history and coadaptation would not have had time to arise. In any statistical procedure it is more difficult to demonstrate an interaction than main effects. More needs to be known about coadaptation; further study may help to predict the likelihood of introgressions of major genes and QTLs being successful, and may help to explain the developmental problems found with transgenic animals. This is of obvious importance for the public acceptance of transgenesis. The question is very often asked, whether transgenesis is essentially the same process as traditional selective breeding and a better understanding of genomic coadaptation may enable this question to be answered in a scientific way.

The growing evidence of coadaptation being a likely, though hard-to-demonstrate phenomenon, confirms that sustainable livestock development strategies should start with locally adapted breeds, and select within them for improved performance (Barker, 2000), rather than assuming that an imported high-performance breed will acquire local adaptation.

Conclusions

The animal breeding theory and practice that have been so successful in increasing production, have achieved this by using genetic models which do not depend on interactions among genes. The main areas of interest have been selection for production traits, and exploitation of heterosis in the F1 generation. If fitness traits become of greater interest and the further development of breeds by introgression of genes from elsewhere become more important, epistatic effects are likely to attract more attention. More theoretical studies are needed, especially in relation to the statistical problems associated with demonstrating the interactions among main factors, and more experimentation, though this will be expensive.

Crossbreeding presents both threats and opportunities to livestock breeds. The most rational short term pattern of utilization of genetic resources is the production of F1 crosses, which often enables heterosis to be exploited. If such a system is run in a sustainable way it can protect the breeds by providing an incentive for continued production of purebreds, otherwise they will lose their identity.

Breed development programmes, introgression and biotechnology change the genotypes of breeds and can enable them to respond to current needs. These are possible routes for the utilization of livestock biodiversity; how they can be reconciled with genetic conservation is considered later.

Chapter 5
Livestock biodiversity and sustainable development

Introduction

Attitudes towards the role of livestock in sustainable development have been changing over the last 20 years. Their role in the developing world is, increasingly, being seen by development agencies as shifting from one of enhancing local or national food production, towards that of contributing to national and regional food security within a global economy. In the developed world, the importance of grazing livestock in maintaining an attractive and diverse landscape is becoming more fully appreciated. Low cost food production is mainly dependent on intensive systems about which many consumers have welfare, environmental and food safety concerns. What are the different opportunities and threats that these trends pose for livestock biodiversity?

Global food security and livestock biodiversity

In 1996, the World Food Summit set a target of reducing the number of hungry people in the world by half – to 400 million, by 2015. When people have access at all times to enough food for an active, healthy life they are termed 'food secure'. The basic causes of food insecurity are insufficient national food availability, and insufficient access to food by households and individuals (Smith *et al.*, 2000). A sensitive measure is the incidence of child malnutrition. Poverty is the most widespread cause of food insecurity. Development programmes need to address both food availability and poverty reduction. Most international development agencies appreciate this and are now emphasizing poverty reduction as a main focus (de Haan *et al.*, 2001).

Are global, regional and national food security policies to be seen as threats or opportunities for livestock biodiversity? (regional, in the international development sense, relates to a group of nearby countries, rather than to an area within a country). There have been many discussions about the relevance of traditional livestock systems to current concepts of food security (for example they are now seen as providing manure, traction, financial security, employment and social cohesion as well as food (Ørskov, 1993). However, with the increasing use of artificial fertilizers and local credit schemes, these ancillary uses of livestock may be attracting less attention in the future (de Haan *et al.*, 2001).

While far too many households remain food-insecure, the income of many others in the developing world is increasing, and it is estimated that demand for meat and milk will become greater in these countries than in more developed countries. The developing countries' share of total world meat consumption will increase from the current 47% to about 63% in the year 2020 (de Haan *et al.*, 2001) – the so-called livestock revolution (Owen *et al.*, 2004), one of whose major challenges is the reconciliation of the need for poverty alleviation with the fact of globalization. National policies therefore have two objectives: to supply the non-poor households, and achieve poverty reduction for the rest. Each of these would be predicted to pose different sets of threats and opportunities for livestock biodiversity (Table 5.1).

Threats predicted from requirement for increased food availability

Intensification following developed world models

It is often assumed that intensification requires the introduction of exotic breeds for crossbreeding and the replacement of traditional breeds, but very many of these introductions have failed. Conelly (1998) reviewed experience with dairy cattle breeding in Kenya during colonial times; while European farmers there used imported Friesians, African farmers were initially discouraged from grading-up their zebu cattle with Friesians. From the 1920s, Sahiwal cattle from India were crossed with their zebus but the crossbreds were said to be not particularly hardy or disease resistant and had poor milk yields.

Table 5.1 Predicted effects (negative and positive) upon livestock biodiversity of available policy initiatives in developing countries aimed at (a) increasing food availability and (b) increasing food security.

	Aim of policy initiative	
	Increasing food availability	**Increasing food security**
Negative effects (threats)	1. Intensification 2. Standardization of marketed products 3. Institutional bias 4. Shift in favoured species	5. Standardization of marketed livestock 6. Social equity and changing employment patterns 7. Provision of credit
Positive effects (opportunities)	8. Preparation for climate change 9. Promotion of sustainable crossbreeding 10. Trade in animal genetic resources 11. Internalization of environmental costs	12. Past experience with livestock projects 13. Lower dependence on veterinary and nutritional inputs 14. Entry points for wider programmes 15. Use of local knowledge 16. Regional purchase of food aid 17. Emphasis on biodiversity conservation

When crossbreeding of African-owned cattle with Friesians did begin to be promoted (in agricultural development projects that were partly politically motivated) lack of extension services, difficulties of disease control and poor availability of feed all retarded progress. Many of the difficulties encountered in this programme have occurred in subsequent projects elsewhere and Conelly (1998) recommends closer study, by today's policy-makers and planners, of previous experience. Lessons that might be learned include the importance of understanding pre-existing agricultural systems, and of appreciating the dangers of concentrating too much on technology and not enough on social matters.

It has been argued that at least in some livestock sectors, the necessary intensification of food production will only be achieved by full adoption of western systems, with abandonment of attempts to utilize local breeds. In principle, improvement could be by selection within the local breed but this would be very slow and also dependent on adequate milk recording procedures, which are not in place (Syrstad & Ruane, 1998). In an exhaustive review of tropical dairying, McDowell *et al.* (1996) concluded that crossbreeding local cows with imported dairy sires was fraught with so many problems that the only viable route to dairy self-sufficiency was 'commercial dairying with high grade or pure dairy breeds'.

Standardization of marketed products

This is particularly associated with meat markets where many purchasers prefer carcases to conform to a standard pattern. As mechanization of the food chain develops this is likely to be accentuated. In many parts of west central Africa, the indigenous humpless shorthorned cattle (the West African Shorthorn breed group, WASH) are being replaced in their traditional range (the humid forest and certain parts of the savanna) through crossbreeding with zebu genotypes which the meat market prefers (Jabbar *et al.*, 1997, 1999; Camus *et al.*, 1981). Although the WASH can tolerate trypanosomosis and, presumably, other local diseases, developments in husbandry and environmental changes are permitting the trypanosusceptible zebu genotypes to colonize these areas (Figure 5.1).

Bias in governments, universities and institutions

Regarding internationally funded research in livestock of the developing world, the major share has gone towards cattle. Research priorities have been 'erroneous' (de Haan *et al.*, 2001), with emphasis on 'modern' production systems and on projects run on research stations. This might be expected from the fact that so many aspiring researchers from the developing world are trained in North America, Australia, New Zealand and the EU, and their research topics will usually be related to developed world production systems.

Animal health programmes have been disease-oriented and have proceeded separately from livestock management. The changing nature and scope of veterinary services is an important issue in developing countries. These have generally followed a western model, with prominent public sector involvement and demarcation

Figure 5.1 Crossbreeding. Somba (left) and Borgou oxen in the savanna zone, Benin, west Africa. Somba cattle are taurine West African Shorthorns and are native to the area, with a reputation for trypanotolerance. The Borgou is derived from a cross between the Somba and zebu cattle.

between animal health and animal production (Kaasschieter *et al.*, 1992). There has also been strict separation of livestock and crops, which has made integrated planning difficult (Conelly, 1998). Budgets have been cut and staffing reduced (by 11% over the last 10 years in 16 sub-Saharan countries; de Haan *et al.*, 2001). There has been very little work on 'microlivestock' – chickens, rabbits, rodents etc. (Figure 5.2) and more research and development are needed here (Blench, 2001).

Shift in favoured species

In Europe, 'red' meat (beef, sheep meat) has been losing ground to 'white' meat (poultry, pig meat). While this might happen in the developing world, it is possible that the shift in species preference in the developed world may increase the world supply of beef which could compete with meat produced locally in the developing world.

Threats arising from requirement for increased food security

Standardization of marketed livestock

Genotypes which are relatively easily marketed, or are suitable as collateral for loans to meet cash flow demands, will be favoured if the priority is poverty reduction. As populations become more urbanized one would expect the market for slaughter stock

Figure 5.2 Rabbit hutch, east Zimbabwe. 'Microlivestock' will in many cases be much more appropriate for household survival strategies than the more conventional species, for social as well as economic reasons.

to become more predictable than that for breeding animals, and this will tend to favour genotypes with good slaughter characteristics over those which possess fertility and viability. Both of these trends will favour currently numerous breeds with a wide distribution, over local, scarce breeds.

Social equity and changing employment patterns

Issues of social equity are intimately bound up with decisions relating to development. An innovation may benefit one sector of society and penalize others. Changes in livestock systems may jeopardize conservation of livestock biodiversity, yet may advance other social aims. In one region of Benin (Hall *et al.*, 1995) the traditional cattle husbandry system, which uses a distinctive local breed (the Somba) of West African Shorthorn type, depends on children to herd the animals. Now that schooling is more accessible to children the herds are entrusted to Fulani herders. As a result of this change the management system has become more extensified, with use of bulls of Fulani breeds and retention of animals in herds rather than marketing of surplus stock. There is a greater risk of environmental degradation under the Fulani herding system, but on the other hand the education of children may be facilitated. This illustrates how livestock biodiversity issues cannot be considered in isolation from considerations of social equity.

Traditional breeds are often associated with traditional ideas on such matters as the role of women and child employment. Food access is only possible if households have a cash income, because self-sufficiency, by which is meant a cashless economy, is probably only a practicable strategy for a small minority of households. Off-farm employment is a key element for many rural households in the developing world, who often send their young men to towns to work and remit funds to support the household – or at the least to avoid being a burden during periods of food shortage. Aspirations towards education reduce the supply of child labour and require households to find cash for school fees.

Economizing on farm labour requirements can therefore be important even for the poorest households, because by doing so they can seek employment off the farm. In some areas the result can be a lack of farm labour at critical times of the year. In the longer term, remittances, whether in cash or as food, can become so much a part of household survival strategy that long-term investment in the landholding is discouraged (Blench, 1999). A side effect will be the loss of traditional livestock husbandry skills, leading to greater dependence on veterinary and nutritional inputs.

Provision of credit

Development programmes which involve providing credit for empowerment of disadvantaged groups will often be biased towards providing a species or breed that is clearly different from the local, traditional breed. Publicity for fund-raising, and auditing of the project, will thereby be easier.

One of the first charities to provide livestock to developing countries was the American charity Heifer Project International which was founded in 1944 (Pelant *et al.*, 1999; http:// www.heifer.org). It provides livestock of many different species to most of the countries of the developing world and both exotic and native breeds are used. In the UK, Send a Cow (http:// www.sendacow.org.uk) was founded by farmers and from 1988 to 1996 (when export was stopped due to BSE), 300 heifers were air-freighted, mainly to east Africa. In the financial year 1999–2000 its total expenditure was £335 585. FARM–Africa, also based in the UK, has concentrated on supplying goats, mainly in Africa. These charities coordinate their activities with each other and with other non-governmental organizations (NGOs) and government agencies.

Some charities focus on specific countries, and some have an agenda which is primarily social reform, for example Grassroots International (www.grassrootsonline.org/pig.html) which has concentrated on providing pigs to peasants in Haiti (see also www.webster.edu/~corbetre/haiti/misctopic/pigs/gaertner.htm and www.netaid.org/hunger/170more.htm).

The strong attachment to exotic breeds shown by these charities might cause concern among conservationists but there is evidence that this is acknowledged. FARM–Africa has sponsored valuable characterizations and comparisons of crossbred and indigenous goats (FARM–Africa, 1996) while Send a Cow has supported research on the broader implications of their programmes.

Opportunities predicted from requirement for increased food availability

Climate change

If distribution patterns of livestock have to change it would clearly be very important to countries to have appropriate livestock biodiversity available. If climate change is rapid, breed replacement would be predicted, while if it is slow, selective breeding could be followed. Thus both breed biodiversity and within-breed variation could be very useful for an appropriate response.

Emphasis on improved management in traditional systems

Livestock development programmes usually involve improved management as well as genetic improvement. Ayalew *et al.* (2003) reported how in Ethiopia, indigenous goats responded with improved performance and made an enhanced contribution to household welfare under improved husbandry conditions. While crossbred goats (produced by mating exotic, high performing goats with local breeds) yielded more milk per female, there was no difference when milk yield was expressed as a proportion of body weight. Benefits to the household took into account purchase of external inputs and contribution of meat and manure as well as milk, and, overall, improved management was of more benefit than the introduction of crossbreds.

International trade in animal genetic resources

The need for increased food supply elsewhere may encourage an export trade in well characterized local stocks of adequate health status. An example is the export of Tuli and Nguni cattle to the northern pastures of Australia (see Chapter 4).

Internalization of environmental costs

Intensified systems will lead to increased environmental damage, though in the developing world it seems unlikely that these costs will be reflected back on the systems themselves. If they were, then ultimately there could be an incentive to extensification. Food production can be increased by farming more land (extensification) or by increasing production per hectare (intensification). In both cases there is a risk of environmental degradation. With intensive systems, pollution is one symptom; with traditional systems, erosion. Selective retention of features of the old system, in parallel with selective adoption of modern innovations, could threaten the sustainability of the system. Can traditional livestock systems, and the breeds associated with them, be developed in such a way as to reduce this risk?

Extensification often means that grazing livestock are forced onto more marginal areas. African pastoralist herds can then lose the freedom of movement, which is fundamental to their survival (Scoones, 1995). Conflicts with wildlife can increase (Bourn & Blench, 1999). At least in some cases, pastoralists increase herd sizes in the hope that as many as possible survive bad years (Roth, 1996). Whether this opportunistic stocking strategy is economically and ecologically optimal is hotly debated (Campbell *et al.*, 2000).

Though many traditional pastoralist strategies have been operating for a long time, today they are encountering new constraints such as pressure on land and new opportunities such as greater availability of water from boreholes (Roth, 1996). Under intensification, crop residues become more freely available for livestock and these can compensate for the loss of grazing land. For example in Burkina Faso, Slingerland *et al.* (1999) found that 1 ha of crops yielded as much biomass in the form of crop residues, as 1 ha of natural vegetation.

Opportunities predicted from requirement for increased food security

Past experience with livestock projects

In developing countries, livestock owners are often seen as the richer members of their communities, and livestock development initiatives may not be seen as aimed at poverty reduction. Indeed, the World Bank funds far fewer livestock-only projects today than it used to (about one per year, instead of ten per year in the 1970s); the emphasis is now said to be on projects in which livestock are a component (de Haan *et al.*, 2001). If there are fewer livestock development projects, this may on balance be good for conservation of livestock biodiversity. In the past, such projects have usually damaged local breeds and this is probably inevitable until a genetic impact assessment procedure (see page 99) becomes routine. Many livestock projects have given disappointing results. An example is the well-publicized poor survival of European dairy breeds in the tropics, compared with genotypes with better local adaptation (de Vaccaro, 1990). 'Hundreds of thousands' of Holstein-Friesian and other dairy breeds have been imported into tropical countries yet survival of calves and of adults has been so bad that in general these imports have been unable to breed their own replacements. Presumably this is connected with there being insufficient economic inducement or management skill to rear heifers. Herd life of imported cows (mean 2.6 calvings) was 24% shorter than that of local and mixed-origin cows (3.1 calvings). Herd life is of course influenced by many factors in addition to breed genotype; it reflects quality of management and perhaps also the market value of cull cows, and there is a need for a full survey of the subject.

Lower dependence on veterinary and nutritional inputs

A policy of reducing dependence on these inputs would favour local breeds because with their local adaptations they would either not require inputs, or would be less sensitive if inputs were not maintained. However, it must be remembered that this is not the only animal health policy that might be adopted. If livestock veterinary services become more effective, through investment and/or privatization, the need to reduce dependence on them diminishes.

The values placed on the attributes of local breeds may depend on household affluence. Scarpa *et al.* (2003) found that the higher disease resistance and lower feed and labour requirement of native pigs in Mexico were more important to poorer than to richer households.

Entry points for wider development programmes

Projects on traditional breeds could be 'entry points for wider development programmes' (Steinfeld & Mack, 1995). This is only likely to be relevant to poverty reduction strategies if these projects focus on the animals kept by poor people, such as chickens and other 'microlivestock'.

Appreciation of value of local and indigenous knowledge

Relationships between livestock biodiversity and human cultural distinctiveness and identity were introduced in Chapter 2. If poverty reduction strategies aim to support local community cohesion, then explicit recognition in the strategies of the value of *indigenous knowledge* would help to secure this aim. Indigenous knowledge is a form of intellectual capital and development projects should consider what use is to be made of it. Projects which value it highly will probably tend to favour local breeds.

Disaster or reconstruction aid

Apart from dried milk, livestock products are not very relevant to disaster or famine relief, but breeding animals are very important for reconstruction programmes. Livestock could be purchased in the region or provided directly by donor countries. The latter approach looks like a way of subsidizing breeders in donor countries, and of increasing dependence on inputs. Programmes which are sensitive to these considerations will clearly favour local breeds.

Biodiversity issues in natural resource management

Traditional livestock and associated husbandry systems are often identified with relatively self-sufficient livelihoods which are not particularly cash-oriented or heavily dependent on trade. If these systems are to contribute more to national food supply they must be intensified and this may involve breed replacement. Policy in the past has generally been in this direction.

However, these systems may provide less easily quantifiable but equally important services to a country, which can be described as ecosystem services (James *et al.*, 1999). This is a broad category, including such aspects as watershed protection, conservation of tree cover in semi-arid areas, maintenance of low-density human populations in peripheral areas, control of bush fires, and maintenance of patchiness of habitat which can enhance floral and faunal biodiversity. The valuation of these services, and how their maintenance can be encouraged through reduction of poverty of the human population, are areas needing further research (Koziell & Saunders,

2001). What is fairly certain is that the local husbandry systems which can help to assure these services are based on traditional breeds.

The relationship between livestock and wildlife can be a difficult one. There is often a need for integrated management and Bourn and Blench (1999) propose the 'co-conservation' of wildlife and local breeds. With appropriate management it should be possible to achieve the peaceful coexistence of traditional livestock systems, with the wildlife which can be so important for tourism. Almost certainly, this coexistence would be more achievable if only traditional breeds are used.

National policies for livestock biodiversity in the developing world

In the developing world, livestock development programmes have, essentially, been either breed improvement schemes or the establishment of self-contained livestock operations. A simple way of promoting conservation of genetic resources would be by requiring all livestock development proposals to be accompanied by a genetic impact assessment (GIA), a quantification of the genetic effects on local livestock. It would be relatively easy to design a standard protocol for a GIA. The concept was first proposed in 1992 (Hall & Bradley, 1995) and has been placed in the context of the Convention on Biodiversity (CBD) by Day *et al.* (1994). In Article 19 of the CBD, there is the 'biosafety stipulation', which states that an imported animal should not have serious environmental impacts. Day *et al.* (1994) raise the point that the introduction of alien genes into local stocks would be such an impact. The CBD therefore reinforces the need for GIA.

Some projects may have a large economic effect, for example chicken production based on importing day-old chicks, but unless animals from the scheme are used for crossing with local stocks the genetic impact would be only indirect. Others, for example open nucleus breeding schemes, would have a moderate effect which would probably be assessed as neutral or benign, others, for example AI schemes, could have a large impact.

The indirect effects on local breeds could be important, for example an advanced pig unit may only use imports, but could lead to eradication of local pigs as possible disease reservoirs, or to a rise in feed prices.

Any importation of semen or embryos should be accompanied by a programme to collect semen of adequate health status in the area. Locally produced semen or embryos, from whatever breed including exotics, would not be subject to this condition (this would foster local AI businesses, and the use of adapted imports). FAO would be the obvious organization to operate an international repository which could store material conforming to various grades of health status.

Environmental aspects and industrialized economies

In the developed world, there are essentially two ways in which livestock farming gives rise to environmental concern. First, there are pollution issues which mainly relate to intensively farmed, housed livestock. Second, there are concerns for the landscape and for floral and faunal biodiversity, arising from intensification of grazing systems. Dealing with these matters is primarily a question of regulation (in relation to pollution) and of policy and subsidy regime (Curry Report, 2002, for the UK perspective) in relation to landscape and biodiversity. However, livestock biodiversity can play a part in (a) developing breeds that cause less pollution and (b) conserving and making available breeds that are compatible with maintenance of landscape and biodiversity.

Ammonia emissions are a problem with intensive poultry, the drier the faeces, the less ammonia is released. Faecal dry matter content is heritable (Preisinger, 2002). It is also correlated with food conversion efficiency. In this case there is probably enough genetic variation within current commercial chicken genotypes for these characters to be selected for, and conserved breeds are unlikely to contribute. However, it is different when landscape conservation is to be achieved.

In 1992 the Common Agricultural Policy (CAP) of the European Union (established in 1957) began to be reformed. This was in response to the political necessity of dealing with overproduction and the high CAP budget and to external pressures to remove production-increasing support, leaving only subventions that are payments for environmental benefits produced or for social development in rural areas. European Union Regulation 2078/92 was devised with a view to supporting farmers while not generating increased production (Bignal, 1998). It has encouraged more extensive production in some countries but its provision for headage payments of 100 Euro per grazing livestock unit for endangered breeds has not yet been implemented in the UK. This may change when the new agri-environment arrangements commence in 2005 (www.defra.gov.uk/erdp/reviews/agrienv/default.htm).

The EU agreed on a new round of CAP reform in 1999, the package of new policies being known as Agenda 2000. It deals with arable, beef and dairy production because these are the sectors where CAP reform was most urgent. The main outcomes were a cut in prices for some commodities with some compensatory direct payments. A new Rural Development Regulation was adopted, supporting diversification and agri-environment schemes. provision was also made for a Mid-Term Review of some CAP finance schemes (Curry Report, 2002). Implications for extensively farmed livestock are being widely discussed (http:/www.macaulay.ac.uk/elpen). Dairy and intensive beef production may become yet more intensive, but in Less Favoured Areas (LFAs) member states will have the option to base support payments to livestock farmers on land area, while previously these have been assessed on headage (i.e. payment per animal). Area-based payments tend to favour extensification while headage payments can lead to overstocking. The reform process is seen as necessary for the accession of new member states in 2004 (Bignal, 2003).

Also in 1992 the EU adopted the Habitats Directive. This extended the scope of the Birds Directive of 1979, to enhance protection for natural habitats and wild fauna and flora through a network of protected areas across the European Community to be known as Natura 2000 (http://europa.eu.int/comm/environment/natura/home/htm). For this network to function, special protection outside these Areas has been acknowledged to be necessary for 'species of Community interest' and for landscape features which may act as ecological corridors. Very many of these habitats to be protected are heavily influenced by farming, especially low-intensity livestock raising. If these traditional systems are intensified or indeed if they are abandoned, the ecological and landscape values of the habitat will be greatly reduced. These questions are discussed in the Forum on Biodiversity and High Nature-Value Farming introduced by Pienkowski (1998).

These initiatives have parallels with national initiatives in EU member states. For example in the UK there are Environmentally Sensitive Area (ESA) schemes and the Countryside Stewardship Scheme (Ovenden *et al.*, 1998). Over 10% of English agricultural land is now under ESA agreements. In Spain the proportion was 2% in 1996, but in Sweden, Austria and Finland it was over 50% (Beaufoy, 1998). Effectiveness of these schemes is discussed in the studies introduced by Ormerod *et al.* (2003).

The full implications of Agenda 2000 for LFAs are not yet clear (Baldock *et al.*, 2000) but the implementing Regulation includes components which enable farmers in LFAs to receive payments compensating for environmental constraints that EC legislation has brought about.

The role of livestock is being reappraised along these lines in many developed countries. In the UK the 2001 outbreak of foot and mouth disease brought into sharp focus the unimportance of much of Britain's livestock farming (especially extensive grazing) as an economic activity in its own right, and its very great importance in maintaining the rural environment as a place for people to visit (Curry Report, 2002). For example, in Scotland, where tourism contributes 5% of GDP compared with agriculture's contribution of 1.4%, the overall loss of gross tourism revenue as a result of FMD in 2001 was £200–250 million. Compensation paid to Scottish farmers for loss of stock was £171 million, and their loss of income was estimated at £60 million (Royal Society of Edinburgh, 2002). Authoritative comparable figures for the rest of the UK do not seem to be available.

Traditional landscapes

Traditional agroecosystems include some very long established areas such as the New Forest in southern England, which includes 19 771 ha of unenclosed common land. Here, management plans trace back to the Orders and Rules of 1537 (Tubbs, 1997) and the area is grazed by an estimated 3343 ponies many of which are of the New Forest breed, 2580 cattle (of mixed breed but often with Galloway affinities) and some pigs, sheep and donkeys (Hickman, 1999). In Spain the holm oak *Quercus ilex* and cork oak *Q. suber* ecosystem known as the dehesa was probably shaped by domestic herbivores over the last 5000 years (Cuartas & Garcia-Gonzalez, 1992). This combines open woodland, arable cropping, cattle and sheep grazing and the rotational grazing of the local pig breed, the Iberian pig (Lopez-Bote, 1998).

In the management of systems like these for conservation, the presumption would usually be that the traditional breed of the area is the one that should be used, because of supposed adaptation to the local vegetation (Hooper, 1998).

The attractiveness of many landscapes often depends on there being a mix of grazing pressures with heavy grazing in some areas promoting grassy swards, and lighter grazing leading to dominance by other non-tree species, and exclusion of grazing from other areas leading to semi-natural tree cover. This mix is very often seen in the UK (Milne, 1996) and the relationships between dominance by heather (of great landscape value and scenic importance in many areas) and grazing regime are summarized in Figure 5.3. It is difficult to derive general principles because habitats are so different from each other. While practical experience is accumulating (Tolhurst & Oates, 2001) on how grazing management can promote floral biodiversity, the relationship between grazing management and faunal biodiversity (especially of mammals and birds) needs further study (Milne, 1996).

For this kind of grazing management, the specific adaptations, if any, of particular breeds to grazing for conservation management need to be considered. This has been reviewed for ruminants and pigs, but not for equines, in Britain by Mercer *et al.* (1997) but general principles are not yet established. The timing of grazing and the stocking rate are very important. Choice of breed would probably be most important when grazing is year-round, but, in practice, managers will probably tend to use whatever genotypes are favoured by local farmers (Grayson, 2000; Laiolo *et al.* 2004).

Under extensive systems livestock may encounter humans only rarely and their fear reactions may be intensified. More research is needed on breed differences in these responses. To assure welfare, potentially stressful operations like handling and transport will need to be particularly carefully managed.

One of the earliest accounts of breed suitability for grazing management schemes was that of Williams *et al.* (1974), who reported that Galloway cattle were well suited to the grazing management of English fenland because of their low selectivity, wide spacing during grazing, tolerance of biting insects and general hardiness. Body weight and bite size affect diet quality, at least in sheep (see Chapter 3) and there are clear breed differences in these respects. Most knowledge in this area is empirical, rather than theory-based. For example from 1980 the Konik horse of Poland began to be studied in semi-natural woodland (Graczyk, 1989) and it is now quite extensively used for conservation grazing especially in the Netherlands. There are several other breeds, such as the Welsh pony and the Exmoor pony, which could probably perform this function, but comparative trials have not been done. In Hungary there is now considerable experience in using traditional Hungarian Grey cattle for grazing management (Allies *et al.*, 1999). While breeds have genetically determined differences in aspects of their social behaviour (Hohenboken, 1986), whether this alters their temporal or spatial patterns of diet selection is not known. There is much other important information without which fully informed choices cannot be made on grazing management for conservation. Research is urgently needed.

It is important to recognize the role of socioeconomic factors in determining patterns of land use and, thereby, the biodiversity of the agroecosystem. In a novel piece of interdisciplinary research, Ellis *et al.* (1999) found that farms where a proportion of income came from off-farm work had more biodiverse swards than those which depended completely on agriculture.

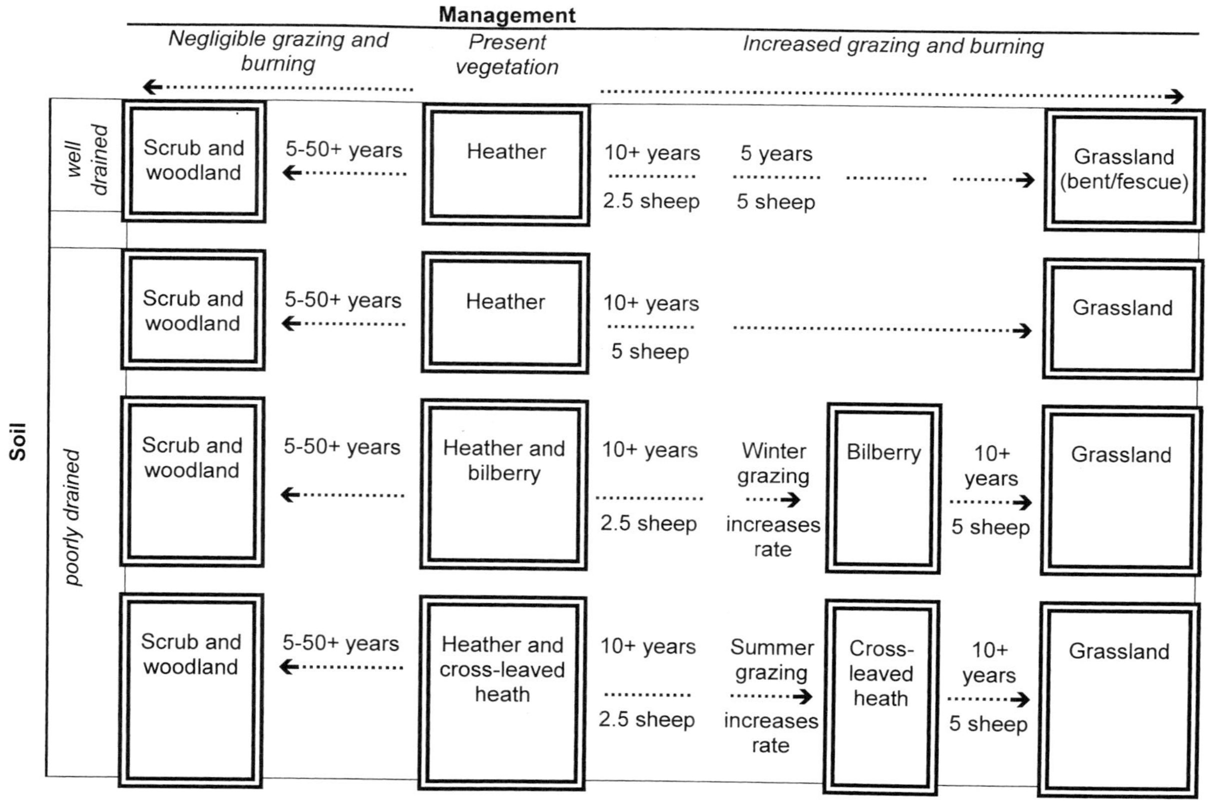

Figure 5.3 Upland ecosystems and grazing/burning regimes. Upland heather ecosystems within the UK: general pattern of probable transitions between vegetation types under different intensities of burning (to encourage heather regrowth) and sheep grazing. (ewes per hectare). (Reproduced from Milne (1996) with permission from Countryside Commission.)

When land is taken out of production the aim might be to restore 'normal functioning multi-(herbivore) species ecosystems' (van Wieren, 1995). In this connection it may be asked why so many European plant species benefit from traditional grazing practices. Vera (2002) suggests that the traditional view that the natural climax vegetation of Europe was closed-canopy forest may be wrong. Relatively few vascular plant species in Europe prefer shade, suggesting the pre-human forest cover was much more open. Gaps in woodlands could arise from 'fire, storms, floods, landslides and beavers'. The true picture of the European Holocene may well be that of an open, wood-pasture-like habitat, owing a great deal to the activities of grazing animals. Supporting this, simulation studies in northern European woodland show that a stocking density of one cow per 100 ha is sufficient to prevent regeneration of trees (Jorritsma *et al.*, 1999).

Wallis de Vries (1995) suggests that it is the requirements of large herbivores that should dictate the design of the new large-scale nature reserves that could be established as a consequence of the removal of land from agriculture in western Europe. As the aurochs and the pre-domestication horse are extinct, these niches would have to be taken by currently existing breeds of cattle and horse. Horses remove more vegetation per unit body weight than cattle and use the more productive plant communities to a greater extent. As they feed closer to the ground, they can maintain structural diversity (Menard *et al.*, 2002). The sheep and goat are not native to western Europe and are therefore sometimes seen as ineligible for use in the restoration of original habitat (van Wieren, 1995).

One genotype of cattle that has been seen as suitable for this purpose is the so-called 'bred-back aurochs'. This was developed in the 1920s, at Berlin and Munich zoos, by the Heck brothers by crossing about fifteen different breeds including Corsican, Camargue and the Spanish fighting cattle (Berlin) and Corsican, Scottish Highland, Hungarian Grey, Friesian, Murnau-Werdenfels and Allgaü (Munich) to produce a synthetic breed resembling pictorial records of the aurochs. These 'Heck-cattle' have been used to stock nature reserves especially in France and the Netherlands (Wallis de Vries, 1995). In an attempt to make them resemble the aurochs more closely they are being crossed with two long-legged breeds, the Chianina and Sayaguesa (Bunzel-Drüke, 2001).

However, Wallis de Vries (1994) found no strong ecological reason for using one breed rather than another in the Netherlands. There might be a presumption that local and rare breeds will be used but differences in grazing and feeding behaviour have often been found to be ascribable to body size rather than to breed. A major issue in management of such livestock is the provision of winter feed. Experience in the Netherlands suggests that cattle may need extensive areas permitting seasonal migration if they are to live as a naturally regulated population. The winter feeding of livestock on localized areas may lead to problems. For example, a herd of Galloway cattle was being fed on an area of 'stony ground' near the Roman Wall in northern England which was of archaeological significance, and as part of an agri-environment scheme partly funded under EU Objective 5b (see page 105), a shed was provided for winter housing (Allies *et al.*, 1999).

Rural development

The rural development policies of the EU are based on regional objectives. These were reformed and strengthened in 1988 soon after Spain and Portugal joined and the need for further promotion of social and economic cohesion became clear. Rural areas that are lagging behind economically, with a GDP less than 75% of the EU average are classified as Objective 1 areas. Those with low levels of socio-economic development, high dependence on agricultural employment, low agricultural incomes and low density or declining populations are known as Objective 5b areas (http://europa.eu.int/comm/dg06/publi.pac2000/rd/rd_en.pdf).

Conservation and sustainable development of traditional agroecosystems in the EU will take place against this policy background, which will vary in important details between countries according to their own national policies. Possible consequences for livestock biodiversity have not been explicitly studied but a body of experience on the utilization of different breeds is growing.

Local livestock breeds clearly have cultural significance but how this can be defined, evaluated and prioritized is not yet clear. Gandini and Villa (2003) suggest local breeds can be assessed regarding their value as 'historical witnesses' and as 'custodians of local traditions'. The Council of Europe is developing concepts of cultural significance of landscapes (http://culture.coe.int/infocentre/pub/eng/epe5.7.html), and if landscape is seen as the marriage of nature and culture then a local breed can be seen as one of the children of this marriage, with history as the midwife. However the policy implications of this vision for breed conservation are not yet developed. The European Landscape Convention (http://culture.coe.int/infocentre/press/eng/2000/express2000.10.20.html) has been signed by eighteen countries and calls for the adoption of policies on landscape. How local breeds fit into such policies needs to be determined.

While the cultural significance of livestock breeds in Europe will mainly be in their landscape associations there will be instances where the association is with cultural or ethnic groups. Here, breeds will be part of the cultural heritage of such groups and whether this qualifies them for government support will depend on national policy towards cultural diversity.

Maintaining or increasing market share

Market demand for animal food products differs considerably from country to country (Bansback, 1995). This can be investigated in three ways: economically, as a proportion of the overall market, sociologically, with reference to the cultural place of different foods in the diet, and behaviourally, by observing direct responses of consumers to different foods. Studies using taste panels (people specially trained to assess the eating quality of meats and other foods) have shown clear differences in taste preferences, for example a British taste panel preferred the tenderness and juiciness of British lamb meat to Spanish, while a Spanish panel preferred the latter (Sanudo *et al.*, 1998).

For many consumers, the place of origin of the food or the breed of the animal that gave rise to it are part of its appeal. These attributes are often capitalized upon, for example the Appellation d'Origine Contrôlée (AOC) system in France. Verrier and Bouffartigue (1993) discuss the possibilities of linking a region, a breed and a product in the context of the French AOC system. This is certainly feasible in limited areas, for example in the northern Alps the local dairy breeds are the Abondance and the Tarentaise and farmers are now producing AOC cheeses. There are some examples, for instance Japanese Black cattle (see page 159) where the breed and the market have probably coevolved over a long period.

In recent years, patterns of meat consumption have become much more volatile in Europe. In a small survey, Holm and Mohl (2000) found that negative attitudes towards meat were becoming prevalent before the BSE crisis. While acknowledging that meat has a central place in many food cultures, many consumers are unhappy about 'modern production and processing of meat, the fact that meat products derive from animals, cultural and social aspects of eating meat, and health'. One response is the redefinition of meat as an ingredient rather than as the central and most highly valued part of the meal. In a post-BSE survey Verbeke and Viaene (2000) found safety and health issues to be especially important, with welfare a consideration regarding pig meat. Perhaps indicating a new trend, in Sweden the use of antibiotics as growth promoters in feeds is now banned (Warriss, 2000). In the case of pork, market research indicates that leanness, healthiness, taste and tenderness should be dealt with before traceability, labelling or marketing (Verbeke *et al.*, 2000).

In western meat livestock there has been strong selection against carcass fat, and the commercially important meat breeds are leaner than the rare and traditional breeds (Table 5.2a, b; Figure 5.4). Higher intramuscular fat is less attractive to most shoppers but it is associated in taste panel trials with higher eating quality (Warriss, 2000). Clearly genetic variation in this trait must be retained in case market preferences change, and according to Notter (1999), '. . . all commercial breeds are believed to still have enough genetic variation to permit restoration of intramuscular fat'.

In the gourmet and restaurant markets, an acceptance of fat as a key factor in flavour may be paralleled by an interest in the health aspects of different kinds of fat. Polyunsaturated fatty acids (PUFA) may be seen as preferable from this angle. While their biosynthesis can be promoted by manipulating the diet (Verbeke *et al.*, 2000) there is growing evidence of breed differences (for example Mezöszentgyörgyi *et al.*, 2001). Welsh Black beef has a higher n-3 PUFA content than that of the Holstein-Friesian (Choi *et al.*, 2000). Explaining the characteristic flavour of Soay lamb meat, Fisher *et al.* (2000) point out that the n-3 and n-6 PUFA content would be liable to peroxidation, giving a venison flavour.

Also very relevant to customer appeal is the colour of meat. In Spain, Insausti *et al.* (1999) found the bright red colour (due to oxymyoglobin) which customers find attractive was most obvious with beef from Morucha cattle, a relatively unimproved and rather fat breed (Mendizabal *et al.*, 1999).

A strong economic argument for improved welfare is the resulting improvement in meat quality, particularly in pigs. Some breeds tend to yield meat that is pale and soft and exudes water (PSE meat: Warriss, 2000), especially if stressed before slaughter.

Table 5.2 Breed differences in fatness of sheep and pigs: British examples
(a) Sheep. From Croston *et al.* (1987), Kempster *et al.* (1987), Taylor *et al.* (1989). Fatness measured in lambs at normal age and condition for slaughter.

Purebred lambs		Crossbred lambs		
Breed	**Fat g/kg carcass**	**Sire breed Fat g/kg carcass**	**Early lambs Fat g/kg carcass**	**Late lambs Fat g/kg carcass**
Soay	142	Texel	254	240
Jacob	247	Wensleydale	263	253
Welsh Mountain	259	Suffolk	267	252
Finnish Landrace	262	Dorset Down	270	267
Wiltshire Horn	280	Oxford Down	273	258
Southdown	290	Southdown	283	273
Oxford Down	317			

(b) Pigs. From Warriss *et al.* (1996). P2 is given (measure of backfat thickness: Warriss, 2000).

Rare breeds		Commercial breeds		Specialist meat breed	
Berkshire British Saddleback Gloucester Old Spot Large Black Oxford Sandy & Black Tamworth	18.0 mm	Duroc Hampshire Landrace Large White	14.2 mm	Piétrain	10.8 mm

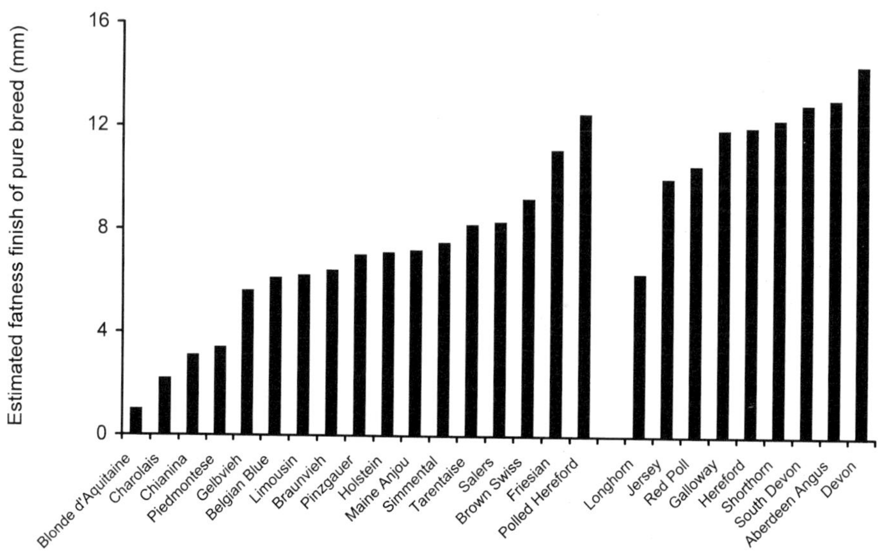

Figure 5.4 Cattle breeds: differences in fatness. Effects of breed of cattle upon fatness finish (mm) of steers. The British breeds (grouped on the right) are generally fatter. Data from Roughsedge *et al.* (2001).

This characteristic is being selected against but a high proportion of meat is of poorer quality than it need be. In the USA the incidence of PSE meat is no lower than it was twenty years ago (Cassens, 2000). Ideally this would be rectified by improving husbandry, transport and slaughter procedures, but if economics prevent this, then the pig industry will be dependent on utilizing genetic variation to improve meat quality. So much effort has gone into developing current high-performance meat breeds that this improvement is more likely to be by within-breed selection than by introgression from other breeds. Thus, livestock biodiversity may be unlikely to have a great part to play in such developments.

Livestock biodiversity and the improvement of animal welfare

The welfare of an animal is the degree to which it can cope with its environment (Broom & Johnson, 1993). Consumers in many countries now expect reassurance that animal products come from high-welfare systems. A general framework for research and practical improvement of welfare has been the list of Five Freedoms which farm animals should be given – this was conceived in 1966 (Appleby, 1999). The Five Freedoms and the possible scope for using livestock biodiversity to secure them (by direct selection or by breed choice) are listed below.

Freedom from hunger and thirst

Breeds which have been selected for high food intake and rapid growth, or for high milk yield, will be more susceptible to hunger and thirst, even under the best management. More slow-growing or lower-yielding breeds would presumably be more easily granted this Freedom but this is unlikely to be compatible with the economic pressure of the mass market, though it could be a selling point in a niche market.

Freedom from discomfort

This Freedom, which relates to adaptation to the physical environment, would be threatened when breeds are farmed in environments to which they are not adapted. For example, traditional breeds of pig would be a better choice than very lean modern genotypes for outdoor systems that claim to be 'welfare-friendly', though formal studies are lacking. Promoting freedom from discomfort for intensively farmed animals is more likely to be achieved by good management than by breed choice. However, this Freedom, like all the others, cannot be treated in isolation and some breeds have behavioural adaptations which would make them less liable to physical discomfort in specific systems.

Freedom from pain, injury and disease

Breeds which are less likely to suffer from production disease (see Chapter 3) are the ones for which this Freedom is more easily provided. In pigs, for example, selection for longevity (defined as time from first farrowing to culling) would improve welfare. Longevity appears to be heritable; it is affected by several management factors, as would be expected, but it can be predicted from the incidence of osteochondrosis (a disorder of the joints leading to leg weakness) which first becomes manifest in recording schemes at the age of six months (Yazdi *et al.*, 2000). Sheep and poultry could also be selected for traits which would help to secure this Freedom (Scobie *et al.*, 1999; Jones & Hocking, 2000; McKay *et al.*, 2000).

When cattle are kept, handled or transported in poor or crowded conditions, horned animals can inflict serious injuries on others. Dehorning raises welfare issues and breeding polled animals would be a benefit (Stookey & Goonewardene, 1996).

Generally, selection (within- or between-breed) for increased longevity and greater resistance to production diseases would promote the first three Freedoms. The remaining two relate to behaviour and raise what are in many ways more complex issues.

Freedom to express normal behaviour

This might be interpreted as (a) enabling animals to exhibit most of the natural behavioural repertoire of the species, or at least those parts of it shown by experiment (Mason *et al.*, 2001) to be particularly important, or (b) developing breeds which only possess in their repertoire behaviours which are compatible with the system. The former is a management issue and is accessible to experiment, while the latter poses some ethical questions.

In the intensive husbandry environment, behavioural adaptation might be assumed to accompany increase in performance (Holmes & Hastings 1995). They carried out behavioural observations on mice that had been selected for up to 50 generations of *ad libitum* feeding to yield four lines, respectively of high and low body weight, and high and low fat content. There were practically no differences in the behavioural traits investigated although the selection had led to massive inherited differences among lines; heavy mice weighed 50 g and light mice 20 g, while fat mice were 20% fat and lean mice, 4% fat. This is evidence that selection for performance in intensive husbandry conditions probably does not lead to correlated changes in behaviour.

It is often possible to select for or against the expression of particular behaviours. Hens of one strain showed restless pacing behaviour around their battery cage before laying, while those of another strain did not pace, but sat instead. These differences were inherited (Mills *et al.*, 1985). Selection against pacing would reduce injuries (such as feather loss and skin damage) but psychological welfare would not necessarily be improved because the pacing behaviour could be a coping mechanism and animals which do not possess it may therefore be deprived of a means of coping.

Le Neindre and Garel (1990) summarize comparative work on Salers and Friesian cattle (beef and dairy breeds respectively). The former normally suckle their calves, while the latter, selected for milk yield, usually do not. Friesian calves are more adapted to artificial rearing, being more active at birth, more ready to drink milk from a bucket and to eat solid food, and having a higher growth rate. Experiments on Salers and Friesian calves under natural and artificial rearing showed this neonatal precocity to be a breed effect. Thus, the Friesian possesses behaviours which are compatible with artificial rearing.

There are many more extreme examples of behaviours whose occurrence suggests poor welfare. Repetitive, usually invariant behaviour patterns ('stereotypies') are quite common in farmed and captive mammals and birds and there is evidence of a genetic component (Price, 2002, Chapter 20). There are at least two hypotheses of their causation; an underlying pathological state, or a mechanism for coping with a stressful environment. Until their causation is better understood, selection against their expression probably cannot be claimed to be a way of improving welfare.

Freedom from fear and distress

These are behavioural manifestations of negative subjective states and as we do not have a full understanding of the latter, the absence of these visible signs cannot be taken as evidence that this Freedom has been secured. Apathy and lack of reactivity both indicate reduced welfare. Taking a pragmatic view, one might assert that this Freedom would tend to follow if the other four have been secured and also that to select for reduced fearfulness would be of benefit in all husbandry systems (Mills *et al.*, 1997). The relative roles of management and genotype on fear and distress have been examined by Goddard *et al.* (2000) who tested immunological response of nine week old lambs (i.e. well before weaning) and found this was lower in Scottish Blackface than in crossbred lambs of lowland type. The Scottish Blackface was developed in extensive systems characterized by a lack of human contact. Plasma cortisol concentration was higher in this genotype at the time of testing, indicating a greater stress response which may have led to immunosuppression. Genotype had more marked effects than management on stress responses in this experiment. Selective breeding and biotechnology could lead to animals that are better or worse at coping with their environment, while replacing one breed with another can also affect welfare. Reactivity to humans can show a genetic basis indicating that selection for reduced fearfulness is possible. Limousin beef cattle have been criticised for problems of temperament and in research aimed at alleviating these, Le Neindre *et al.* (1995) put 905 heifers sired by 34 bulls through a 2.5 min test of reactivity towards a human and found that docility scores had a heritability of about 0.22.

In chickens, there are between-breed differences in behaviour and one of the main responses studied has been tonic immobility (TI). If a chicken is placed on its back and restrained for about 15 seconds, it will remain in this position for up to 2 min before righting itself. A long duration of TI is thought to indicate a heightened fear reaction; heritability of TI duration is high (Ferrante *et al.*, 2001).

Genetic bases for many of the behavioural responses indicative of stress seem clearly established (Tarricone *et al.* 1995) and in some cases the underlying physiology can be inferred. For example, Hérens cattle are less fearful than the French Brown and the former breed has higher plasma testosterone (Plusquellec & Bouissou, 2001).

General considerations

Welfare can also be assessed in terms of the degrees to which an animal's different needs are met, and this concept can be extended into a framework of breeding goals (Kanis *et al.*, 2004), leading to production of genotypes that can tolerate poor husbandry conditions. Many people would find this ethically suspect, unless husbandry conditions were, at the very least, made no worse than they are at present. Arguably an ethical appraisal system for livestock breeding programmes is needed (Lawrence *et al.*, 2000).

If there is selection for genes that lead to better welfare, it must be remembered that welfare would be promoted not merely by increasing the mean of any quantitative characters that are proportional to adaptation, but also by reducing their variance, so that fewer individual animals are so far from the population mean that they have difficulty coping with the environment. Increasing the mean tends to decrease the genetic variance and the heritability, while the environmental component of variance increases (see Chapter 1). This indicates a greater environmental contribution to the phenotype, so a high quality of management would be even more necessary. Recent work on mouse behaviour has shown the importance of genotype–environment interactions (behaviour of a given strain may differ between laboratories) (Crabbe *et al.*, 1999; Nevison *et al.*, 1999). This has practical implications as any genetically-based improvement in welfare should not be environment-specific.

Conclusions

In the developing world, livestock are an extremely important component of the economy, while in the industrialized economies they are less so. In both, livestock are not merely providers of food – they serve very many functions and their conservation and development must take account of this. Traditional breeds are of cultural significance in both; and in the developing world their local adaptations will contribute to the sustainability of the systems of which they are part. In the industrialized world this is not yet of critical importance, because management, nutritional and veterinary inputs are freely available, but environmental and other concerns may change this situation.

In the developed world, there are pressures for change in livestock systems, as reflections of concerns about landscape conservation, animal welfare, and sustainable rural development. Livestock biodiversity whether in the form of local breeds or as within-breed variation is indispensable for these developments, especially as they relate to extensification and to niche marketing.

Part 3
The Assessment of Livestock Biodiversity

Where is the world's livestock biodiversity to be found and what determines this distribution? Livestock are important in all continents but there must be differences in cultural status and historical dimensions of livestock, at least between Asia, Europe, Africa and the other continents. Although there was husbandry in South America before the Conquest began in 1492, this was almost eradicated and the time depth of continuous husbandry is, effectively, only 500 years in the Americas while it is nearly 10 000 years in Asia, Europe and Africa.

Are there more breeds in some regions than in others and if so, why? How many breeds are extinct or at risk? Much work has been done on characterizing the performance of breeds, but until recently there has been a lack of objective information on their evolutionary relationships. New molecular methods have in many cases confirmed historical accounts. Much variation exists within breeds and conserving this variation ensures that the breeds retain evolutionary potential. While the importance of conserving breeds has been appreciated for a long time, now the recognition is growing that within-breed variation is being lost through the unguarded use of modern reproductive technology, and must be carefully protected.

Chapter 6
Global status of livestock biodiversity

Introduction

How many breeds of livestock are there in the world and where are they to be found? What proportion are rare and how many have gone extinct? Do numbers of breeds relate to sizes of livestock populations? How reliable are the data and is it possible to deduce trends and predict future patterns? Is it possible to make these inferences at global or continental level, or is it only possible for individual countries?

Feral livestock populations are widely distributed and in some cases quite large. What do they have to offer? Some feral populations, especially of pigs and goats, have a bad reputation for environmental destructiveness, but are there circumstances in which they could be seen as a valuable component of biodiversity? Given that they were once dependent on people, do feral animals have a special ethical status?

Numbers of breeds

In 1951 the first edition of I.L. Mason's work *A Dictionary of Breeds Types and Varieties of Livestock* appeared, in which the 'livestock names which may be encountered in the literature' are listed with summary information on geographical distribution, numerical status and basic characteristics. Now in its fifth edition (Porter, 2002), *Mason's Dictionary*, though not intended as a database, can be used for the deduction of historical, organizational and geographical patterns of breed diversity.

The FAO database (www.fao.org/dad-is) includes more specific population data and characterizations and is based on Mason's dictionary. It gives an indispensable current picture of world livestock biodiversity, and it is highly reliable for some countries and breeds and less so for others. While Mason's dictionary deals with the seven main mammalian livestock species (ass, water buffalo, cattle, goat, horse, pig and sheep) the FAO database includes all other livestock species.

Following the proprietorial definition of 'breed' (see Chapter 2) the FAO database shows considerable duplication of breeds – the same breed is often listed under several countries. This makes it difficult to deduce true world and national numbers of breeds and their degree of endangerment. In several cases breeds are described as 'endangered' in one country while they are very numerous in others; this can distort priority lists for conservation. In Mason's dictionary the consensus on whether such breeds are truly different in different countries can be inferred.

Using Mason's dictionary (Porter, 2002) it is deduced that of 5535 breeds or breed varieties of the seven major mammalian livestock species (ass, water buffalo, cattle, goat, horse, pig and sheep) described in the literature from about 1900 to 2002, 775 (14%) are estimated to have become extinct either recently or in the past, and, of the 4760 existing breeds, 1012 (21%) are defined as rare. These data are summarized in Table 6.1. A corresponding table from the third edition (Mason, 1988) is given by Hall and Ruane (1993).

The FAO database (Scherf, 2000, Sections 2.1.1 and 2.1.3) yields a global total of 5219 breeds of these species of which 704 are extinct, that is 4515 breeds still in existence. To these can be added 111 breeds of other mammalian species and 1049 breeds of farmed birds. These data are in Table 6.2. If the duplicated breeds are removed the number according to FAO is reduced by 446, to 4069 existing breeds. After these removals, the number of 'rare' breeds is calculated as 722 (16% of existing breeds). With numbers deduced from Porter (2002) in brackets, the rare breeds of each species number are: ass 13 (21); water buffalo 11 (7); cattle 209 (269); goat 73 (104); horse 149 (231); pig 103 (141); sheep 164 (239); totals 722 (1012).

Information on the minor livestock species is beginning to accumulate. For example there may be 48 breeds of dromedary (single-humped camel; Blanc & Ennesser, 1989). There are over 100 breeds of rabbit (Joly *et al.*, 1994). Regarding poultry, in addition to farmed breeds, inbred lines are kept for research at several institutes (Abplanalp, 1992). Valuable recent breed inventories include those of Pisenti *et al.* (1999) and Romanov *et al.* (1996). For comparison, worldwide there are at least 4327 recognized mammalian species and 9672 species of bird (Groombridge, 1992).

Recent changes in numbers of breeds

There are new entries that have been added to successive editions of Mason's dictionary and these are summarized in Table 6.3. From 1988 to 1996 there was a net gain of 999 breeds and a further 705 were added in the 2002 edition. It seems that between 1988 and 1996 85 breeds were found to have become extinct at some time and 217 breeds whose numerical status was not commented upon in 1988 were noted as 'rare' in 1996. Presumably most of these newly listed breeds had been in existence for some time but had not previously been described. In the 1996 edition the major single group to be added was 111 pig breeds from China (Table 6.4).

There is no absolute way of knowing how many more breeds remain to be distinguished. Hall and Ruane (1993) calculated for each continent and each of the main seven mammalian livestock species the ratio of population to number of breeds, using an earlier edition (Mason, 1988). A low ratio indicated a relatively large number of breeds and thus a fairly complete differentiation of breeds. Livestock population data are collated by FAO and can be found in several locations including CAB International (2002). The same general pattern observed in 1988 still remains

Table 6.1 Numbers of breeds classified according to numerical status, compiled from Porter (2002).

		Donkey	Water buffalo	Cattle	Goat	Horse	Pig	Sheep	Totals
Africa	Rare			25	5	9		10	49
	Extinct			36	1	5		4	46
	Common	27	5	218	130	38	17	161	596
	Total								**691**
Asia	Rare			18	6	30	30	12	98
	Extinct			13	1	5	17	6	42
	Common	29	74	244	213	131	285	275	1251
	Total								**1391**
Europe	Rare	18	3	182	68	131	84	174	660
	Extinct	7		186	30	73	93	114	503
	Common	11	7	227	128	151	142	385	1051
	Total								**2214**
N&C America	Rare	1		14	5	27	10	14	71
	Extinct			4	1	4	24	15	48
	Common	7	1	94	21	50	37	38	248
	Total								**367**
S America	Rare	2	2	15	14	4	9	3	49
	Extinct			21		1	2		24
	Common	3	4	64	30	29	18	24	172
	Total								**245**
Oceania	Rare			2	1	2	1	12	18
	Extinct			3		1	1	4	9
	Common		1	32	11	3	7	51	105
	Total								**132**
Ex-USSR	Rare			13	5	28	7	14	67
	Extinct			24	6	19	21	33	103
	Common	15	1	66	15	42	54	132	325
	Total								**495**
World totals	Rare	21	7	269	104	231	141	239	1012
	Extinct	7		287	39	108	158	176	775
	Common	92	93	945	548	444	560	1066	3748
	Total	**120**	**100**	**1501**	**691**	**783**	**859**	**1481**	**5535**

(Table 6.5); when the 2002 data are considered the lowest ratio is obtained for the horse (implying this is the species most completely differentiated into breeds).

If this ratio had decreased markedly from 1988 to 2002, it could be inferred that many more breeds have been discovered. The decrease has indeed happened for pigs but not for water buffalo. Either breeds of water buffalo remain to be discovered or this species is not very differentiated into breeds.

Numerical changes in livestock populations in relation to trade

Changes over the last ten years in world populations are given in Table 6.5. The most dramatic changes are in bird species. The increase of 100 million in goat numbers is

Table 6.2 Numbers of livestock breeds in FAO database (Scherf, 2000).

FAO category	Donkey	Water buffalo	Cattle	Goat	Horse	Pig	Sheep	Other mammals	Birds	Totals
Unknown	62	20	295	166	179	111	391			**1224**
Critical maintained[1]	0	0	30	3	12	11	6			**62**
Endangered-maintained[2]	1	2	66	10	34	21	43			**177**
Critical[3]	5	3	76	28	115	47	62			**336**
Endangered[4]	15	6	127	60	144	85	156			**593**
Not at risk	14	55	630	303	242	223	656			**2123**
Extinct	6	0	255	17	94	151	181			**704**
Totals	**103**	**86**	**1479**	**587**	**820**	**649**	**1495**	**111**	**1049**	**6379**

[1,2] These are breeds for which conservation programmes are in operation.
[3,4] To take account of evident duplications (when the same breed is listed for more than one country), the following numbers of breeds can be removed from the 'critical' and 'endangered' categories.

	Donkey	Water buffalo	Cattle	Goat	Horse	Pig	Sheep
Critical	0	0	35	7	75	18	29
Endangered	8	0	55	21	81	43	74

There is probably also scope for removing breeds from the 'unknown' and 'not at risk' categories.

Revised world totals:

	Donkey	Water buffalo	Cattle	Goat	Horse	Pig	Sheep	Totals
Rare	13	11	209	73	149	103	164	722
Extinct	6	0	255	17	94	151	181	704
Common	76	75	925	469	421	334	1047	3347
Total	**95**	**86**	**1389**	**559**	**664**	**588**	**1392**	**4773**
Total extant breeds	**89**	**86**	**1134**	**542**	**570**	**437**	**1211**	**4069**

also striking. It is almost entirely explained by increases in China (from 95–157 million), India (115–123 million), Pakistan (39–49 million) and Sudan (23–39 million). World trade in goats is small (1.8 million head in 1992, 2.9 million head in 1999) so the increase in goats is leading to increased national herds and production rather than to international supply.

The world pig population has increased by 7% while international trade has increased by 13% to 16 million head in the same period. This reflects increasing globalization and perhaps also a global increase in production efficiency. Globaliza-

Table 6.3 Changes in the breeds listed in consecutive editions (1988, 1996, 2002) of I.L. Mason's dictionary of world livestock breeds.

	Donkey	Water buffalo	Cattle	Goat	Horse	Pig	Sheep	Totals
Existing in 1988, extinct in 1996	0	0	21	4	26	11	23	85
Not rare in 1988, rare in 1996	3	2	53	17	47	34	61	217
Newly added as extinct breed in 1996	0	0	22	1	11	16	9	59
Newly added as rare breed in 1996	6	2	41	26	56	27	33	191
Newly added as not rare breed in 1996	16	15	144	106	126	195	160	762
Removal of breed as not valid		−4			−3	−4	−2	−13
Net change in 1996	+22	+13	+207	+133	+190	+234	+200	+999
Newly added as extinct breed in 1996	2	0	36	7	2	9	5	61
Newly added as rare breed in 1996	1	1	52	26	33	37	17	167
Newly added as not rare breed in 1996	15	12	121	125	43	98	63	477
Total additions in 2002	+18	+13	+209	+158	+78	+144	+85	+705

tion is given as one reason for the five fold increase in Brazilian beef exports between 1997 and 2003, along with low land prices, currency devaluation and reduced sales in many markets of poultry meat (Pearce, 2004).

The increase in cattle numbers of 4.4% is partly due to increased demand for milk, at least in the developing countries where the number of milking animals is increasing. In developed countries, yields are increasing rather than numbers of animals. Only about 7% of global dairy product output is traded internationally (FAO, 1999). International trade in cattle increased by 3.4% to 9.5 million head in 1999, but this is still a very small proportion of the global population.

Sheep have shown complex patterns of change. Of the ten countries with the highest numbers of sheep in 1992, four have shown an increase and the others a decrease, while Sudan and Kazakhstan have reported dramatic changes in opposite directions (Table 6.6). National livestock numbers often change independently of those in neighbouring countries. For example sheep numbers have declined in Pakistan and have increased in neighbouring India; likewise, a decline in Turkey has accompanied an increase in Iran.

The FAO statistics enable the major horse breeding and exporting countries to be identified (Figure 6.1). Numbers have increased in the high income countries, while

Table 6.4 Numbers of rare and uncensused breeds added to I.L. Mason's dictionary in 1996 (Mason, 1996) and in 2002 (Porter, 2002), indicating which countries made the main contributions.

	Added in 1996	Main countries	Added in 2002	Main countries
Africa	87	Sudan 15, Morocco 7	136	Ethiopia 20, South Africa 34
Asia	352	China 219, Mongolia 18, India 17	157	China 53. India 22
Europe	282	UK 33, France 28, Spain 24	236	France 31, Germany 23, Spain 30
North/Central America	97	USA 66, Canada 8	29	USA 16
Oceania	31	Australia 19, New Zealand 10	14	Australia 7, New Zealand 5
South America	53	Brazil 35	52	Brazil 35
Ex-USSR	51	Russia 20	20	Russia 6, Moldova 5
	953		644	

Table 6.5 World livestock populations and numbers of breeds, 1992 and 2001. Census data from CAB International (2002); breed numbers from Table 6.1.

	Population 1992	Population 2001	Percentage change	Ratio of livestock numbers (10^3) to number of breeds worldwide 1992	Ratio of livestock numbers (10^3) to number of breeds worldwide 2001
Donkey	43.7×10^6	42.68×10^6	−2.29	545	356
Water buffalo	153.1×10^6	167.6×10^6	+9.50	1902	1676
Cattle	1302.8×10^6	1360.1×10^6	+4.41	1179	906
Goat	597.6×10^6	693.5×10^6	+16.0	1393	1014
Horse	60.3×10^6	58.3×10^6	−3.40	118	74
Pig	869.6×10^6	927.7×10^6	+6.68	1781	1080
Sheep	1161.3×10^6	1059.1×10^6	−9.80	995	715
Camels	18.2×10^6	19.1×10^6	+5.17		
Other camelids	5.41×10^6	6.20×10^6	+14.5		
Mules	15.1×10^6	13.5×10^6	−10.4		
Rabbits	519.7×10^6	481.6×10^6	−7.33		
Rodents	22.9×10^6	14.9×10^6	−34.9		
Chickens	11.5×10^9	14.7×10^9	+27.3		
Ducks	652.1×10^6	915.4×10^6	+40.4		
Turkeys	223.4×10^6	241.4×10^6	+8.04		
Geese	163.5×10^6	240.6×10^6	+47.2		
Beehives	57.9×10^6	58.9×10^6	+1.68		

Table 6.6 Principal sheep populations and changes from 1992 to 2001 (FAO data from CAB International, 2002). The overall world decline in this period was about 102 million.

1992	Numbers of sheep	2001	Numbers of sheep	Percentage change
Australia	148.2×10^6	China	133.2×10^6	+23
China	110.9×10^6	Australia	120.0×10^6	−28
New Zealand	52.6×10^6	India	57.9×10^6	+8
India	50.8×10^6	Iran	53.0×10^6	+7
Iran	46.0×10^6	Sudan	48.0×10^6	+22
UK[1]	43.9×10^6	New Zealand	43.9×10^6	−8
Turkey	40.4×10^6	UK	42.0×10^6	−2
Kazakhstan	33.9×10^6	Turkey	29.4×10^6	−11
South Africa	30.9×10^6	South Africa	28.8×10^6	−2
Pakistan	26.9×10^6	Pakistan	24.2×10^6	−6
Sudan	26.5×10^6	Kazakhstan	10.4×10^6	−24

[1] This number conflicts with that given in Fig. 4.1; here, the figure includes the year's lamb crop.

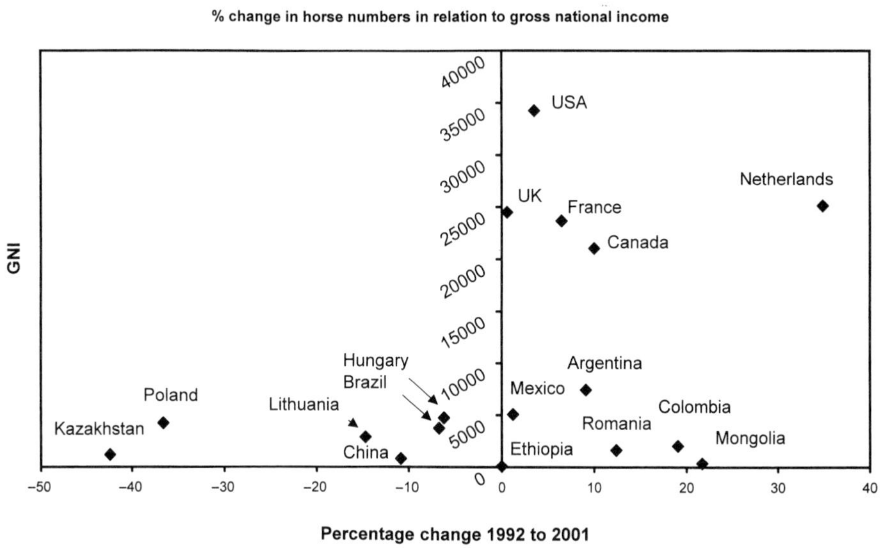

Figure 6.1 Relationship between national wealth and horse numbers. Relationships between gross national income (GNI) and percentage change in horse numbers from 1992 to 2001, for the major horse breeding and exporting countries (FAO data from CAB International, 2002).

in some middle and low income countries numbers have increased and in others, decreased. This reflects national differences in the role of horses in the economy. While Romania is exporting large numbers, the population is increasing but Poland appears to be destocking with high exports and rapid decline in numbers.

Global patterns of trade in live animals and their products are highly complex, changing rapidly in response to exchange rate fluctuations and market developments. Some very poor countries (Table 6.7) depend greatly on exports of live animals while some wealthy countries are big exporters and importers, reflecting local patterns of trade. Exports of live animals around the EU, especially, are of high cash value.

Effects of changes in livestock populations upon livestock biodiversity

Increased national wealth and a drive for greater productivity are likely to threaten livestock biodiversity. Loss of breeds in Europe has been greater in the richer countries (Hall & Ruane, 1993). It is not clear what the effects of overall change in livestock numbers would be on livestock biodiversity. Considering Europe, Asia, Africa and Oceania, there is a non-significant negative correlation ($r = -0.22$) between change in numbers from 1992 to 2001, and change in the proportion of breeds that are defined as rare or extinct, from 1988 to 2002 (Figure 6.2).

Table 6.7 Countries with the highest dependence on live animal exports, FAO data from 1999 (CAB International, 2002).

	Total merchandise exports ($)	Live animal exports ($)	As percentage
Somalia	150×10^6	62×10^6	41
Chad	211×10^6	33×10^6	16
Sudan	596×10^6	87×10^6	15
Mali	665×10^6	95×10^6	14
Niger	276×10^6	29×10^6	11
Burkina Faso	255×10^6	20×10^6	8
Mauritania	448×10^6	33×10^6	7
Namibia	1246×10^6	83×10^6	7

The world's main exporters of live animals (by value: US$, data source as above).

France	1.6×10^9
Canada	1.0×10^9
USA	622×10^6
Germany	511×10^6
Netherlands	496×10^6
UK	488×10^6
Australia	446×10^6
China	383×10^6
Ireland	314×10^6
Mexico	292×10^6

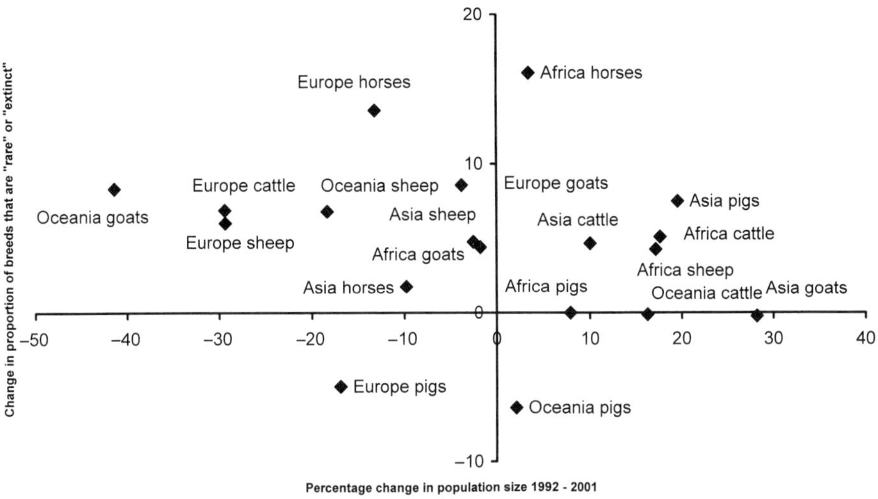

Figure 6.2 Effects of changes in livestock populations upon proportions of breeds that are 'rare' or 'extinct'. Changes from 1992 to 2001 in population size of each species on each continent do not seem to have affected the proportions of breeds in each continent that are listed as 'rare' or 'extinct'. Data from CAB International (2002), Mason (1988) and Porter (2002).

A significant negative correlation would imply that as populations decline, the proportion of breeds that are rare or extinct will increase. A more detailed country by country study would be rewarding. Effects on breed biodiversity are also likely to be fairly country-specific.

Where there is heavy dependence on the export trade one might expect selection or crossbreeding for particular characteristics such as conformation or growth rate. Disease resistance, fertility and longevity would not be so favoured.

When there is increase in numbers and only a small dependence on exports, one would expect local adaptation to be selected, and this would tend to favour local breeds.

Historical and geographic background

Though continental-scale livestock population processes are hard to use for understanding current changes in the status of national breeds, there are some continental-scale generalizations that can be made.

Europe including former USSR

Livestock development has been rapid in Europe and so has the rate of extinction of local breeds (Hall & Ruane, 1993). Against this, public interest and the perception of the value of conserving local breeds in their accustomed habitat is relatively strong here but government support has been patchy. In eastern Europe and the former USSR, privatization of agriculture and access to western technology and finance have placed local breeds under threat (Grunenfelder, 1992; Romanov *et al.*, 1996).

Probably few if any local and traditional European breeds are still undescribed or unreported. Many of the new breed entries in Mason (1996) and Porter (2002) relate to new synthetic commercial breeds which, in conservation terms, are less significant than the local breeds which are at risk of extinction. The amount of recent government-funded work on local breeds has been variable; EC funding in genetic resource research has been biased towards molecular characterization work. This is against a background of a considerable body of work on performance of commercial breeds conducted mainly between about 1950 and 1990.

Asia and Africa

Asia (excluding the former USSR) and Africa both cover about the same proportion (20% each) of the world's land surface area, but Asia has about 60% and Africa 13% of the world's people (Crystal, 1997). Africa is much better endowed with land than Asia (7.6 km^2 per 100 adults south of the Sahara, compared with 0.9 km^2 in east Asia and 0.6 km^2 in south Asia; Wood, in press) but most of it is rather remote from the coast and from large river systems, making trade difficult. Africa's relatively infertile soils make cropping less rewarding and trypanosomosis is a major constraint on livestock and a threat to people. One third of Africa's area is infested with tsetse flies, with today about 50 million people and 46 million cattle at risk (Kristjanson *et al.*, 1999). One might therefore expect livestock biodiversity in Africa to reflect ecological constraints, while that in Asia might reflect human geography. There has been much more domestication and between-species hybridization in Asia (see Chapter 1) and these processes, together with the much greater human population density perhaps explain why Asia (excluding the former USSR) has 1483 breeds of the seven main mammalian species and Africa has 677. In both continents the richness of human cultural diversity and the time-depth of livestock history make human–livestock relationships more complex and subtle than they are in Europe and the Americas.

North and South America

While one of the most potent livestock images – the cowboy – is firmly associated with North America, the time depth of human–livestock interactions is a fraction of that in the old world. Perhaps surprisingly there is no evidence of New World domestication of dogs (Vilà *et al.*, 1999), though several present-day breeds show evidence of introgression from the grey wolf. Although the turkey originated in North America, it appears that it was actually domesticated in Europe having been taken there around 1500 (Brant, 1998). Horses evolved in North America but died out at the end of the ice ages, leaving the Americas horseless until the Spanish introductions after 1492 (Clutton-Brock, 1992b). While in Britain, for example, there is a set of breeds associated with almost every county, this is far from the case in North America. For example, each rare breed of livestock in the USA has a very specific set of historical associations (Christman *et al.*, 1997), rather than there being a general inventory of region-specific indigenous breeds as in Europe.

In South America the pre-conquest biodiversity of domesticated animals persists, with guinea pigs and camelids being components of local farming and smallholding systems, but knowledge of their importance, genetic diversity, the threats that face them and the opportunities for their development is incomplete (Kyle, 1987; Scherf, 1995).

Oceania

The history of the livestock breeds of Australia and New Zealand goes back no further than 1788 and about 1840, respectively (Crystal, 1997), so the present day complement of breeds owes far less to long term cultural processes than it does in other continents. Yet with the large numbers of animals involved, the often harsh weather conditions (especially in Australia), the need to respond to world markets, and at least in more recent years two of the strongest livestock research and development communities in the world, the conditions have been right for divergence of breeds. In many islands of the Pacific, pigs both husbanded and feral are highly important, though not exhaustively documented.

Feral livestock

Feral animals are those that live in a self-sustained population after a history of domestication (Clutton-Brock, 1999). Bixby *et al.* (1994) expand on this:

'A feral animal is one that was once domesticated, or with domesticated ancestors, but now lives as a wild creature. It is not under the effective ownership of humans, and does not receive protection, care, or food as a deliberate gift from man.'

This definition would admit free-living animals that are managed, for example, by culling or castration. There are also feral populations that receive supplementary winter feeding, for example the Chillingham cattle (Figure 6.3; Hall & Clutton-Brock, 1988). Rudge (1986b) points out how some animals acknowledged as feral are managed to some extent, while for others there can be a spectrum from wild, through feral, to domesticated (for example, pigs on some Asian islands). World-wide, wild and feral domesticated livestock are highly important members of many faunas (Long, 2003).

In total, 72 feral populations have become sufficiently distinct to have acquired breed status (Figure 6.4). Feral populations generally tend to be found on islands, but feral horses live in a wide range of environments 'in North America, South America, Europe, Asia and Australasia, from the equator to the temperate-boreal frontier, in deserts and high-rainfall regions, and at low altitudes on river deltas and islands or in the high-altitude mountainous regions of central continents' (Linklater, 2000).

Foundation and survival of feral populations

An important factor in the success of establishment of an introduced species may be the numbers introduced – the larger the founder group, the more likely the survival of

Figure 6.3 The Chillingham cattle are (along with the Soay sheep) the only British livestock to have escaped selective breeding during recorded history. Morphologically they are medieval cattle; genetically their homozygosity reflects their history of inbreeding and genetic bottlenecks.

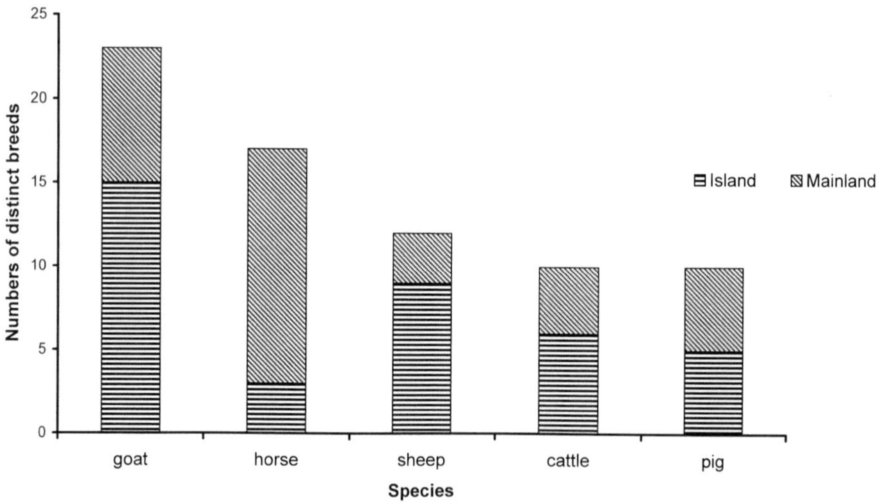

Figure 6.4 Habitats of feral breeds. Classification according to habitat of the feral livestock breeds listed as distinct by Porter (2002). Apart from horses, most are found on islands.

the population (Green, 1997; Fraser *et al.*, 2000). Experience suggests this may be less important with feral animals. Some populations are from very small founder groups – though whether domestication confers upon animals a genotype that is suited to expansion of numbers from a small founder group has not been explicitly studied. In 1788, two bulls and four cows (imported from the Cape of Good Hope) escaped at Port Jackson in Australia. By 1801 this herd totalled 500–600 (Daly, 1981). In the 1950s one male and two female goats were liberated on Pinta (Galápagos Islands), by the 1970s there were 10 000 goats present (Hamann, 1993). Five cattle founded the Amsterdam Island herd (see page 131).

Presumably it is for reasons of ecological preadaptation that some feral populations have expanded enormously while others have not. For example, in northern Australia, water buffalo have been very widespread and numerous while, in contrast, feral banteng, which were censused at around 3000 have not spread very far (Freeland & Boulton, 1990; Choquenot, 1993).

Goats are both physiologically and behaviourally preadapted to the feral state, especially in the absence of predation, as there has been selection under domestication for sociability and fertility (Parkes, 1993). Rudge (1970) found that feral goats hide their young kids, unlike domestic stock – this implies the latter exhibit a telescoped version of the natural behaviour. They also show an interesting lack of propensity to extend their range and most feral populations remain geographically localized.

The strict seasonality of many breeds of sheep constrains the success of some of their feral populations. For example the Bizet sheep, a hardy hill breed, reproduces all year round in its native France but a feral flock now established for 40 years on Kerguelen (subantarctic island) tends to lamb in the austral winter, which is unfavourable for survival of offspring. This appears to be because ewes are in better body condition and more likely to conceive in summer. Thus, reproductive seasonality has not yet been optimized in this feral population (Réale *et al.*, 2000a). Feral sheep are less widespread than feral goats, possibly because unless accumulation of wool around the perineal region is controlled, sheep are prone to flystrike and to problems with mating (Rudge, 1986a).

Feral horses often show high foal mortality (Greger & Romney, 1999). At the same time, populations often appear to be out of control, prompting concern for welfare and suggestions for such management practices as contraception. Generally, feral ungulates show a pattern of density-dependent adult mortality and density-independent juvenile mortality (Chillingham cattle: Hall & Hall, 1988; Soay sheep: Clutton-Brock & Pemberton, 2004). Perhaps genes inhibiting breeding under poor or crowded conditions were selected against under domestication.

Physical appearance

The wild relatives of domesticated mammals normally have cryptic coloration and under domestication, variant colour phenotypes are usually selected for (Clutton-Brock, 1999). One might expect, therefore, that when livestock go feral the wild type would be selected for, but this has not been the case in cattle or sheep and the position is complicated in pigs.

The feral pigs at the Savannah River Plant in South Carolina, USA, showed a decline in frequency of solid black animals and an increase of spotted animals from 1968 to 1984, while a solid white phenotype persisted (Mayer *et al.*, 1989). There was no obvious reason for this change. In South Island, New Zealand, there was local variation in colour that probably reflected which breeds went feral in different areas (Clarke & Dzieciolowski, 1991b). Overall, black was the major coat colour (77% of 2604 pigs shot during 18 months) and grey/blue with a woolly winter undercoat the next most common (12.7%), which was probably derived from wild boar introduced from India to be hunted (Clarke & Dzieciolowski, 1991a). Pigs of the Large Black breed were known to have gone feral in one area, but no trace of their lop ears remains. Other characters to have been lost include the razor-back, mane, long snouts, high skulls, short legs, curly tails and fine hair.

The general picture therefore is that feral pigs develop a phenotype that to a great extent depends on the breed from which they were derived and there is no reversion to wild type unless there has been introgression from wild boar.

Lessons to be drawn from feral livestock

With the growing interest in extensification of grazing and in the importance to animal welfare of natural behaviours (see Chapter 5), it might be thought that study of the behaviour and ecology of feral populations could help with the design of sustainable management systems. Feral animals may be thought particularly well adapted to the environment but selection regimes may vary from year to year and chance demographic events may occur, so exposure to the natural environment is not the most effective way of selecting animals to be well adapted to rigorous conditions (Simm *et al.*, 1996). However, if the regime is maintained for long enough, a feral population could well have a high frequency of genes that would be adaptive under some other environment. The best documented example of feral animals being found to be a genetic resource is feral goats in New Zealand, Australia and the UK which still possess the cashmere undercoat and which have been crossed with Angora goats to build up herds of fine-fibre goats (Russel, 1991). In the 1980s the feral goats of New Zealand, then around 300 000 in number, were found to be suitable for crossbreeding with Angora goats for fibre production (Parkes, 1993). The new venture was not, however, economically viable and a lot of the goats were released back into the wild.

However, there may be genetic insights to be obtained. So far, studies of feral populations have shown that inbreeding is commonplace. The Chillingham herd is the most homozygous population of cattle in the world, only segregating at one of the 25 microsatellite loci for which it has been tested (Visscher *et al.*, 2001). This is explained by their history of isolation rather than their feral state. They have probably been inbreeding for at least 67 generations; for comparison, all the classical mouse inbred strains have been inbred for at least 60 generations (Silver, 1995), though usually based on a single pair. This makes the Chillingham herd a unique genetic resource, perhaps of value as much because of its lack of heterozygosity as for any traits of interest it may have. As an inbred strain it may be capitalized upon for genomic mapping of traits. The current long-term studies of the Soay sheep of

St. Kilda (Clutton-Brock & Pemberton, 2004) are yielding much insight into mammalian ecological genetics, population dynamics and evolutionary biology.

Several other feral populations also show only limited genetic variability. In Australian feral goats and many traditional, extensively farmed goat breeds in south-east Asia this is apparently due to uncontrolled mating (Barker *et al.*, 2001). The feral horses which inhabit a Namibian desert are genetically very uniform and also inhabit a very harsh environment (Cothran *et al.*, 2001). Continued viability of these highly inbred populations suggests that their low levels of genetic variation are not a handicap so long as the environment remains unchanged. However, it seems likely that they would be very vulnerable to disease, making biosecurity a prime concern for their conservation.

Regarding behaviour, there may be insights from feral animals into how their wild ancestors behaved and also some feral populations such as Soay sheep, the Chillingham cattle and the Mostrenca cattle of the Doñana National Park in Spain (Lazo, 1994) and feral horses generally, have shown how livestock animals are likely to behave if free of human control. Thus they serve as ethological resources, of potential value for understanding of animal welfare and of human–animal interactions. For example, the feral cattle of the Chihuahuan Desert in Mexico (Hernández *et al.*, 1996, 1999) number about 1000. They have a flight distance of up to 2 km in response to a man on horseback. This suggests that cattle that are not exposed to people during a sensitive period of their social development as young calves remain wary of people throughout life.

In winter once grazing has been exhausted pastured cattle will tell the farmer it is time to provide feed by gathering at the gate and waiting, more or less quietly. Is this behaviour innate or learned? It is striking how the feral cattle of Swona, Orkney Islands, Scotland, which have not been fed in winter since about 1978 (Hall & Moore, 1986), forage as a single herd when the herbage has all been grazed. The alternative strategy would be to forage individually. If this grouping behaviour was a characteristic of the ancestral aurochs, it could have been a contributing factor to the development of human–cattle relationships and of domestication.

Genetic management of feral populations will need to be designed in a sensitive manner. As conservation of these populations would have as a goal the maintenance of their full behavioural repertoire, it would not be appropriate to try and manipulate reproduction. The strategy might be to permit the population to operate its own sexual selection regime and to maintain reserve breeding groups and practise cryo-conservation. This is the policy for the Chillingham herd; there is a reserve population in northeast Scotland and semen and embryo stores are being established.

Conservation significance of individual feral populations

In the USA, when feral populations are to be removed or eradicated the American Livestock Breeds Conservancy considers whether the stock should be conserved elsewhere (Brisbin, 1990). Their criteria for conservation are:

- Historical evidence of existence in the feral state for a substantial period of time, usually 100 years, without introgression

- Possession of unique traits of scientific interest or value
- Commercial value, for example for recreation or food
- Importance to local communities or cultures.

Accordingly, Brisbin (1990) considered the Andaman Islands pig (Bay of Bengal) and the Ossabaw Island pig (Georgia, USA) as particularly worthy of conservation. A conserved herd derived from feral Australian goats was found by Barker *et al.* (2001) to represent a large proportion of the genetic variation in south-east Asian goats. If it were to become extinct, 14% of the microsatellite diversity (see Chapter 7) would be lost. These issues of valuation and prioritization are considered for breeds generally in Chapter 8.

Feral populations can be expensive to manage *in situ*. A feral horse population in Wyoming (Bastian *et al.*, 1999), which has numbered up to 500, has led to reduced populations of wildlife, range deterioration and reduced cattle stocking rates. Management is impeded by objections by activists to rounding up the horses and by the difficulty of disposing of them in ways that satisfy public opinion. At high population numbers, the opportunity cost of the horses (the wildlife and cattle income lost) is $1900 per animal per year.

A very distinctive feature of the livestock biodiversity of Australasia is the coexistence of domestic and feral stocks. In Australia alone there are feral banteng, water buffalo, cattle, dromedary, horses, goats and pigs, while the dingo is a feral dog. Feral pigs are widespread in New Zealand (Fraser *et al.*, 2000) but the most common introduced ungulates there are deer. In the Indonesian and Papuan archipelagos there is a remarkable diversity of 'native, introduced, feral, domestic and hybrid pig populations' (Oliver & Brisbin, 1993). The first pigs were introduced here 6000 years ago and the subsequent history of pigs in the region is highly complex. These stocks include many possibly valuable genetic resources and many of them have profound cultural significance for particular human societies (Hertrampf, 1990).

In 1770 Captain Cook introduced pigs to Australia – as a result, Australian feral pigs are known as 'Captain Cookers'. Pigs might have been introduced into the Cape York Peninsula across the Torres Strait from New Guinea well before then by aboriginal peoples (Baldwin, 1983). The evidence is biological, cultural and linguistic. A nematode parasite in the local feral pigs is common in New Guinea but absent elsewhere in Australia – the adult pigs are of distinctly small body size while the newborn piglets have longitudinal body stripes. Known patterns of aboriginal trade are consistent with an introduction from New Guinea and pigs are depicted in Cape York rock paintings. Finally, the word for pig is similar in at least one Cape York language to that in the language used on some of the Torres Strait islands – in other aboriginal languages the word is clearly derived from English.

Ethical and legal issues

Feral animals presumably receive different degrees of legal protection in different countries, although no international survey exists. For some purposes wild animal law might apply, for others domestic and farm animal law and food safety law may also be involved. Ethically and legally the most interesting situation is probably the deliberate release of farmed animals into the environment. This is relevant to the use of animals for habitat enhancement (see Chapter 5) and it has been discussed in the Dutch context (van Liere, 1999; Kampf, 2000). Here, the concept of 'de-domestication' has been developed and some breeds are seen as intrinsically suitable for this process, such as the Konik pony and Heck cattle. This suitability is due partly to the adaptations they are claimed to have to the de-domesticated state and partly to their appearance which is very different from those of familiar husbanded breeds. Perhaps the main issue in the public mind is what is to be done if animals are injured or diseased. The general approach is to leave them alone if they are expected to recover of their own accord, or to euthanase them if they are suffering badly.

Feral livestock, sensitive habitats, human culture and animal disease

Amsterdam Island (Micol & Jouventin, 1995) is one of the world's most remote islands, being 3200 km from Australia, 4200 km from South Africa and 3300 km from Antarctica. It is 55 km^2 in area (Figure 6.5). Five cattle were introduced in 1871 from La Réunion and these founders were of French taurine type. By 1988 they numbered 2000 and had done serious damage to the vegetation and were threatening the breeding success of the endemic Amsterdam albatross. However, their interest as one of the world's very few herds of feral cattle and potential value as a distinct adapted population (the island is very short of water for part of the year) led to a management plan being devised which protected the albatross and conserved the

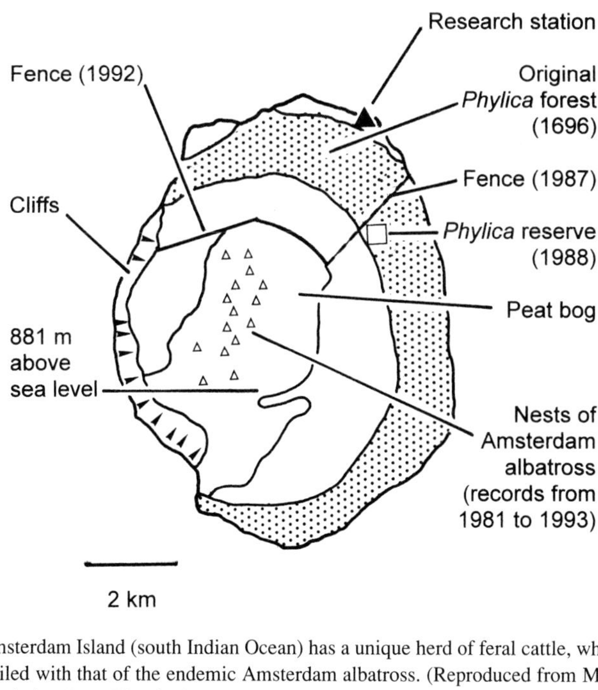

Figure 6.5 Amsterdam Island (south Indian Ocean) has a unique herd of feral cattle, whose conservation has been reconciled with that of the endemic Amsterdam albatross. (Reproduced from Micol & Jouventin (1995) with permission from Elsevier.)

cattle. A fence was erected across the island to keep the cattle off areas dedicated to the recovery of the native flora and the herd heavily culled (the population in May 1993 was 532 animals).

Recovery of the vegetation on one of the floristically richest small islands of the Galápagos group, following elimination of feral goats, has been documented by Hamann (1993).

In contrast, feral populations may sometimes benefit the environment. Carcasses of large animals support rare carrion-feeders and provide, through local enrichment, niches for rare plants (Kampf, 2000; Towne, 2000). The management plan prepared by Scottish Natural Heritage to protect rare seabirds on Swona stipulates no change in the grazing regime which depends on the small feral herd of cattle. Effects of feral animals are not uniform across the natural flora and fauna – grazing by feral horses on the saltmarshes of Shackleford Banks, North Carolina, leads to increased diversity of foraging birds and densities of crabs, but increased predation on fishes (Levin *et al.*, 2002).

In New Zealand there is a National Feral Goat Control Plan which began in 1990 and which, it was anticipated, would lead to about half the feral goat populations on land that is reserved for the indigenous biota being eradicated or controlled within ten years (Parkes, 1993). Thus, there is a whole range of approaches to management of feral livestock, from complete eradication, sometimes with 're-domestication', through range restriction, to appreciation of their positive role, and to the establishment of new feral populations.

Feral pigs can be numerous. For example the population in Australia was estimated as between 3.5 and 23.5 million (95% confidence limits; Hone, 1990). There are about three million farmed pigs in Australia (Porter, 1993). Feral pigs predate lambs and could be a reservoir of livestock disease (Choquenot *et al.*, 1997). While they can cause highly visible damage to the environment and to native flora and fauna, this has not often been compared to that due to other environmental factors (Pavlov *et al.*, 1992). That feral animals could be a reservoir for disease is a common worry. Feral bees are a potential reservoir for the varroa mite (Cobey, 2001) while many populations of feral ungulates in New Zealand carry bovine tuberculosis (Fraser *et al.*, 2000). The large-scale programme of eradication of feral livestock in northern Australia was aimed at eliminating or getting into husbandry the population of feral water buffalo because they were considered a reservoir for brucellosis and bovine tuberculosis, which were eradicated from Australia in 1989 and 1997, respectively (Cousins & Roberts, 2001).

Conclusions

The number of known breeds continues to increase and now stands at 5535 for the seven main mammalian livestock species of which 775 are relatively recently extinct and 1012 are known to be rare. There are big differences between the continents in nature of this livestock biodiversity, which reflect differences in human and cultural history. In the developed world the main threats to livestock biodiversity are related to agricultural development, while in the developing world there is a lack of documentation of breeds, which could easily be lost before they are recorded.

Feral animals can give insights into the behaviour of husbanded animals and in principle they could serve as genetic resources. Whether they receive conservation support should probably be determined on a case-by-case basis. In many cases feral populations are detrimental to wild fauna and flora, while in others they are a source of enrichment. Scientifically, they have contributed to understanding of ecology and population dynamics, but their contribution to understanding the behaviour of husbanded animals has been relatively limited. Ethical issues of responsibility for feral and free-ranging animals have not been addressed in depth yet and this requires further thought.

Chapter 7

Characterization of livestock biodiversity

Introduction

The foundation of floral and faunal conservation biology is taxonomy. Not only does this lead to the species inventories that show which habitats are most worthy of conservation, it also gives an objective basis for deciding which species are most important. From one perspective, the most important species to conserve are those which, collectively, represent the maximum amount of evolutionary history or distinctiveness. Does this paradigm also apply to livestock biodiversity? While distinctiveness can be determined through the deduction of genetic distances, contributing to a rational basis for prioritization for conservation, distinctiveness in agricultural function and in aspects of performance are clearly relevant. Do taxonomies based on these different criteria relate to each other? Within breeds, there can be much individual variation. How is this to be characterized, and what is its significance?

Breed characterizations: phenotypic variation

Many, perhaps most, breeds, are defined as distinctive from their close relatives on the basis of a small number of genes. One of the most obvious ways to characterize breeds is by their coat colour. In mammals, six main loci are involved (Nicholas, 1996a, Chapter 12) and certain other loci are important in particular species. In pigs, the molecular genetic events leading to distinct coat colour patterns have been defined (Table 7.1). Sometimes breed characteristics that appear trivial in comparison with performance attributes can be seen as quite important. For example, in Canada, Black and White Holstein and Red and White Holstein dairy cattle have similar performance (Sullivan *et al.*, 1988) but the latter may be preferred for crossing with the Simmental (a brown and white breed) because it would not threaten the uniformity of the latter's colour pattern, which is important for breeders.

In many breeds, a feature may arise independently and in time it may become a distinguishing attribute. For example several European beef breeds exhibit double muscling (muscular hyperplasia, an increase of about 20% in muscle mass, caused by proliferation of muscle fibres). In Belgian Blue cattle, this is the result of an 11

Table 7.1 Molecular genetics of pig coat colours.

Phenotype (ref.)	Breed	Genetic basis	Operation of gene	Nature of mutation
Dominant white Marklund *et al.* (1998)	Large White, Landrace[1]	*KIT* locus: 3 alleles recessive *I* – wild type (in wild boar) semidominant I^p fully dominant *I* (dominant black)	Encodes growth factor controlling migration and survival of melanocyte precursors	*I* seems to have arisen from a gene duplication and a single (G → A base substitution leading to dysfunctional gene product
White belt Giuffra *et al.* (1999)	Hampshire[2]	4th allele found at *KIT* locus		
Dominant black Kijas *et al.* (1998)	Hampshire, Large Black[3] and Meishan[4]	*MC1R* locus (same as *Extension* locus which is well known in mammalian coat colour genetics) 4 alleles *MC1R*1* (wild type), *MC1R*2* (Large Black and Meishan), *MC1R*3* (Hampshire), *MC1R*4* (recessive red colour of Duroc)	Encodes melanocortin 1 receptor, regulates eumelanin (black/ brown) and phaeomelanin (red/ yellow) synthesis within melanocyte	Base substitution
Black spots Kijas *et al.* (2001)	Gloucester Old Spot[5], Berkshire[6] and others	5th allele found at *MC1R* locus.	Mutation causes recessive red colour; black spots result from somatic reversion events that restore *MC1R* function. These reversions are likely to be under separate genetic control and have presumably been selected in different breeds	

[1] Major commercial breeds worldwide. *I/I* homozygotes may have reduced white blood cell count. Fairly high frequency of I^p in commercial herds, in spite of ineligibility of spotted pigs of these breeds for pedigree registration
[2] American breed
[3] British breed, with history linking it to eighteenth century imports from China
[4] Chinese breed
[5] White British breed with a few black spots
[6] Completely black British breed with 'six white points' – feet, blaze of nose, tail tip. The Berkshire is to be seen as a white pig with very extended black spotting

base-pair deletion in the gene coding for the hormone myostatin, but double-muscled Maine-Anjou cattle do not carry this deletion, implying that a number of different alleles produce this phenotype (Grobet *et al.*, 1997). Double muscling is also seen in some sheep and pig breeds with a different mode of inheritance and pattern of development. True muscular hypertrophy, caused entirely by enlargement of existing cells rather than by proliferation, is also a genetically determined condition in some sheep breeds where it is called callipyge (from the Greek words for beautiful and buttocks; Cockett *et al.*, 1999). Particular breeds of pig are known to have specific characteristics (Warriss *et al.*, 1990, 1996). For example, the Hampshire pig (an American breed) has a high frequency of a dominant mutation denoted RN^- which leads to elevated muscle glycogen (and thus to improved meat quality)–the mutation evidently took place within the breed (Milan *et al.*, 2000).

Some breeds have a phenotype that is due to a single distinctive allele. For example the 'Sauteur Alfort' rabbit has a recessive mutant causing it to do 'handstands'. This is associated with an ocular cataract (Joly *et al.*, 1994). Another breed defined by a single gene is the Polish morph of the mute swan (Scherner, 2000). The *immutabilis* allele makes birds look 12 months older and such swans have been bred as ornamental birds, while other varieties have been bred for meat.

Examples of characterizations can be found in FAO DAD-IS, and in Simon and Buchenauer (1993). These databases provide the fundamental information for breed inventories. An obvious use of this information is for making comparisons among breeds, in phenotypic features of economic value. The simplest way to make such comparisons is to give figures on what performance is expected under standard conditions (Nix & Hill, 2001 for British dairy cattle). These comparisons have been arrived at by trials (Nicholas, 1987, Chapter 15), conducted by government or university research organizations, or breed societies.

The first step in characterization is to identify the homogenous groups (the breeds) within the heterogenous assemblage (the livestock population) and to establish how the breeds are themselves grouped. In the developed world this is straightforward as the breeds have breed societies. In a far more challenging environment, FARM-Africa (1996) surveyed the goats of Ethiopia and Eritrea by recording 21 qualitative and quantitative variables from approximately 500 goats at each of 103 sampling sites. They screened these variables for variability and discriminating power, calculating certain ratios among them, and performed principal component analysis on the set of 103 observations to generate a hierarchical classification tree of 14 clusters, each of which was accepted as a breed. These clusters themselves fall into four groups (Figure 7.1a,b). The general description, herd demography, physical characteristics, products and productivity of each breed were then summarized. The sample sizes were large (altogether 41 447 goats were recorded) enabling means for growth traits, which are known to have high coefficients of variation to be calculated with small standard errors. The survey provides a solid foundation for the further characterization and rational utilization of each breed. The morphological clusters in Figures 7.1a,b show some correspondence with the distribution map; this could imply some ecological adaptation or natural selection, as mentioned in connection with goats in Chapter 1.

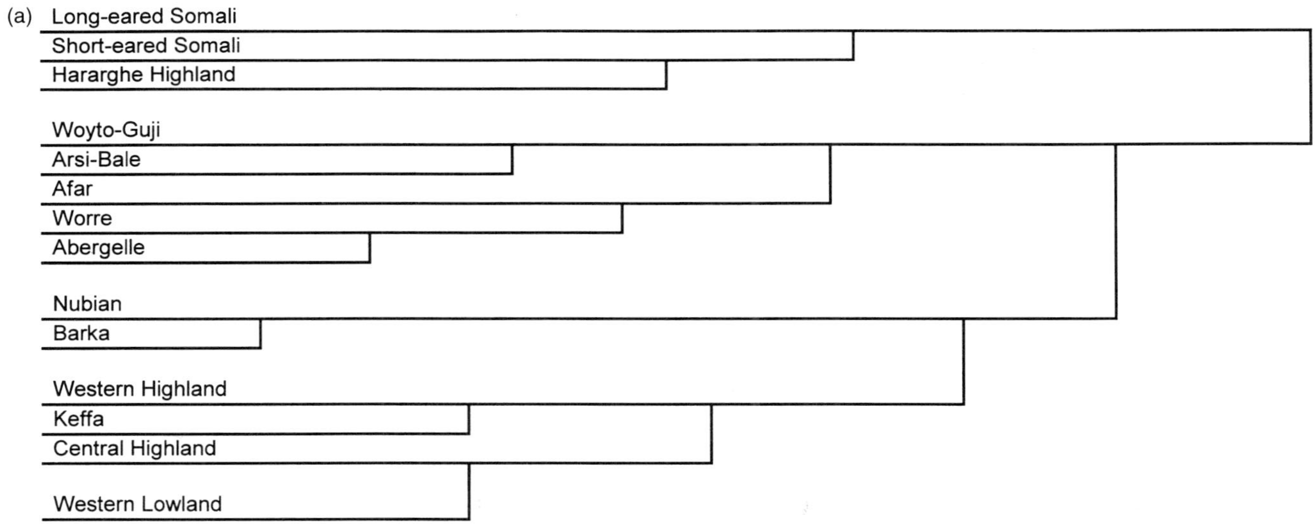

Figure 7.1 The 14 clusters into which Ethiopian goat breeds can be classified on morphological grounds, and their geographical distribution. (Reproduced with permission from FARM-Africa.)

(b)

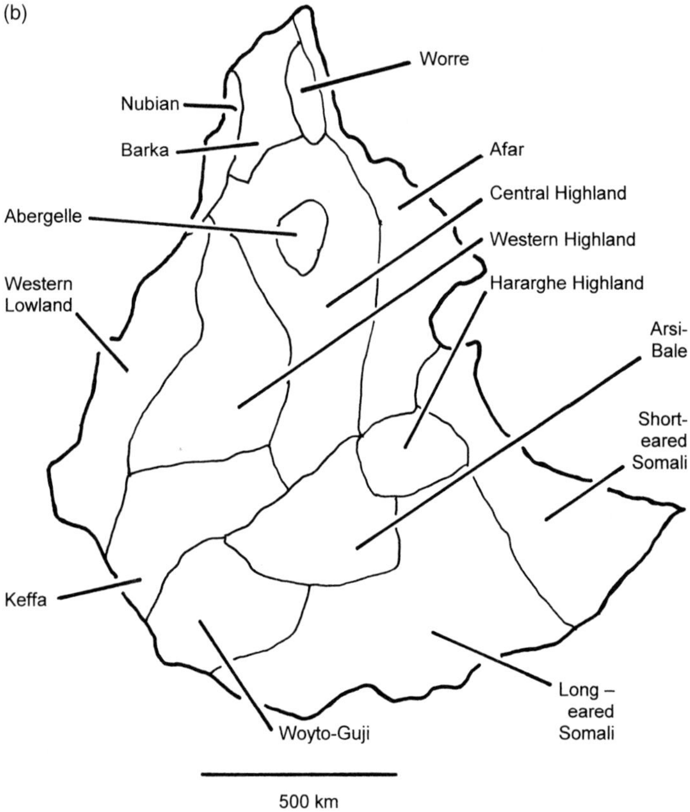

Figure 7.1 (*Continued*)

Results are not always so clear cut. Using microsatellites Wimmers *et al.* (1999) grouped individual chickens from six local populations in Nigeria into clusters, but the new groupings did not correspond to the sampled populations indicating a lack of geographical differentiation and making it impossible to consider them as distinct breeds.

Pre-existing data can be used to characterize breeds. Roughsedge *et al.* (2001) used genetic data on puberty and fertility, preweaning, postweaning and milk yield traits, to classify 17 commercially important taurine cattle breeds into two groups – the larger terminal sire breeds and the dairy and small-bodied maternal breeds. Blanc and Ennesser (1989) used a matrix of published data on morphology, biology, habitat and geographical distribution of 48 populations of dromedary to classify them subjectively into three groups and eight subgroups.

Sometimes these phenotypic classifications can overlook differences which have cultural or biological reality. Camels in India provide an example. Pre-existing classifications, introduced by British colonial officials, have been essentially ecological or functional, using terms such as Hills camels, Plains (Desert or Riverine) camels or Baggage and Riding camels (Kohler-Rollefson, 1993). With increasing knowledge of the husbandry and breeding practices of the camel-owning ethnic groups and an appreciation of the segregation that exists between camel populations, it is becoming clear that camel breeds can be defined, though not usually according to western preconceptions about what constitutes a breed (see Chapter 2). For example Kohler-Rollefson and Rathore (1995) describe how until recently only two Indian camel breeds were generally accepted as distinct, but information is now available about several more, notably the Malvi, which numbers about 2500–3000 animals in 13 villages in Madhya Pradesh. The Malvi is said to be a good milking animal and is also distinctive by virtue of its rather small size and white colour. Many other livestock populations may also have a distinct though unacknowledged identity (Kohler-Rollefson, 1997). However, this kind of classification might well not be supported by molecular genetics – Mburu *et al.* (2003) found Kenyan dromedaries to comprise just two separate genetic entities rather than the four previously recognized.

Barker *et al.* (1997a,b) reported genetic studies on 17 populations of water buffalo in south-east Asia in a study which is of great methodological interest. It showed that populations which were morphologically very similar and which do not have acknowledged breed identity are found, once genetic distance measures have been applied, to be as distinctive as breeds of cattle, pigs, goats and sheep. These findings are also important because it can easily be assumed that local and previously undescribed livestock are not of any particular breed and thus not worthy of special attention or recognition. For example in India and Pakistan the term 'desi' ('native, indigenous, and hence nondescript, breedless, unimproved'; Mason, 1996) has been widely applied, and distinct local varieties might thus have remained unappreciated.

The phenotypic features that are most useful for classifying breeds would be the ones which are most heritable, although this has not been explicitly studied. Some traits are the visible expression of single alleles. Machado *et al.* (2000) compared Mediterranean and Brazilian goats using loci coding for presence/absence of such features as long hair, reduced ears, horns, wattles (these are small fleshy protuberances hanging from the neck), beard, roan colour, brown eumelanin and eumelanic standard pigmentation to establish breed relationships. These characters are completely heritable ($h^2 = 1$). Some quantitative traits are highly heritable, for example cattle shoulder height and mature female body weight (0.48 and 0.50 for Holstein-Friesian and for beef cattle, respectively; Simm, 1998). Others are less so (rump width in Holstein-Friesian 0.22) or else the heritability is unknown.

Lauvergne *et al.* (1993) developed the idea of 'index of primarity' and have applied it to goats. This is based on the assumption that a wide range of phenotypes in a population is indicative of a lack of directional selection and thus of antiquity. It provides a quantitative means of assessing the degree to which a population has lost its supposed original characteristics, for example through introgression or selection. This technique may well prove useful in investigating large-scale processes but is clearly sensitive to population size and inbreeding, though there are examples of numerically small breeds that have considerable morphological variation (Folch & Jordana, 1997). Nevertheless morphological methods of characterizing breeds would be much cheaper than molecular techniques (Figure 7.2).

Figure 7.2 Morphological characterization. Recording a Kuri cow, Lake Chad, northern Nigeria.

Measuring animals is not a technically advanced activity but as a means of characterizing breeds it has certain advantages. It measures aspects of the phenotype (size, conformation) that traditional breeders probably include in their selection criteria. It can easily be combined with an interview-based survey. However, some countries are becoming sensitive about the export of genetic material, which might make molecular work difficult.

Characterization by performance

Traditionally, breed characterization has been based on a comparative approach, the study aiming to detect a given degree of difference between breeds with a given degree of error (Figure 7.3). On this basis it can be decided whether to practise selection within a breed, or to replace a breed with a better breed. Bruns (1992) tabulates the numbers of animals it is necessary to record, for given degrees of accuracy and under different assumptions. For example, in an on-farm study, 200 unrelated animals per breed from a total of 100–200 randomly selected flocks (i.e. not closely related animals) would suffice to detect differences between breeds of about 10% of the mean. If breeds are kept together at a research station, so there is no between-farm variance, then 200 animals per breed, the progeny of 20 sires, would suffice. In their meta-analysis of performance, Roughsedge *et al.* (2001) used data from trials where at least ten sires per breed had been used. With relatively minor changes to the number of animals and the design, heritabilities can be estimated too.

Whether this kind of study would ever be done for any rare breed or for breeds in the developing world generally, is highly unlikely. Taylor (1976) pointed out how

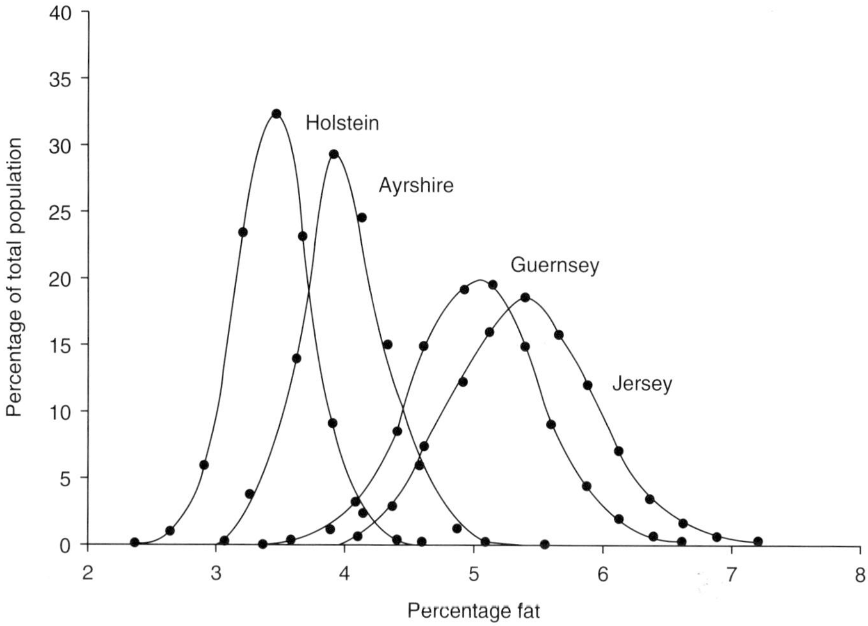

Figure 7.3 Within- and between-breed differences in dairy cattle. A classic illustration (from Lush, 1946) of how breeds can overlap in a quantitative character, but the differences between their means are real.

this approach aims to answer such questions as 'what is the difference between breed A and breed B in growth rate and feed efficiency'. He wrote that it may be more helpful to answer more general questions like 'what is the extent of between-breed variation in growth rate and feed efficiency'. 'Multibreed' designs can achieve this (Thiessen *et al.*, 1984, 1985). If within-breed genetic variation is found to be more important than between-breed genetic variation, then breed replacement is not justified. Taylor (1976) reanalysed earlier work on sheep; for ewe weight and litter weight, a multibreed design with a total of 120 ewes and 30 breeds with four ewes per breed can give as accurate results on between-breed variation as a design with a total of 6000 ewes and 13 breeds with an average of 500 ewes per breed.

Heritability is a property not only of a character but also of the population and environment in which it is measured and of the way the phenotype is measured (Falconer & Mackay 1996, p.161; Simm, 1998, p.270). In populations where genes have become fixed, heritabilities will tend to be lower and in variable environments they will be lower than in more stable environments. It is therefore very difficult to compare heritabilities between breeds. Generally traits associated with reproduction, behaviour, disease resistance and survival have low heritabilities; with milk production and early body size traits, medium; and with adult size and general quality traits, high. In practical animal breeding, response to selection can only be predicted if heritabilities are known (preferably using data from the population to be selected), but if this information is not available 'a published estimate for a similar kind of population kept under similar conditions' will generally suffice (Wiener, 1994).

A very important aspect of characterization in the developing world is village-level monitoring. This can be done using individually identifiable animals. The French overseas programme on animal production included a database of 490 000 individual animals (Matheron & Planchenault, 1992). Many studies are based on one-off surveys, like the Ethiopian/Eritrean project previously described. Female case histories are relatively easily collated and analysed (RIM, 1994; Hall, 1998). Farmers are asked to describe the breeding careers of particular animals in their flocks and herds, and the fates of these offspring are summarized. Offtake rates and other herd statistics can be calculated, but there are no publications on how to validate the findings.

Monitoring and survey programmes like these are usually intended to identify constraints on productivity of traditional systems or to determine how novel systems and innovations can be disseminated. It is important that traditional systems should be described and characterized themselves, including such aspects as the use of ethnoveterinary medicine and surgery and patterns of ownership, transfer and marketing. Here social science researchers have played a major part (for example RIM, 1994). There is great scope for such work in local livestock systems all over the world. Mariner and Roeder (2003) give an excellent example of the value of interviewing herders about the incidence of disease; traditional laboratory-based approaches would have been inadequate in the remote areas where they worked.

Characterization by genotype

Genetic characterization of a breed can take place in three ways. First, the quantitative genetics underlying aspects of performance could be characterized, second genes, which may not have detectable effects on performance could be scored for presence or absence. In the first case, for each breed of interest the heritability of characteristics of interest would be calculated and genetic correlations with other characteristics established. In the second case, presence or absence of particular alleles would be used to calculate genetic distances between breeds and to make inferences about their history. The first two approaches are tenuously linked, in that it is possible that genetic distance between breeds may predict the heterosis to be expected when they are crossed. In the third approach, breeds could be scored for presence or absence of single genes of large effect (major genes) which could in principle be available for use as a genetic resource elsewhere. Genetic distance measurements could be useful in planning crossbreeding strategies (see Chapter 4). However, there has been very little progress in characterizing the quantitative genetics of any breeds other than those which are most important commercially. Heritabilities depend on environment, characterization work done on minor breeds would be just as expensive as that done on commercially important breeds where a financial return is less uncertain, and the relevance of work done on experimental stations to the commercial environment is not clear, especially in the developing world. Consequently this situation is unlikely to change.

Functional genes

In livestock, many phenotypes are coded in a simple Mendelian way and the existence of different alleles can be deduced. For example, immunogenetic studies began in cattle in the late 1930s and the availability of blood grouping led to tests for the validation of parentage data. Biochemical polymorphisms in livestock began to be investigated in the 1950s and presence of particular alleles was inferred from the activity of the different enzymes they produced.

Molecular techniques mean alleles can be detected that differ by a single base substitution (Nicholas, 1996a, p.58) and the phenotypic consequences can be observed. For example, in Merino sheep, animals with particular variants of a gene connected with keratin synthesis had different wool fibre diameters (Parsons *et al.*, 1994).

Microsatellites and minisatellites

Microsatellites and their applications in conservation genetics are described by Frankham *et al.* (2002). They are currently the first choice as indicators of evolution in livestock. Some studies (for example Arranz *et al.*, 1996) have compared them explicitly (and favourably) with other polymorphic systems. A big advantage of microsatellites is their high heterozygosity. Assumptions of neutrality are easier to justify for microsatellites than for biochemical polymorphisms. While some microsatellite loci may be linked with loci that are under selection (Kantanen *et al.*, 2000) this is less of a problem given the large number of microsatellite loci that are available for study. Most genetic distance models assume that any allele can mutate to any other (the infinite alleles model). In microsatellites, mutations are often stepwise, an allele gaining or losing a single repeat unit, though large jumps in repeat number are also possible. Mode of mutation is relevant to the calculation of expected distributions of allele frequencies, so microsatellite studies usually include discussion of which evolutionary model has been used.

Minisatellite analysis can give individual-specific profiles (DNA fingerprints, Frankham *et al.*, 2002) but population genetic inferences are not straightforward because the different bands are not ascribable to particular loci.

Amplified fragment length polymorphisms (AFLPs)

To produce AFLPs (Mueller & Wolfenbarger, 1999; Herbergs *et al.*, 2000), genomic DNA is incubated with two restriction enzymes, one which acts at cleavage sites which are known to occur frequently and the other at sites which are rarer. This produces large numbers of fragments. Adapters of known sequence are then attached by DNA ligase to the ends of the fragments. The next step is a PCR using two primers, which are complementary to the two adapters. This generates many copies of a large number of fragments of varied length from all over the genome. What these fragments have in common is that at one end they were cleft by the frequently-acting restriction enzyme and at the other end, by the other enzyme. PCR is then carried out again with radio-labelled primers and the fragments can be examined after electrophoresis. The very large numbers of different fragments generated must be reduced or else analysis is too complex. This is achieved by modifying one of the PCR primers to an appropriate extent. Extending the primer by one base (A, C, G or T) into the unknown part of the fragment beyond the restriction site reduces the number of fragments the primer binds to by a factor of 4; by two or three bases, 16 and 64, respectively. The modified primer is radiolabelled so a fingerprint pattern is visualized after polyacrylamide gel electrophoresis. The patterns are analysed by special software.

Polymorphisms are caused by a mutation either in the restriction site or in the region of the selective nucleotide(s) or an insertion/deletion in the amplified fragment. This gives a presence/absence matrix. Ovilo *et al.* (2000) describe AFLP on Iberian pigs leading to 1733 amplification products ranging from 70–700 nucleotides in length of which 26 were strain-specific (present in all individuals of one strain and absent in the others). AFLPs are clearly valuable for understanding within-breed diversity but because they are based on 'presence' or 'absence' until recently it was impossible to tell whether an individual is heterozygous or homozygous for a given fragment (Ajmone-Marsan *et al.*, 1997).

Centromeric satellite DNA

DNA which is apparently non-functional is found in the centromeres of chromosomes – in cattle it makes up 15–20% of the total genome. One bovine satellite consists of 23 base pair subrepeats – restriction enzymes cut out the satellite, which is probed with a labelled oligonucleotide. Nijman *et al.* (1999) found differences between zebu and taurine cattle in numbers of subrepeats, enabling degree of zebu introgression into African breeds to be quantified.

Mitochondrial DNA

Mitochondria are self-reproducing units found in all eukaryotic cells. They have their own DNA, which has been fully or partly sequenced in the main livestock species. The mitochondrial genome has a 'control region' and in vertebrates this includes a displacement loop (D-loop) structure, which functions in replication. As the mtDNA genome is usually maternally inherited (Frankham *et al.*, 2002) there is no recombination so mtDNA can be seen as lineages transmitted in the female line. Mitochondrial DNA can evolve very rapidly. Giuffra *et al.* (2000) found the region which includes the cytochrome B gene evolved at different rates in different lineages of pigs and suggested causes may be differences in efficiency of DNA repair, in generation time and in metabolic rate. Mitochondrial effects may account for up to 5% of the total variation in dairy performance of cattle (Simm, 1998).

Summarizing and interpreting genetic differences between breeds

When breeds are formed, what was previously a single population becomes subdivided and gene frequencies begin to change in the new subpopulations. To quantify this differentiation F_{ST} can be calculated which is

$$\frac{\text{variance of gene frequency among subpopulations}}{\text{mean gene frequency (1 - mean gene frequency)}}$$

F_{ST} was developed by Sewall Wright (Cavalli-Sforza *et al.*, 1994; Frankham *et al.*, 2002, Chapter 13). It can take values between 0 and 1. When a locus is being considered with only two alleles their frequencies in a set of populations follow the binomial distribution. Here, the variance is equal to the mean so if two populations are being considered the numerator in the above formula becomes the difference in allelic frequencies in the two populations. This formula gives a measure of the genetic distance between two populations and describes genetic differentiation. It can be seen as the proportion of genetic variation that is ascribable to differences between breeds. In natural species, if F_{ST} is over 0.15, subpopulations are considered to be 'significantly' differentiated (Frankham *et al.*, 2002, Chapter 13). If there is random mating, and the genes studied are selectively neutral, then the only population genetic process operating is random genetic drift, and F_{ST} can be used to establish an evolutionary timescale. F_{ST} is equal to

$$1 - e_z^{-t/2N}$$

where t is the number of generations since the population split, e is the base of natural logarithms (2.718) and N is the effective population size.

With appropriate assumptions, F_{ST} can provide statistics relating to population processes. It is approximately equal to $1/(4Nm + 1)$ where m is the proportion of migrants and N is the effective population size (Frankham *et al.*, 2002, Chapter 13). Nm is the effective number of migrants per generation, i.e. the number of individuals whose transfer between the population would counter the effects of genetic drift in driving them apart. In principle this could provide guidance to breed societies proposing to accept introgression from other breeds yet wishing to maintain breed identity. Effective number of migrants per generation has been calculated for some groups of breeds, for example in seven Spanish horse breeds (plus the Thoroughbred) these numbers range between 1.4 and 9.5 (Cañon *et al.*, 2000) for different pairs of populations.

There are two more F statistics, (Frankham *et al.*, 2002, Chapter 13):

F_{IS} measures genetic differentiation within a single subpopulation
F_{IT} measures genetic drift within the population as a whole
and they are related thus:

$$1 - F_{IT} = (1 - F_{IS})(1 - F_{ST})$$

Falconer and Mackay (1996, p.96) point out how in random mating populations with gene frequencies only being affected by genetic drift the F statistics are the same as inbreeding coefficients. F_{ST} is the average inbreeding of the subpopulation relative to the whole population, F_{IS} the inbreeding of the individual relative to the subpopulation and F_{IT} the inbreeding of the individual relative to the whole population. For genetic distance calculations among breeds other formulae are often used and those

implies that these (respectively) Spanish and British breeds separated about 2000 years ago. Buchanan *et al.* (1994) inferred a time of divergence of 1094 years ago, suggesting that microsatellites gave more reliable estimates of the times since breeds diverged. Manwell and Baker (1977) had recalculated the genetic distance, using only the polymorphisms where there were low allele frequencies (< 0.05). They suggested that these polymorphisms were the ones which may have arisen by mutation after divergence. This yielded a more realistic time since divergence of 1470 years. They also suggested that polymorphisms with high allele frequencies (between 0.2 and 0.5) represented selection and the estimate of 69 700 years since divergence was partly a measure of divergence under different selection regimes in the two breeds. These contrasting results are a reminder that evolutionary divergence between taxa can be a consequence of time having elapsed since there was a common ancestor, or of there having been different selection histories.

The large-scale genetic distance projects that were carried out in the 1990s have generated very many papers. In a 'preliminary literature review' Barker (1994) found over 100 papers on genetic distances in livestock. However these studies have not led directly to any proposals or rationale for allocation of conservation efforts (Ruane, 1999) but they have helped to confirm the place of animal genetic resources as an important component of biodiversity. They have added objectivity to the belief that today's livestock biodiversity represents summed evolutionary history that is worthy of conservation. New interdisciplinary initiatives (Blench & MacDonald, 2000) are demonstrating how cross-fertilization between these hitherto rather separate sciences, can radically improve our understanding of the place of domesticated animals in human culture.

The results of these studies are not always clear-cut. Some breeds have been shown to have genetic affinity on the basis of microsatellites which are clearly very distinct in function (the Charolais appears close to the Friesian and the Jersey to the Aberdeen Angus, MacHugh *et al.*, 1994, 1998). In Canadian dairy cattle, even though the Canadienne is firmly believed to be distinct and has been given high priority for conservation, it was found to have strong microsatellite affinities to the Holstein, Brown Swiss and Jersey (Hansen *et al.*, 2002).

In pigs there have also been problems producing reliable microsatellite phylogenies of breeds (Laval *et al.*, 2000). Worryingly, there is not yet a clear set of genetic distances for Chinese breeds of pig, many of which are under threat and the available data do not clearly link genetic distances with either geographical distribution or production characteristics (Wu *et al.*, 1999). New developments in the study of human genetic variation, e.g. single nucleotide polymorphisms (Frankham *et al.*, 2002) are being applied, revealing for example that several pig breeds that have been selected for high meat production (Piétrain, Large White, Landrace and Hampshire) have an allele at a QTL that differs from that in breeds which have not been thus selected (Berkshire and Gloucester Old Spot; Van Laere *et al.*, 2003).

Microsatellites have been effective in confirming relationships among inbred lines of chicken, in many of which they show high homozygosity and also in showing that lines descended from the jungle fowl have variation that has been lost from commercial and other inbred lines (Romanov & Weigend, 2001).

As gene mapping and sequencing proceed, it may be that breeds are investigated for possession of specific alleles of practical use. At present this is mainly being done for pigs. Ciobanu *et al.* (2001) describe how traditional breeds can be searched for alleles for meat quality, prolificacy, growth and leanness, and disease resistance.

Use of genetic distance studies to deduce conservation priorities

The Weitzman approach to biodiversity (Thaon d'Arnoldi *et al.*, 1998) is a mathematical procedure, which computes what would be the effects upon total diversity of the extinction of a particular taxon. Examples of its application include Asian and Australian goats (Barker *et al.*, 2001), where the hypothetical loss of the Ujung Pandang goats of Indonesia is calculated to reduce genetic diversity of this group of breeds by 18% and 18 European cattle breeds (Cañon *et al.*, 2001) where loss of the Mirandesa breed would reduce genetic diversity of the sampled group of breeds by 11% (Figure 7.5), and 49 African cattle breeds, where it was combined with a novel assessment of extinction probability (Reist-Marti *et al.*, 2003).

The technique is currently being developed with refinements, including ways of linking it with deductions of probability of extinction and with likely economic benefits of conservation, as well as modifications to take account of variation within breeds (Bruford, 2004; Caballero & Toro, 2002).

Molecular approaches to breed history

Very soon after biochemical and immunological polymorphisms were first applied to livestock breeding in the late 1950s, they began to be used to investigate breed

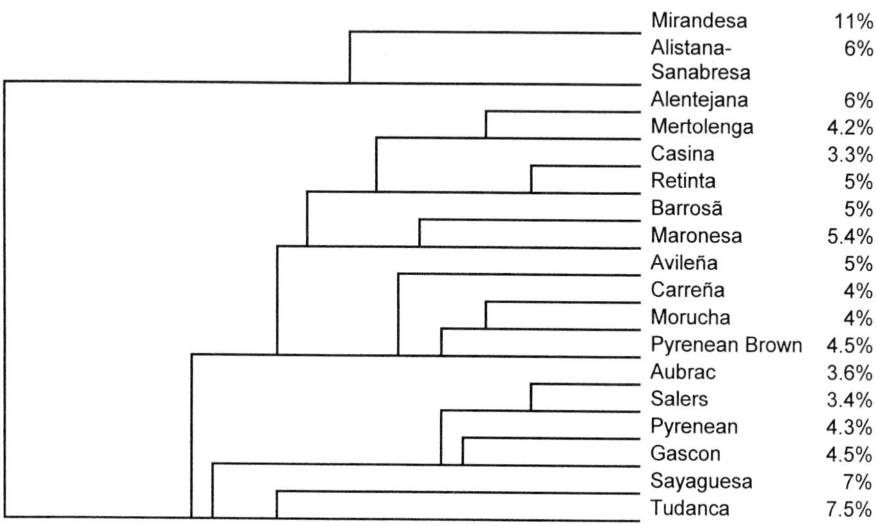

Figure 7.5 Relationships among southwest European cattle breeds. Tree of relationships of ten Spanish, five Portuguese and three French cattle breeds. The values in brackets represent the loss of diversity that would be caused by the hypothetical extinction of a breed or group of breeds. (Reproduced from Cañon, Alexandrino, Bessa *et al.* (2001) with permission from EDP Sciences.)

history. The new molecular techniques allow these questions to be investigated again in much greater detail and a whole new range of hypotheses to be tested. For example, patterns of variation in the genotypes of today's breeds can in principle provide 'genetic signatures' of historical events such as genetic bottlenecks, geographical spread and population expansion (Bruford, 2004).

The studies of individual breeds that have been made have tended to use molecular methods to verify historical accounts. For example Cunningham *et al.* (2001) confirmed that the modern Thoroughbred horse is derived from a very small number of ancestors (Figure 7.6) while Visscher *et al.* (2001) demonstrated extreme microsatellite homozygosity in the highly inbred Chillingham cattle. In Holstein-Friesian cattle, patterns of linkage disequilibrium imply that effective population size has decreased from about 1000 to 100 over the last 700 years (Goddard, 2003).

Within-breed compared with between-breed variation

Within a population phenotypic variance can be divided into environmental, additive genetic, dominance and epistatic components. When the population is subdivided into breeds, variation can be partitioned into within-breed and between-breed components. The proportions of the genetic variation explained by the latter component can be quite high, especially when variation in quantitative characters is considered (Table 7.2).

Within-breed variation provides the flexibility that breeds need to have if they are to respond to changing conditions. Its presence also means that an individual animal will be heterozygous at enough of its loci to be able to cope with the challenges of life. When a high proportion of variation in a useful character is within-breed it

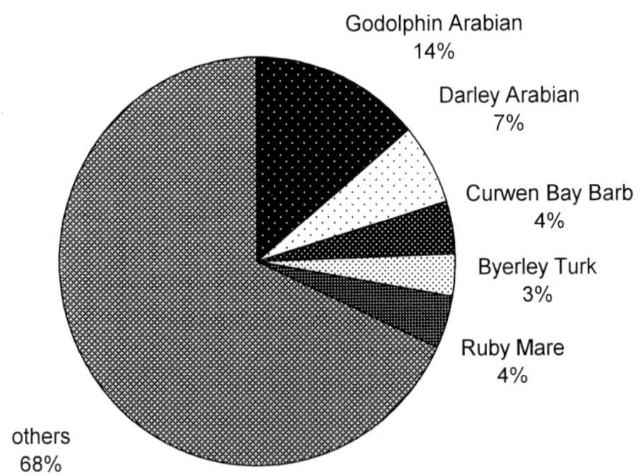

Figure 7.6 Contribution of named ancestors to present-day Thoroughbred horses A large proportion of genes in today's Thoroughbred population (over 300 000 horses worldwide) comes from just a small proportion of the founders of the breed (Cunningham *et al.*, 2001). 78% of alleles trace back to 30 founders of which 27 were male. The Darley Arabian is responsible for 95% of paternal lineages.

Table 7.2 Some estimates of the proportions of variation that are ascribable to variation within-breed and between-breed.

	% variation between breeds	% variation within breeds	Variant	Reference
Cattle	7	93	bovine microsatellites	Cañon *et al.* (2001)
Cattle	10–11	89–90	bovine microsatellites	MacHugh *et al.* (1998)
Dogs	9.9	90.1	biochemical polymorphisms	Jordana *et al.* (1992)
Donkeys	4	96	horse microsatellites	Aranguren-Méndez *et al.* (2001)
Goats	11	89	AFLP	Ajmone-Marsan *et al.* (2001)
Goats	17	83	bovine microsatellites	Saitbekova *et al.* (1999)
Goats	8.8	91.2	AFLP	Crepaldi *et al.* (2001)
Horses	8	92	horse microsatellites	Cañon *et al.* (2000)
Pigs (23 Chinese breeds)	31	69	blood proteins	Wu *et al.* (1999)
Cattle	28	72	food conversion efficiency	Thiessen *et al.* (1984, 1985)
	25	75	relative growth rate	
	70	30	body weight	
	60	40	cumulated voluntary feed intake	

means that within-breed selection is appropriate. If a high proportion is between-breed, it is quicker to change to a different breed.

The nature of within-breed variation

When there are two or more alleles at a locus and the frequency of the scarcest one is too high to be due to recurrent mutation, the locus is said to be polymorphic. Genetic diversity is usually measured as the proportion of loci that are polymorphic, the proportion of loci that are heterozygous in an average individual or the average number of alleles per locus (Frankham *et al.*, 2002, Chapter 3).

These are the main underlying causes of presence of polymorphism:

- The heterozygote may be advantageous. In Hereford cattle in the USA between 1945 and 1955 there was a high frequency of a recessive gene (producing 'snorter dwarfs') that in its heterozygous state seemed to produce a compact, blocky carcase that was favoured by the market (Nicholas, 1987, p.175).

- The heterozygote may be disadvantageous. In pigs a neonatal diarrhoea is caused by strains of *E. coli* bacteria with a cell surface antigen called K88 (Nicholas, 1987, p.177). This antigen attaches to a receptor in the pig's intestinal mucosa which is coded by a dominant gene *S* (the recessive gene *s* codes for absence of receptor). Sows which are *ss* do not attach the K88 positive bacteria so do not produce antibodies; if their piglets are sired by a boar which has the *S* gene, therefore, the *Ss* piglets will attach the bacterium but will not have received the antibodies from their dam and will be likely to be affected by diarrhoea.
- The physical environment may differ from time to time, so each of the alleles may have particular advantages and the gene frequencies reflect the balance of these advantages. Long-term studies in Scottish Blackface (1954–1974) and Welsh Mountain (1966–1972) sheep (Hall & Purser, 1979) showed that ewes homozygous for haemoglobin type B produced significantly more twin pairs (18% for Blackface and 38% for Welsh) than those homozygous for type A (8% and 30% respectively). However, overall, BB ewes lost a significantly higher proportion of their lambs during the years of greatest environmental stress (Blackface: 22% of lambs of BB ewes lost, 13% for AA; Welsh: 18% of lambs of BB ewes lost, 9% for AA).
- Population density may play a part. In Soay sheep, individuals with a wider bite survive the winter better but are perhaps at a disadvantage when population density is low, which is when selectivity in grazing would be advantageous (Illius *et al.*, 1995).
- Countervailing selection could maintain polymorphism, favouring different genotypes according to which fitness component is being considered. A possible example comes from pigs. About 10% of Swedish Landrace and Swedish Large White pigs carry the partially dominant *Ip* allele at the dominant white spotting locus. As homozygotes these would have black spots and would not be eligible for pedigree breeding. This is a high frequency for an apparently undesirable allele and this implies a countervailing selective advantage, perhaps in some aspect of fertility or the production of white blood cells (Marklund *et al.*, 1998).
- The observed polymorphism may be a transitional stage in the evolutionary replacement of one allele by another. It may reflect selection at a closely linked locus. This replacement is sometimes called a selective sweep; the newly favoured allele increases in frequency in the population and so do any other closely linked alleles. Selective sweeps have been demonstrated in *Drosophila* (Nurminsky *et al.*, 1998) and in maize (Wang R.-L. *et al.*, 1999). They are probably more evident in livestock than in other animals, thanks to the rapid genetic change that often accompanies the heavy use through AI of small numbers of sires.

Gain and loss of within-breed variation

Within-breed genetic variation of these kinds is clearly essential for a breed to retain viability and adaptability. However, the population processes that inevitably operate within breeds (inbreeding, genetic drift and selection) will all tend to reduce within-breed genetic variation. To some extent this variation may be restored by mutation.

Conservation geneticists are interested in mutation mainly because, with the action of inbreeding, harmful mutations will lead to depressed fitness. This is likely to be relevant to rare livestock breeds. In contrast and in spite of the likelihood that only about 10–12% of mutations are neutral or beneficial, in highly numerous breeds there can, in principle, arise through mutation useful genetic variation that is accessible to selection (Hill, 1982). In practice, though, the actual value of this variation for livestock breeding is not known (Notter, 1999).

How within-breed variation persists is explicable by population genetics theory, which is founded on the behaviour of an idealized population (Frankham *et al.*, 2002, p.188). This idealized population is one in which:

- There is no migration
- Generations do not overlap
- Breeding population size remains the same
- All individuals are self-fertile and hermaphrodite
- Gametes unite at random
- There is no selection or mutation
- Number of offspring per adult (family size) has mean and variance of 1

If genetic processes in a real population of size N, such as loss of genetic variation, take place at the same rates as those which would be observed in an idealized population of size n, then n is said to be the effective population size (usually termed N_e). N_e can be a tiny fraction of N (see Chapter 9). When population size fluctuates from generation to generation the long-term N_e is close to the smallest N_e observed (Frankham *et al.*, 2002, p.247).

There are many ways in which the actual population might differ from the ideal, so there are several reasons for N_e being lower than N. In wildlife (Frankham *et al.*, 2002, p.241) the main cause is fluctuations in population size, then variance in family size and finally variation in sex ratio. In livestock breeds because of the general effect of only a few males having the majority of the matings, variance in family size is probably the main cause, at least in the short term. This is considered further in Chapter 9.

Small population size

The expected loss of heterozygosity over t generations from a population whose effective size is N_e, is as follows:

$$H_t = H_0(1 - 1/2N_e)^t$$

where H_t and H_0 are heterozygosity at time t and time 0 respectively (Frankham *et al.*, 2002, Chapter 10). Thus an animal population with a generation length of three years, at an effective population size of 50 for 200 years would lose 49% of its original heterozygosity.

The mechanism for this loss of heterozygosity is inbreeding, which changes genotypic frequencies, but not gene frequencies (Nicholas, 1987, p.370). Inbreeding thus moves alleles out of heterozygotes into homozygotes. For the inbred individual, this can have two direct effects – deleterious recessive alleles become expressed and heterozygous allele pairs are lost. The results are that genetic defects are likely to be expressed and performance is likely to suffer too, because if there has been heterosis (heterozygotes performing better than homozygotes, see Chapter 4) this is lost. For the breeder, inbreeding exposes recessive alleles and selection can be applied to make the population uniform for some easily selected trait of simple genetic background such as coat colour, possession or absence of horns, etc. Once the breed is established and selection is proceeding for quantitative traits, inbreeding must be restricted or viability will suffer.

In principle, inbreeding reduces heterozygosity of neutral alleles in proportion to the inbreeding coefficient and without inbreeding their genotypic frequencies would be as predicted by the Hardy-Weinberg equilibrium (Nicholas, 1987). The average inbreeding of a population can be calculated by finding the deviation of genotypic frequencies of neutral characteristics from Hardy-Weinberg equilibrium.

Selection

Inbreeding does not change gene frequencies, but selection does. After many generations of selection for a particular characteristic a population would be expected to be homozygous for favourable genes (or at an equilibrium, if overdominance is involved, see Chapter 4) and no further genetic improvement would be possible unless new genetic variation is brought in. The population is then said to be at a selection limit. There are several examples of this in experimental animal populations (Falconer & Mackay, 1996, p.215). In one study mice were selected for increased body weight while another group from the same base population was selected for lighter weight; when the populations ceased to respond to selection the heavy mice were 2.5 times the weight of the light mice. In contrast, most livestock breeds seem to have been much less inhibited by selection limits (Simm, 1998; Falconer & Mackay, 1996). Many of the traits being selected nowadays have clearly not been heavily selected in the past (as shown by their high heritabilities) and rapid progress is still possible.

Some studies suggest that within-breed genetic variation is very resistant to depletion even under apparently intensive selection. For example, Thoroughbred racehorses are selected purely for racing performance and are derived from a very narrow genetic base (see Figure 7.6). The stallion 'Darley Arabian', who began to be used in 1704 (30 generations ago) accounts for 95% of the paternal lineages and ten founder females account for 72% of maternal lineages. Yet in spite of this narrow pedigree base there is still considerable microsatellite variation, and evidence (Gaffney & Cunningham, 1988) that additive genetic variation for racing performance persists in the breed, because about one third of the differences between horses in track performance can be attributed to genetic causes. However, winning times are not decreasing so it appears that racing performance may have reached a plateau. There are several possible reasons for the apparent selection limit – current selection methods may not be exploiting the available variation, training methods may not be ideal (Hill, 1988) or there may be a physiological limit. Racing performance may involve so many anatomical and physiological adaptations and their interactions, that gains in one component may impose penalties in others (Eckhardt *et al.*, 1988).

Bottlenecks

When a population is founded by a small number of individuals, gene frequencies in the founder group are usually different from those in the original population (Nicholas, 1987, Chapter 5). This can be accompanied by geographic fragmentation and the result is genetic differentiation among the separated populations. This is the founder effect or 'genetic bottleneck' and it is a special case of genetic drift (Frankham *et al.*, 2002, Chapter 13). A population size bottleneck of N individuals is predicted to reduce heterozygosity by the proportion of 1/2N.

Considering quantitative characters a bottleneck would be expected to cause the same proportionate reduction in additive genetic variance. Several experiments have tested this expectation (Meffert, 1999; Wolf *et al.*, 2000) and, surprisingly, evidence is accumulating that genetic variance can actually increase following a genetic bottleneck. In a two year study involving the breeding of over 10 000 animals, Cheverud *et al.* (1999) put experimental mouse populations through a genetic bottleneck (two male-female pairs through four generations); 55 populations were maintained thus, and their adult body weight compared with that of control non-bottlenecked populations. Genetic variance was calculated within each strain in the F2 generation and was found to be 75% greater than would be expected on a purely additive model.

There are at least two possible mechanisms for this. A genetic bottleneck can mean that rare recessives become common, and then variance due to dominance increases (Frankham *et al.*, 2002, Chapter 8). Secondly, there has been debate on what effect epistasis would have, and while it could lead to an increase of additive genetic variance after a bottleneck (Goodnight, 2000a) it has also been argued that inbreeding depression would reduce the accessibility of this variation to selection (López-Fanjul *et al.*, 2002).

Whitlock and Fowler (1999) found experimentally that additive genetic variance in morphological characters did not increase after a bottleneck, rather it declined with considerable variation among inbred lines. Evidently the change in additive genetic variance for a given population is unpredictable, even though the average change is calculable. Clearly founder effects on quantitative traits are complex, and may depend on whether the traits are morphological or fitness-related.

These findings are particularly interesting in the context of domestication. Current models (Clutton-Brock, 1999) seem to embody a contradiction in that it is accepted that only small numbers of wild animals were domesticated, but there was enough variation, much of it seldom if ever seen in the wild, for humans to exercise their preferences upon. These models of increase in heritable variation as the result of a bottleneck may demonstrate that there is, indeed, no contradiction.

Genetic bottlenecks can also be viewed from the perspective of effects on single genes and DNA sequence data can in principle show whether a population has undergone a selective sweep or a genetic bottleneck. Both processes will reduce the average number of alleles per locus (allelic diversity) but a selective sweep will not have uniform effects across the entire genome while a genetic bottleneck would be expected to lead to evolutionary change at all tested loci (Galtier *et al.*, 2000).

In a bottleneck, rare alleles are lost, so allelic diversity declines but heterozygosity is not reduced in the same proportion because these rare alleles contribute little to it. If this difference is statistically significant in a population this can be taken as indicating a bottleneck (evident in Thoroughbreds, but not in Turkish and Egyptian horses and Shetland ponies; Cunningham *et al.*, 2001).

Genetic bottlenecks are at the basis of breed formation in that they lead to the fixation of genes that identify and distinguish the breed. Many distinguishing features of this kind, such as coat colour in pigs are inherited as single genes and breeds are homozygous at the relevant loci.

Inbreeding depression

Inbreeding depression is a very frequent observation in livestock, but the actual effect of a given degree of inbreeding can be quite variable. For example, in beef cattle the weaning weight of female calves shows greater depression at a given degree of inbreeding than does weaning weight of males, while the converse is true for weight at about one year of age (Burrow, 1993). It can also vary between lineages within species (Frankham *et al.*, 2002, Chapter 12). For example, there are two sections in the American Standardbred horse breed, pacers and trotters. Cothran *et al.* (1986) found that with increased inbreeding pacers tend to show a decrease in conception and foaling rates, while trotters tend to show an increase. Generally, though, performance deteriorates with inbreeding (Figures 7.7a,b,c,d). Considering Holstein dairy cattle, Smith *et al.* (1998) found that inbreeding had little effect on conformation traits (udder shape, general body form) but did affect performance and reproduction. A 1% increase in inbreeding led to an increase of 0.6 days at age of first calving, decrease of six days in total productive life, and reduced lifetime production of milk by 177 kg.

The overdominance model of inbreeding depression says that heterozygotes have the highest fitness, and the partial dominance model says that inbreeding depression results from increase in frequency of deleterious recessive or partial recessive alleles (see Chapter 4). The weight of experimental evidence favours the partial dominance model (Roff, 1997, p.338; Bijlsma *et al.*, 1999) and this leads to the important prediction that deleterious alleles can be purged. It is also clear (Lande, 1998) from experiments with *Drosophila* that populations are more likely to be able to purge recessive lethal mutations as they become expressed in homozygotes, but 'it is difficult or impossible for inbreeding to purge the more nearly additive mildly deleterious mutations'. It is conjectured that half of the inbreeding depression is due to rare recessive lethal mutations and half to a large number of mildly harmful mutations. Indeed the slightly harmful alleles may become fixed rather than being purged by inbreeding (Bijlsma *et al.*, 1999).

Bijlsma *et al.* (1999) found that *Drosophila* populations which had been artificially purged of harmful recessives exhibited inbreeding depression when kept under novel environmental stresses – inbreeding depression may therefore be due to different loci in different environments.

In wild animals, inbreeding depression has greater effects on life history traits (related to fitness) than on morphological traits (Frankham *et al.*, 2002, Chapter 12). A possible reason is that while the mutations in fitness-related traits are normally

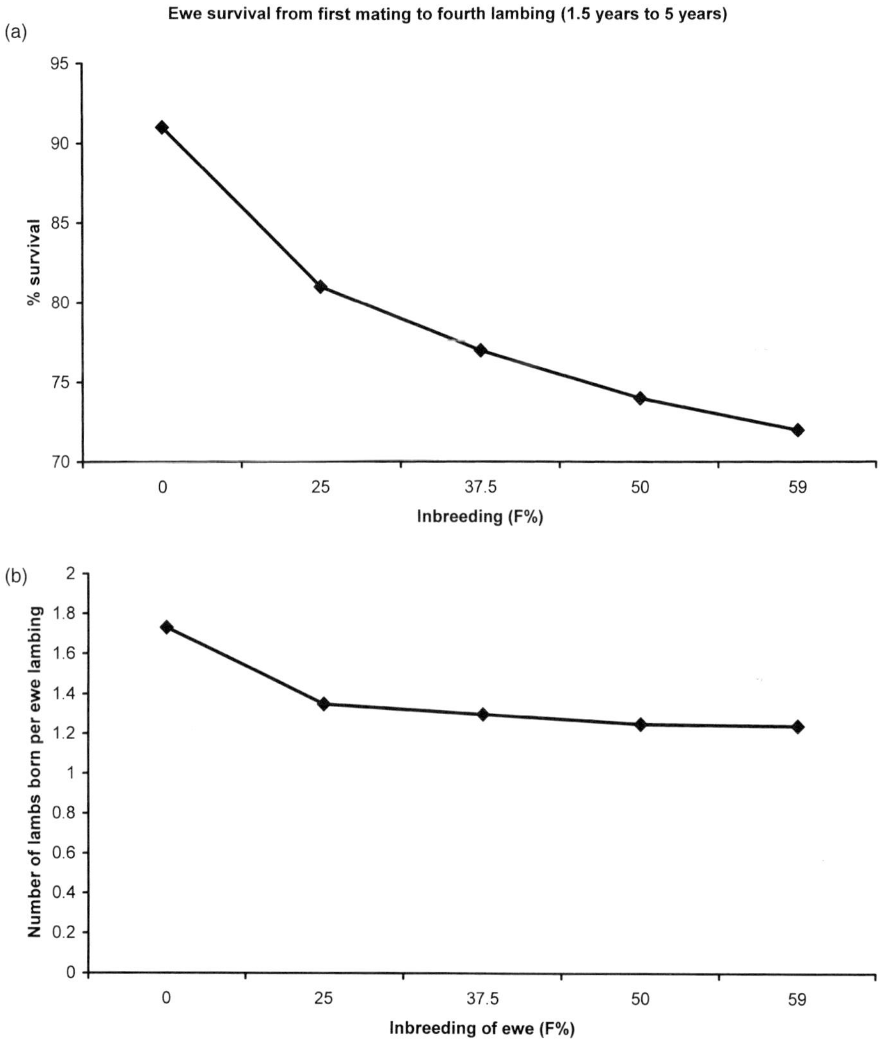

Figure 7.7 Experimental rapid inbreeding of sheep. Effects of rapid inbreeding (parent–offspring mating) in sheep with deliberate avoidance of selection (Wiener, 1994). The main effects were: (a) decline in ewe survival from about 92% to 74%, (b) reduction in lambs born per ewe lambing from 1.73 to 1.26, (c) reduction in lamb survival – inbreeding of lamb has a greater effect than inbreeding of dam, (d) overall a decrease of about 50% in the total weight of lamb weaned for each ewe originally mated. (Reproduced from Wiener (1994) with permission from Macmillan Education.)

(c)

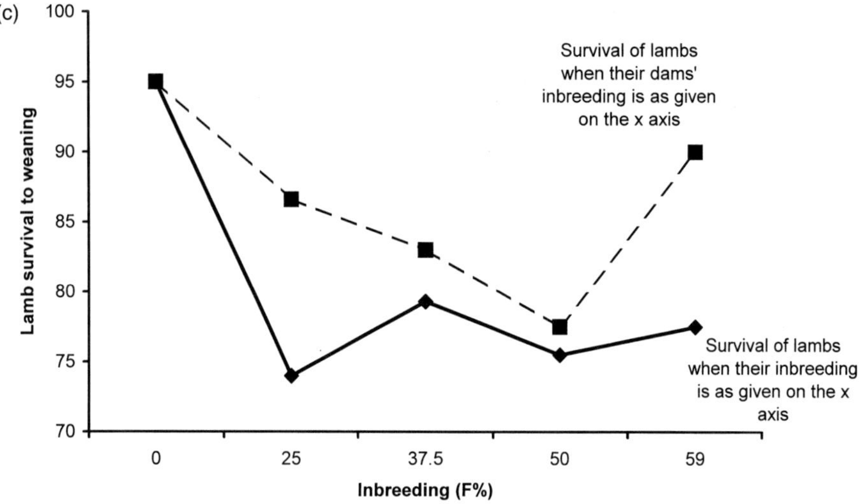

(d)

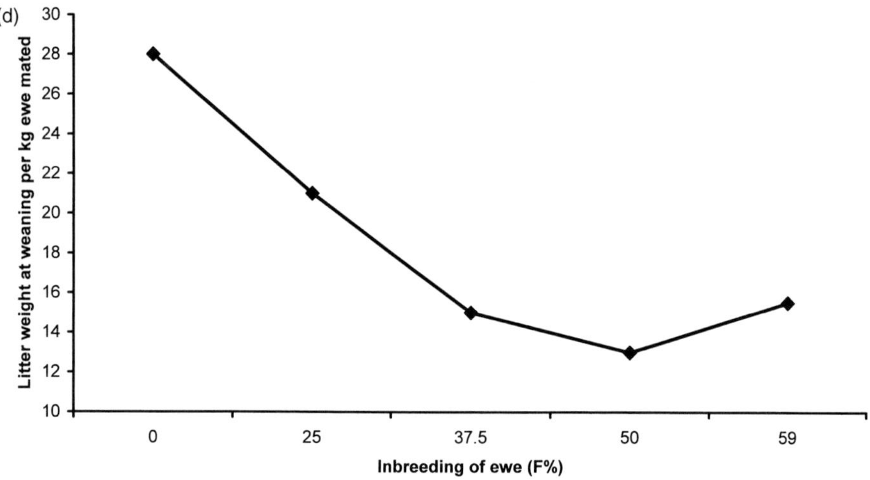

Figure 7.7 *(Continued)*

deleterious, those in morphological traits will tend to push the trait to one side or the other of an optimum and, on average, these effects will even out. The observations on dairy cattle previously mentioned are consistent with this.

Inbreeding: rates and thresholds

If a given level of inbreeding has been reached quickly, i.e. if the rate of inbreeding has been high, then the degree of inbreeding depression can be rather greater (Frankham *et al.*, 2002, Chapter 12). There is some evidence of thresholds of inbreeding beyond which inbreeding depression becomes accentuated. Age at calving for Holstein and Jersey cattle may become increased when inbreeding passes 10% (Thompson *et al.*, 2000a,b) and in Japanese Black cattle breeding life is curtailed when inbreeding exceeds 11.2% (Mukai *et al.*, 1990).

Experiments on inbred lines have often shown that only a minority of such lines survive for any reasonable length of time. Wiener *et al.* (1992b) found only 44 out of 159 sheep lines (28%) survived rapid inbreeding. The importance of rate of inbreeding is shown by contrasting these results with those of Ercanbrack and Knight (1981). They describe 56 lines of sheep, developed by slow inbreeding (1% increase per year) from 1927 to 1969. None died out through lack of viability. Falconer and Mackay (1996, p.253) describe how only 3 out of 20 lines of mice remained after 6 generations of experimental inbreeding. So breed differentiation involving prolonged and relatively complete reproductive isolation will succeed in only 5–10% of cases. In practice breeders will not continue to breed pure if there is the prospect of extinction or of substantially reduced performance, through genetic drift. Wright (1977, Chapter 16) describes how new individuals were introduced three times into Thomas Bates's famous Duchess strain of Shorthorn cattle, during the seventeen years following thirteen years of close inbreeding. Indeed, all the early Shorthorn breeders practised outbreeding to some extent (Darwin, 1868, vol.2, p.119).

Inbreeding will cause reduced performance and when it is combined with use of small numbers of sires it can also lead to high incidences of genetic defects. Japanese Black cattle have been heavily selected for meat quality with intense inbreeding and today they exhibit a high incidence of Mendelian recessive genetic defects including renal tubular dysplasia, achondroplasic dwarfing, Chediak-Higashi syndrome (Nicholas, 1987, p.83) and a red cell membrane defect (Ohba *et al.*, 2000). These defects are widespread as a result of selection for correlated traits and of genetic drift due to the use of small numbers of sires.

Mutation and breed divergence

Can mutation generate enough new variation to enable newly-established breeds to diverge from each other? While it is easy to envisage new breeds acquiring distinct identities through selection for a few alleles with major visible effects, e.g. coat colour or presence/absence of horns, what are the prospects for divergence in quantitative characters such as those relating to performance? It has been found that genetic variance due to mutations (mutational variance, V_M) is about 10^{-3} of the environmental variance V_E (Falconer & Mackay, 1996). This leads to the conclusion that mutation has negligible effects on variance within inbred lines, but between-line variance increases due to mutation. If there are populations with N_e of 50, that had a common ancestor 20 generations previously, then the variance between means attributable to mutations since the populations separated would be about 4% of the phenotypic variance within the populations (Falconer & Mackay, 1996, p.270).

Conclusions

Many breeds are distinctive in appearance, with high degree of uniformity, while others are much less so. Genetic distances calculated between breeds are very often consistent with known history. Once defined, breeds can be characterized by their performance. Genetic variation within breeds is depleted by genetic drift and inbreeding, and by side effects of selection for particular characteristics.

Part 4
The Conservation of Livestock Biodiversity

Livestock biodiversity and floral and faunal biodiversity are both genetic resources but they are conserved in very different ways. Legislation and regulation are fundamental to how floral and faunal biodiversity has customarily been conserved, while the main policy instruments for conservation of livestock biodiversity are incentives and subsidies.

These differences arise because livestock and crops are almost always private property, easily traded and not necessarily owned by the same entity as the land they are on. Wild terrestrial animals and plants usually belong to the owner of the land where they are taken, although they can be state property or subject to state protection which overrides property rights. When livestock biodiversity requires active conservation measures, this implies interference with normal market processes.

Essentially, successful conservation of livestock biodiversity requires attention to two aims, first the maintenance of numbers of individual animals of the breeds in question and second the maintenance of genetic variation within those breeds. These are clearly both vital for rare breeds, but it is now evident that for some very important breeds, notably the Holstein, the latter is becoming highly significant.

Techniques for conserving livestock biodiversity come from several branches of genetics and from commercial reproductive technology. Often an attempt at valuation of a genetic resource must be made before a conservation programme can be put into effect. Valuation of biodiversity is a developing area of conservation biology and ecological economics. Its application to livestock biodiversity is in its infancy. In contrast, reproductive biology and the genetics of livestock production are well understood. The challenge is in the application of this knowledge to the breeds themselves.

Chapter 8
Management, organization and policy

Introduction

Conservation of livestock biodiversity is, in principle, supported by activities at local, national, regional and international levels. How can these activities be integrated? For over 50 years, FAO has taken an interest in the subject and over the last decade especially this interest has given birth to practical activities that can be very supportive of national initiatives. How can countries and groups within them be encouraged to work for conservation of livestock biodiversity?

History of concern for livestock biodiversity: the example of the UK

In Britain between 1799 and 1814, George Garrard published detailed drawings with measurements of cattle and in the 1840s Professor David Low commissioned paintings (half life size for cattle and horses, life size for sheep, pigs and goats) and produced descriptions of the native breeds with assessments of how valuable they were likely to be in improved agriculture (Hall & Clutton-Brock, 1988). Some breeds he was prepared to see decline to extinction, such as the Old Wiltshire and the Ryeland sheep, while he also wrote that 'several entire breeds have been lost which ought to have been preserved.' He pointed out that in the course of intensification it was much easier to replace a breed by one better adapted to the new system, than to apply selection to the pre-existing one. His reason for regretting the loss of certain breeds was because valuable raw material for farming was being lost. For example, he wrote of the Falkland, a Scottish breed of dairy cattle:

'It is much to be regretted that the former breeders of Fifeshire should have been too careless of the preservation of a breed so much superior to the mixed varieties that have succeeded to it. Had the Falkland Breed been cultivated with care during a period when artificial food could have been supplied in the requisite quantity, it is probable that Fifeshire would now have been possessed of a breed combining, in a degree not surpassed by any other in the kingdom, the properties of grazing and yielding milk.'

The story of systematic conservation of livestock biodiversity in Britain began apparently by accident in the early 1950s when the potential of sheep farming to improve the meat and wool supply attracted government attention, at the same time as it became clear that many breeds were becoming rarer (Zuckerman, 1994). Interest extended to the other species and in 1961 a 'gene library' was established at Whipsnade Park (owned by the Zoological Society of London). In February 1964 there were two breeds of cattle, seven of sheep and nine of poultry at Whipsnade. The financial and practical commitment was considerable and in 1968 the stock began to be transferred to another non-governmental organization (NGO), the Royal Agricultural Society of England (RASE). Animals from the RASE stocks were used to start many privately owned flocks and herds around the time the Rare Breeds Survival Trust (RBST) was founded, in 1973.

The RBST renews its priority list every spring. A copy is lodged in the House of Commons library, effectively giving it official government status. Most of the breeds under the RBST's care have increased in numbers, some dramatically (Table 8.1), since the pioneering survey of British rare breeds by Bowman and Aindow (1973).

In Britain, breed societies are well established and, though perpetually short of money, command much support and loyalty, as well as partisanship and politics. One of the functions of the RBST has been to act as a breed society for those rare breeds that lack this framework. Measures of its success include the increase in numbers of most of the breeds under its care and the fact that since its foundation no British breed has gone extinct, though some breeds must be seen as still marginal (Figure 8.1).

Minority breeds are in a more difficult position. Their 'rarity value' is naturally less than that of the rare breeds, so the profile they present to the general public is less sharply defined. Their rather stronger numbers make them more likely to serve as a genetic resource for mainstream agriculture and arguably they would be a good direct or indirect investment of public funds.

Breeds as the basis for conservation

Conservation of livestock biodiversity will almost certainly continue to be based on the maintenance of pure breeds. There are other approaches but none have found favour. In a 'gene pool' approach a mixed population derived from many breeds is kept. This could be helpful when it is clear how the genes are to be utilized, for example tropically adapted breeds could be crossbred to form a composite or synthetic population which could then give rise to new breeds (Notter, 1999). Alternatively, Smith (1985) suggested choosing aspects of the performance of each species of livestock and then breeding groups of animals that express these aspects to extremes. For example, in pigs divergent stocks could be selected for extreme fatness and for extreme leanness. This idea has not been taken up, even though the costs would be small in relation to the potential return if the conserved stocks proved to be of value. However, when future needs are not yet clear, as explained by Barker (2000), '...because about 50% of the total genetic variation within a species is between breeds (Hammond & Leitch, 1995) and because between-breed variation

Table 8.1 Changes over 30 years in populations of the rare British breeds censused in 1974 by Bowman & Aindow (1974); 2002 data from DEFRA (2002). (For further information about specific breeds, see Appendix).

	1973	2002
Cattle		
Blue Albion	220	95
British White	168	1368
Dexter	260	3000
Gloucester	75	389
Longhorn	191	1500
Shetland	220	300
White Park	66	525
Sheep		
Black Welsh Mountain	1200	7200
Cotswold	98	712
Hebridean	319	2298
Jacob	3000	No data
Lincoln Longwool	592	1345 (probably overestimate)
Lleyn	218	129 643
Manx Loghtan	66	1253
North Ronaldsay – island	2–3000	3756 (L. Whistance, pers. comm)
North Ronaldsay – mainland	54	666
Portland	81	1333
Rhiw	Believed extinct	No evidence of presence
Shetland	120 000 (island)	2500 (mainland)
Soay – mainland	386	1107
Wensleydale	415	1624
White Face Dartmoor	3000 (estimate)	2094
Whitefaced Woodland	466	656
Wiltshire Horn	689	3000
Pigs		
Berkshire	49	391
British Lop	58	162
Gloucester Old Spot	620	628
Middle White	65	233
Tamworth	117	214

is more accessible the primary focus in the conservation of domesticated animal diversity is on the conservation of breeds.'

How breeds are organized

In the developed world, the management framework of each established breed is its breed society. These vary from country to country in the degree of support they receive from national governments. In the developing world, few national institutions for breed conservation and sustainable development exist while 'many development agencies... are yet to be convinced of the importance of conserving indigenous breeds' (Blench, 2001). There is an urgent need for studies on what kinds of breeders' organizations might be appropriate and sustainable in developing countries.

Figure 8.1 One of Britain's rarest goats with about 80 breeding females, the Bagot goat is not particularly hardy and its long-term survival will not be easy to assure.

Breed societies

A breed society is an organization which promotes a specific breed and advances the interests of its breeders. They operate a registration programme which leads to the centralized documentation of parentage information for individual animals. Animals will be refused registration if they do not conform to the breed standard, which can be defined however the society wishes, by phenotype, performance or pedigree. Some breed societies register the name of the breed as a trademark (for example, the Southdown Sheep Society; Hall, 1989b).

Breed societies have had to reconcile two aims which sometimes run counter to each other – they are dedicated to maintenance of the breed standard and at the same time to promotion of selective breeding and improvement. It is not always easy to achieve the right balance. 'Show points' such as commercially irrelevant aspects of coloration or conformation have often been overemphasized (for example in nineteenth century British pigs; Figure 8.2; Hall & Clutton-Brock, 1988) while if breed societies are to exist at all, some agreement on the general appearance of the animal is necessary.

There is probably no breed society which has as its aim the maintenance of the breed in an unchanged state. Even societies concerned with what might be thought to be relatively unchanging breeds have permitted selection. In the Chillingham herd, in the nineteenth century there was culling and castration meaning that selection had in effect been practised though the herd was kept pure (Hall & Hall, 1988). The first object for which the Chillingham Wild Cattle Association was established in 1939 was 'to maintain, preserve and improve the herd of white

Figure 8.2 One of Britain's rarest breeds with 211 registered sows in 2002, the Middle White was developed as an early-maturing pork breed. It is difficult to see a major commercial application for this breed, at least in the short or medium term, though it has a niche market appeal. From Spencer (1897).

cattle . . . in their natural surroundings in Chillingham Park, Northumberland, and to keep and maintain the same in their indigenous condition, and in particular maintain their purity of breeding . . . '.

Although some breed societies date back many years, new societies continue to be formed (Figure 8.3). Some are intended to advance the interests of new breeds, while others are founded in order to protect local breeds which have not previously had a breed society. Worldwide, 1047 of the 4760 extant breeds of the world have a breed society (Table 8.2 a,b). Of these, 177 have a breed society in the country of origin and at least one 'daughter' breed society in some other country where the breed has also been adopted – there are 904 such daughter breed societies in total (Table 8.3).

Breed societies are much more common in Europe and North America than elsewhere. Here, they are the dominant form of institutional framework for livestock development and conservation and existence of a breed society may be sufficient for a group of animals to be considered as a breed. Lerner and Donald (1966, Chapter 7) give an entertaining account of breed societies, written at the time when many of the trends which led to today's status of livestock biodiversity in the developed world, were taking shape. Today pedigree data are essential for rational conservation planning and as a framework for the use of data from relatives in calculating breeding values. However, few breed societies have formal programmes to monitor founder representation, trends in inbreeding and the like.

Administrative costs may force smaller societies to amalgamate, most probably with those for similar breeds, but this does not mean the breeds will amalgamate.

Rising costs of data input and checking could lead to flock- and herdbooks being produced as a sideline to performance-recording information systems (Barton, 1984).

Although the theory of livestock genetic conservation is well understood and manuals are available (FAO, 1999) there is still a major need for computer packages and expert systems to help breed societies to design and manage conservation measures (Woolliams & Meuwissen, 1999). Breed structure studies can alert managers to conservation problems, either by enabling comparisons with other breeds or with the same breed some time in its history. However these are time consuming (Boichard *et al.*, 1996) and dependent on good pedigree data. A review of conservation programmes of French cattle which had been running for 20 years (Danchin-Burge & Avon, 2000) was hampered by the fact that because of the long generation interval of cattle many of the pedigrees only went back two or three generations, making it difficult to calculate inbreeding rates.

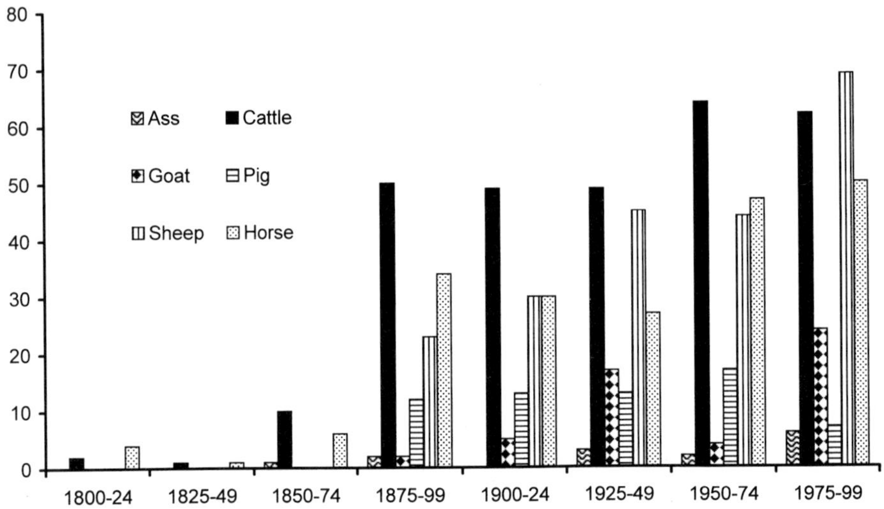

Figure 8.3 Dates of foundation of herd books and breed societies. Some breed societies are old while new ones continue to be formed. Some of these relate to new breeds, while others represent the taking of responsibility for breeds which have not previously had a breed society. Compiled from Porter (2002).

Table 8.2 Breed societies and registration authorities (herd, flock or stud books) for the seven major mammalian livestock species; breeds stated to be 'rare' or 'nearly extinct' are tabulated as 'rare' (compiled from Porter, 2002).
(a) Numbers

Total breed societies and/or registration authorities in country of origin of breed	1047
Total breeds for which date of foundation of above is known	826 (79%)
Total breed societies with at least one 'daughter' breed society or registration authority in another country	183 (17%)
Total such 'daughters'	904
Total breed societies and/or registration authorities whose breed is defined as 'rare'	352 (34%)
Total 'daughters' whose breed is defined as 'rare' in country of origin	183 (20%)

Table 8.2 *(Continued)*
(b) Classification of breed society/registration authority according to species, location, and whether the breed is listed as rare.

	Africa		Asia		Europe		N. America		S. & Central America		Former USSR		Oceania		Totals		
	R	C	R	C	R	C	R	C	R	C	R	C	R	C	Rare	Not rare	
Donkey					12	0	1	14		1					13	5	**18**
Water buffalo										3					0	3	**3**
Cattle	2	11	2	17	73	132	9	32	3	26	4	22	1	15	94	255	**349**
Goat		5			19	35	6	2				1			25	43	**68**
Horse	3	6	10	6	69	70	13	36	2	14	1	7	1	2	99	141	**240**
Pig		1			23	31	6	11			1	7	1		31	50	**81**
Sheep	4	12		2	71	130	13	12		7		14	2	21	90	198	**288**
Totals	9	35	12	25	267	398	48	97	5	51	6	51	5	38	352	695	**1047**

Table 8.3 Classification of daughter organization according to species, location, and whether the breed is defined as rare in its original country. Data from Porter (2002).

	Africa		**Asia**		**Europe**		**N. America**		**S. & C. America**		**Former USSR**		**Oceania**		**Total**		
	R	C	R	C	R	C	R	C	R	C	R	C	R	C	Rare	Not rare	
Cattle	2	2	2	5	45	285	8	23	2	3	2	4	2	5	63	327	**390**
Goat	1	2			3	12	3	9							7	23	**30**
Horse			1	4	35	173	7	12	1	1					44	190	**234**
Pig					8	28	7	22					1	1	16	51	**67**
Sheep	1	1			40	108	4	4					8	17	53	130	**183**

Another need is for a set of simple validated tests to be performed on registration, field survey or questionnaire data for rapid assessment of the conservation status and population viability of a breed. For example, the proportion of flocks or herds that supplies sires for the breed might be an indicator of conservation status. If this is low, the breed has a hierarchical structure which can mean it is dominated by the breeding goals of an élite and this can be antagonistic to genetic conservation. For example in British rare and minority sheep and pigs, between 29% and 52% of breeders supply sires to other breeders (Hall, 1986, 1989a,b). In Galloway cattle, which are declining in numbers in their native Scotland, about 23% of herds supply sires; in the Lincoln Red, an English beef breed with about a third of the numbers of the Galloway, the proportion is about 50% (Wright *et al.*, 2002). Maybe genetic variation is being more effectively conserved in the Lincoln Red.

National activities

Around the world concern for livestock breeds was building up in the early 1970s. The dedication of the individual breeders who had kept old breeds going, was achieving recognition. The Rare Breeds Survival Trust was founded as a culmination of various private and NGO activities that had been going on for many decades with some support from NGOs like the RASE and the Zoological Society of London (Alderson, 1981). Following the pioneering plant breeding work of N.I. Vavilov, the concept of a 'gene fund' as a national asset was formulated in the 1920s in the USSR and in 1970 the eastern bloc countries agreed on multilateral cooperation in genetic conservation (Romanov *et al.*, 1996). In the USA, the American Livestock Breeds Conservancy (www.albc-usa.org) was founded in 1977 (Christman *et al.*, 1997).

In France, livestock conservation could be seen as having gone through two generations (Audiot *et al.*, 1993). In the first, which took place during the 1970s, the aim was to conserve genes or combinations of genes that could be of use in the future. Since 1976 the Ministry of Agriculture made annual funding available to the extent of 0.5% of that available for national programmes for the genetic improvement of livestock and certain regional authorities have also supported their local breeds. Several different programmes were started with considerable involvement of government agricultural research institutes. However, concern has been expressed as to the sustainability of these conservation efforts. Second generation programmes have attempted to gain economic benefit from the conserved genetic resources, and the local affinities of breeds have been capitalized upon.

The Gascon pig of France (Molenat *et al.*, 1993) was reduced to 2 boars and 30 sows in 1982 and had grown to 81 boars and 311 sows by 1993 with 120 breeders. The breed has been characterized to a limited extent with a small number of published reports and dissertations and a long-standing involvement of government scientists. Recently studies under the supervision of the Toulouse Veterinary School on the husbandry and meat processing appropriate for the Gascon pig have been financed by the Chambre Régionale d'Agriculture. Since 1981 conservation of the breed has been overseen by the Institut Technique du Porc. Also in France, Regional Natural Parks are committed to conservation of local livestock breeds as well as of wild flora and fauna and in the west the Parc Naturel du Marais Poitevin has been heavily involved in collaboration with the French national stud in conserving the Poitou donkey (Philippe, 1995; Camac, 1997). There is a striking contrast between this approach to conservation of rare breeds and that evident in Britain. In France, regional and national governmental and academic institutions have contributed sustained effort over many years while in Britain leadership in this activity has been almost entirely by the Rare Breeds Survival Trust and the breed societies.

Although the rare French cattle breeds, for example, have increased in numbers (Table 8.4) this has not been at the same rate as the British rare cattle (Table 8.1). The effectiveness of the different approaches has not been compared; the French system is much more centralized and formal, which may have retarded growth of numbers, but inbreeding may have been more effectively controlled. Strikingly the Breton Black Pied whose conservation has been reviewed by Colleau *et al.* (2002), though not, unfortunately, in comparison with that of other breeds, clearly retains its dairy attributes as indicated by the proportion of breeders who practise milk recording (Figure 8.4a,b,c,d). Few herds of the British rare cattle breeds are being milked commercially.

In the USA a National Animal Germplasm Program is being put into effect, based on collaboration between institutions. In Germany, NGOs and a federal research centre are collaborating to screen and evaluate poultry breeds (Weigend *et al.*, 1995). The activities of many other breed conservation organizations were reviewed by Grunenfelder (1992).

Particularly in poultry there may exist stocks that were developed in government, commercial or university research departments. Often these are under threat because

Table 8.4 French rare cattle breeds: changes in numbers, from Danchin-Burge and Avon (2000). Breed names are from Porter (2002).

	Date of first census	Population at first census	Female population in 1998
Armorican	1981	47	76
Aure et Saint Girons	1982	73	163
Béarnais	1982	70	112
Ferrandais	1981	248	594
Froment du Léon	1981	42	168
Gascon Aréolé	1985	135	395
Lourdais	1982	28	108
Maraîchin	1986	75	281
Nantais	1986	59	204
Villard-de-Lans	1979	60	228

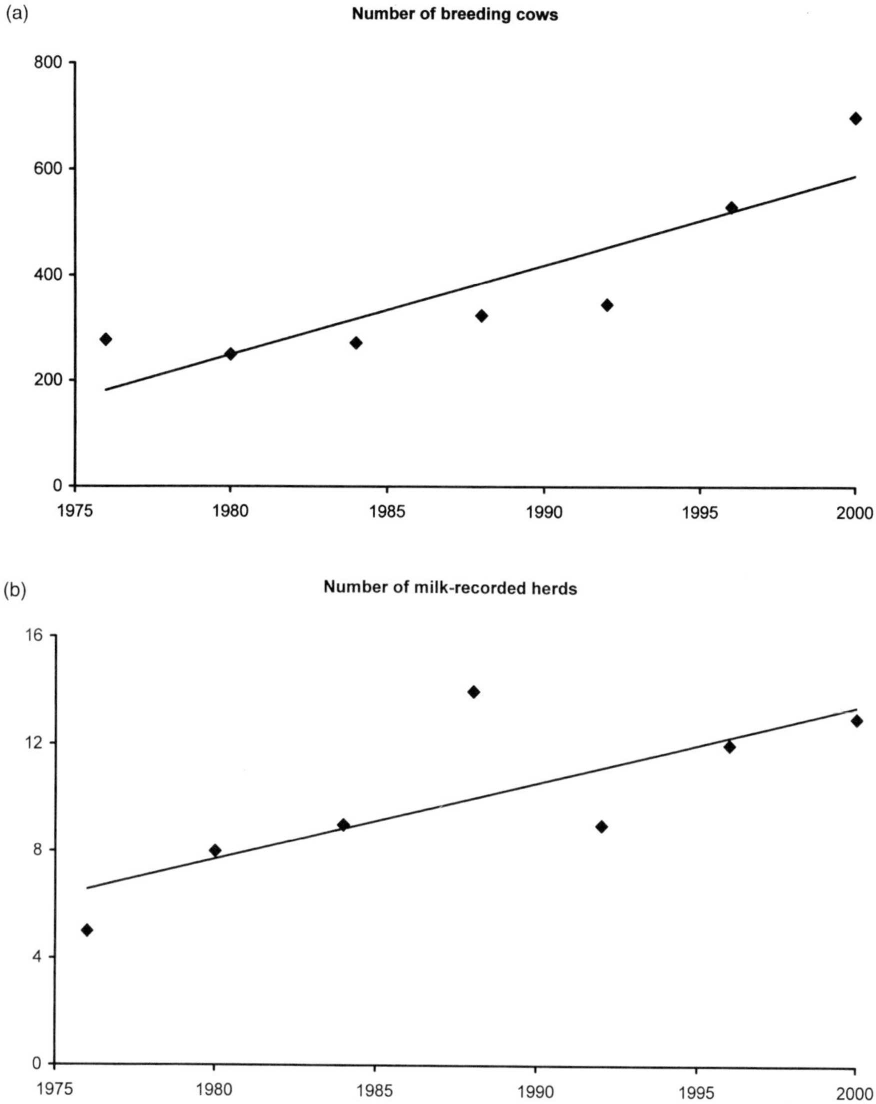

Figure 8.4 Conservation of French cattle: the example of the Breton Black Pied. From 500 000 in 1900, the BPN numbered 15 000 in 1975 (a) and was clearly about to disappear when a conservation programme was started, involving 46 breeders with a total of 311 cows (Colleau *et al.*, 2002). Since then there has been an increase in numbers of cattle and of breeders (c), milk recording (b) has been continued, and the annual increase of inbreeding has been restricted to 0.22% (d).

of financial pressures. Pisenti *et al.* (1999), reported that in the USA and Canada, over 200 mutant, inbred and selected stocks of poultry have been destroyed. There remain 268 chicken, 20 turkey, 65 Japanese quail and 8 waterfowl or gamebird stocks, a third of which are at risk. An Avian Genetic Resources System is envisaged which would also protect commercially developed hybrid and selected lines and

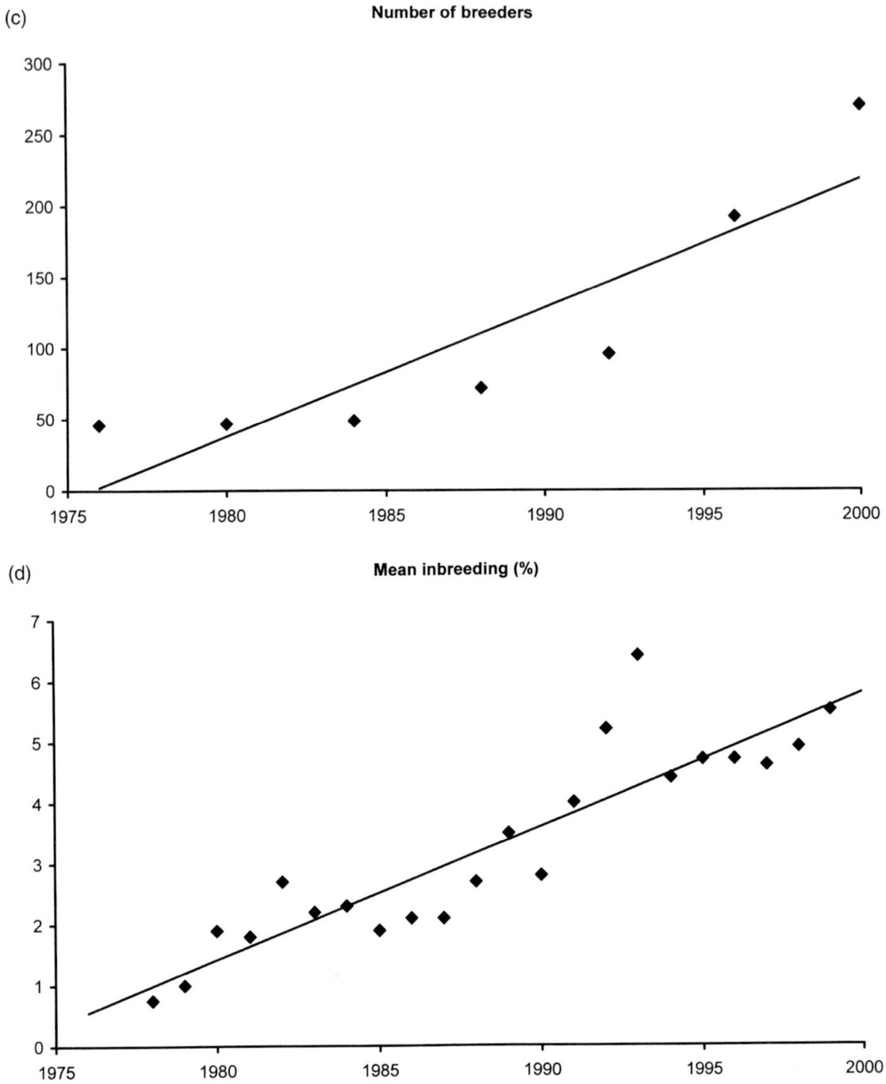

Figure 8.4 (*Continued*)

breeds of a primitive nature from elsewhere in the world. 'Mid-level production and fancy breeds' would also be included although it is acknowledged that systems for their conservation are already in existence.

These national activities take place against a background of declining public investment in agricultural research. During the 1960s and 1970s funding for agricultural research increased in real terms in most parts of the world but since then the rate of increase has declined. In 1981, 51% of agricultural R&D in 22 OECD countries (the EU and other advanced economies including the USA and Japan) was privately funded, in 1991 it was 58% and over the same period the proportion of public R&D

Table 8.5 Reductions in grassland, upland, or other pastoral-based research in selected countries (Hodgson, 2001).

England & Wales	Fall of 59% between 1990–91 and 2000–01
Scotland	Fall of 9% between 1993–94 and 1999–2000
New Zealand	Fall of 12% between 1992–93 and 1998–99
Japan	Fall of 48% in support of main research institute from 1989–1998
China	Fall of 32% in Ministry funding from 1991–95 to 1996–2000, following rise of 66% between 1986–91 and 1991–95

funds allocated to agriculture declined from 8.9% to 7.5%. Some areas of agricultural research have been particularly hard hit (Table 8.5).

International and regional activities

International interest in the conservation of livestock biodiversity was first formalized at a meeting in Copenhagen in 1946 (Barker, 1994), which led to the newly founded FAO (as a secretariat of the member states of the United Nations) being given the duty of cataloguing, maintaining and utilizing livestock biodiversity (Phillips, 1981). Useful publications were produced, beginning with catalogues of the cattle of India, Pakistan and Africa in the 1950s. In 1966 a consultation was held to investigate the problems of conserving and using these stocks and further meetings were held on cattle (in 1968), pigs (in 1970) and poultry (in 1973). Concurrently, international concern for the environment led to the United Nations Conference on the Environment in Stockholm in 1972 and the establishment of the United Nations Environment Programme (UNEP) which funded several FAO field projects, desk studies and meetings which were reported mainly in the form of paperback books called *FAO Animal Production and Health Papers*.

The European Association of Animal Production (EAAP) set up a working party in 1980 and its findings, along with earlier work, were summarized by (Maijala *et al.*, 1984). FAO held its first technical consultation on the subject in the same year, in partnership with UNEP (FAO/UNEP, 1981). The EAAP data bank started at Hannover in 1988 was transferred to FAO in Rome in 1992–93, published as a book (Simon & Buchenauer, 1993) and expanded into a global data bank (the FAO/DAD-IS system; Scherf, 2000). Early initiatives elsewhere in the world are summarized by FAO/UNEP (1981).

On 29 December 1993, the Convention on Biological Diversity came into force. Its main concern is with non-domesticated species and ecosystems but FAO was involved in the negotiations and this led to recognition that biodiversity includes domesticated plants and animals.

From 1995 the EU directorate DGXII Biotechnology funded research projects in European cattle, sheep and goats, chickens and pigs. In 1997 EU support in the form of a Concerted Action was given to enhance the EAAP inventory and subsequently to establish a permanent DNA archive and database for horse breeds. The support was from DGVI Agriculture which also supported projects for genetic characterization in rabbits, pigs and cattle.

The importance of animal genetic resources was also acknowledged at the 1996 World Food Summit in Rome and at subsequent Conferences of the Parties to the Convention on Biodiversity. In parallel, FAO has organized expert consultations, many of which have been published. FAO's strategy for 2000–2015 explicitly mentions genetic resources at several points, though not perhaps as forcefully as many conservationists would like, as well as emphasizing sustainable intensification and environmental protection.

Livestock biodiversity has begun to attract more attention from conservation biologists. Two standard works produced by biodiversity scientists as their contribution to explaining the problems facing the world's biodiversity (Groombridge, 1992; Heywood & Watson, 1995) include accounts of the genetic resources represented by livestock breeds and plant varieties. This demonstrates how agricultural genetic resources are accepted as elements of biodiversity by the International Union for the Conservation of Nature (IUCN) and the United Nations Environment Programme (UNEP). The Convention has three principal objectives – conservation, sustainable use and equitable sharing of the benefits, of biodiversity. Signatories to the Convention undertake to have plans for conservation of biodiversity and to carry out specific tasks, including identification and monitoring and *in situ* and *ex situ* conservation, which together with activities of a policy or administrative nature, operate to achieve the objectives of the Convention.

Since then, FAO has continued to promote the cause of livestock biodiversity with the establishment of a global database (Scherf, 2000; Weigend & Romanov, 2002) that is now on the internet, and the publication of the journal *Animal Genetic Resources International*. An extremely important initiative is the FAO *State of the World's Animal Genetic Resources* which is based on country reports and which is scheduled for publication in 2006. It will stimulate the formulation of national policies and an international FAO strategy, all of which will be expected to be consistent with the Convention on Biological Diversity.

However, substantial funding for a global programme has not been forthcoming although regional initiatives have started, including one in 12 countries in Asia and the Pacific (Bhutan, China, India, Indonesia, Laos, Malaysia, Myanmar, Nepal, Pakistan, Philippines, Thailand, Vietnam) supported by the Japanese government (Steane, 2002) from 1993. These countries hold 85% of the region's breeds and 26% of the global total. Another regional initiative is supported by the International Livestock Research Institute (ILRI) in Africa (www.cgiar.org/ilri).

As an international secretariat, FAO can only accept data for the database that have come through the national coordinators (one for each country) who are government appointees. It is easy to find fault with some of the information in the database and livestock specialists should contribute to refining the database via their national coordinators.

Man-made and natural disasters

Wars, invasion and civic collapse can lead to the loss of livestock which can have a strong effect on local breeds. This was most obvious in South America after the Spanish conquest. Understanding of livestock history has been helped by the peculiar features of archaeological preservation there, permitting survival of desiccated remains of carcases. At the time of the Spanish conquest large herds of llama and alpaca were maintained but then there was a breakdown in specialist breeding and distinct breeds of fine-fibre producing llamas and alpacas disappeared. A great deal still remains to be discovered about the history and current status of these systems. The role of other livestock species is becoming better appreciated (Archetti, 1997) but the camelids are the most distinctive features of South America's livestock biodiversity (Kadwell *et al.*, 2001).

It is claimed (references in Cothran & van Dyk, 1998) that during the Boer war in South Africa (1899–1902) the British made a 'concerted effort to eliminate' the Cape Horse (or Boerperd). This was because it was better adapted to the environment and more resistant to African horse sickness virus than the imported British army horses. Poultry in eastern Europe were much affected during World War II (Romanov *et al.*, 1996). In Nigeria the main reason for the decline in West African Shorthorn cattle given during a 1990 survey (RIM, 1994) was the destruction of these animals during the Biafran war (1967–70). In Bosnia 95% of all livestock was destroyed during the fighting in the early 1990s. Friend (1999) describes how a World Bank project is supplying replacement animals: '9100 heifers of Holstein-Friesian, Milking Simmental, Brown Swiss and the Tyrol breed from the Austrian/Italian border... 10 000 Pramenka ewes, 500 Wurttemburg/Texel rams, 1500 goats and 250 breeding pigs'. The Buša cattle, probably the area's most distinctive cattle breed (Medjugorac *et al.*, 1994), and 'minimally endangered' before the war, according to Simon and Buchenauer (1993) has probably become endangered as a result of the combination of war and subsequent reconstruction (Arapovic, 1997). The Turopolje pig of Croatia might also have been endangered through war (Grunenfelder *et al.*, 1994; Wilson, 1995). After the Falklands war of 1982, British livestock including several rare breeds were sent to make up for losses (Joss & Mason, 1983).

The international community is becoming more interested in predicting emergencies involving livestock (electronic conference in 1998: www.fao.org/ag/aga/agap/lps/drought1.htm). East African outbreaks of the mosquito-transmitted viral disease Rift Valley fever can be predicted two to five months in advance (Linthicum *et al.*, 1999). This permits emergency vaccination of people and cattle. During the 2001 FMD outbreak in Britain, emergency collection of sheep semen was carried out by the RBST (Mansbridge, 2001) and by the newly-formed Sheep Trust (Bowles *et al.*, 2003) but generally in the face of incipient disaster, it is very difficult to conduct emergency biodiversity rescue programmes.

Capturing the economic benefits of livestock biodiversity

The Convention on Biological Diversity

The Preamble to the Rio Convention of 1992 specifically mentions the conservation of crop and livestock biodiversity. The Convention aims to conserve world biodiversity and to use it sustainably, with fair sharing of its benefits with appropriate access to genetic resources and by appropriate transfer of relevant technologies while taking account of national and farmers' rights.

Before the Convention, biodiversity was generally accepted to be the common heritage of humankind and states were seen to have a responsibility to preserve it for everyone. However, in response to concerns of developing nations (which arose from the history of utilization of crop genetic resources), a new concept was developed whereby national sovereignty over biodiversity is accepted. Signatories to the Convention are expected to develop national strategies for conservation and sustainable use of biodiversity and there is financial support for this.

Under the Convention, states have sovereign rights over any genetic resources transferred in accordance with the Convention. Any benefits from the utilization of plant genetic material are supposed to be shared with local communities, following principles (including that of farmers' rights) established in the FAO Undertaking on Plant Genetic Resources but this does not apply to animal genetic resources (Day *et al.*, 1994). The taking of genetic resources from a developing country, their incorporation into varieties, frequently accompanied by the insertion of a genetic marker and the establishment of patent rights, followed by the sale of the new variety back to the developing country, has involved plants rather than animals. This is the process that has sometimes been referred to as 'biopiracy'. In 1997 it was alleged that genes from the dwarf Vechur cow of Kerala, India, had been patented by a British research institute, but this was false (Bulfield, 1998). Partly reflecting differences in the speed of developments, the Uruguay Round's Agreement on Trade-Related Aspects of Intellectual Property Rights has led to plant varieties but not animal varieties being likely to receive intellectual property protection (Cardelinno, 2003; www.to.org/english/docs_e/legal_e/27-trips.pdf). Genes can be patented as can plant varieties, but animals apparently cannot (Day *et al.*, 1994). A gene can be patented and incorporated into an animal but once having purchased breeding stock farmers are free to propagate the new animal variety. Local breeds may carry genes that could be of economic value and patentable once defined and engineered, but when the animals are exported, the country of origin has no further rights in the genotype.

In plants, the sale of the material goods (generally seeds) necessarily conveys the immaterial goods (knowledge, information) so intellectual property rights of the latter need to be defined. In contrast, it would be relatively easy for animal breeders to protect their intellectual property, perhaps by only selling animals of one sex (for example egg laying chickens or breeding female pigs) or by only trading within their own vertically integrated industry.

Export and import prohibitions and quotas

Merino sheep used to be closely guarded by the guild of sheep farmers, the Mesta, in their native Spain (Hall & Clutton-Brock, 1988). There are many more recent examples of protectionism in livestock biodiversity. The export of Danish Landrace pigs used to be prohibited (King, 1991). Currently, breeding Merino ewes and embryos cannot be exported from Australia and a quota is set on the export of Merino rams and semen. The quota is 900 rams for the entire world market apart from New Zealand which is permitted to import without restriction (www.merinos.com.au/whatwedo.htm). There are many examples of breeds being moved from one country to another – between 1970 and 1996 at least 39 breeds from 18 countries were imported into the USA for breed development (Notter, 1999).

Commercial confidentiality means that it is not usually possible to find out how much was paid for breeding animals that were taken from one country to form new breeds in others. Examples would include Chinese pigs, Finnish Landrace sheep and the Boran and Tuli bovine embryos exported from east Africa to Australia (see Chapter 5). The available indications are that the price paid in the exporting country was the local market price for breeding stock.

Kingwell *et al.* (1995) and Lightfoot (1987) describe the establishment of the Awassi fat-tailed sheep breed in Australia to set up a meat sheep industry particularly aimed at the Middle-East export trade. The project began in 1983 when sheep were imported from Cyprus as embryos. Australia has very tough restrictions on the import of sheep – the new lambs were screened for disease and kept in quarantine for seven years to confirm absence of scrapie. During this time they were multiplied by embryo transfer and were ready for widespread adoption by farmers within 12 years (Kingwell *et al.*, 1995). The expense and risk of these procedures and their sensitivity to changes in national and international legislation make them rather unattractive for private companies and in general they may only prove feasible for state organizations or multinational companies.

Economic valuation of livestock biodiversity

The principles of valuing biodiversity are well understood (Swanson, 1995; Primack, 2002) but methodologies are not yet adequate for specific animal or plant genetic resources (Drucker *et al.*, 2001). This is perhaps surprising in view of the advanced state of development of the application of economics to animal breeding programmes. This should have made it relatively easy to establish the value of conserving genetic resources for use in future breeding programmes. The pioneering work of Smith (1984, 1985) provided the overall economic justification for conservation.

Biodiversity can be valued in several ways – essentially in relation to the use that can be made of it and in relation to the value that is placed on its continued existence without reference to its use. Both these general approaches are commonly used for floral and faunal biodiversity and both are applicable to livestock breeds.

If a country loses a breed, through disaster or by grading-up to some other genotype, it loses the opportunity to use the breed's special characteristics. Future needs are unpredictable so the value of this loss can only be guessed (Smith, 1984). The replacement value of the breed is calculable, provided a similar breed is available elsewhere. For example the British cattle breed most threatened by foot and mouth disease in 2001 was the Belted Galloway, which lost 21% of its approximately 1100 breeding cows (Townsend *et al.*, 2002). If it had been eradicated it could in principle have been restored to some extent as there are 'daughter' populations in the USA, Germany, New Zealand and Australia which could be used, with appropriate reproductive technology and genotyping. Crudely, the project cost could be deduced from the need to purchase or breed 1100 cows – maybe £600 000 at 2003 prices. However, if the Hill Radnor sheep breed had been lost (with 23% of its breeding ewe population of about 1500 having been culled, this was the worst hit sheep breed) replacement would not have been possible as there are no foreign counterparts. There are some British breeds to which it is closely related so its replacement value might be some function of its genetic distance from other breeds, or might be calculable from the proportion of genetic variation that it harbours (following the Weitzman approach: see Chapter 7).

In at least some countries, rare breeds have a clear existence value in that people will pay in order to keep the breeds in existence. Probably at least 43% of the 7864 members of the Rare Breeds Survival Trust do not keep livestock (Evans & Yarwood, 2000) yet still feel a desire to support breed conservation work, while the Chillingham Wild Cattle Association has over 300 members. However, valuation approaches based on use values will be of more general application.

Costs of maintenance of conservation populations can be deduced from standard animal husbandry texts. The Chillingham cattle are perhaps an extreme example. Here, annual costs of running the herd in their native 134 ha parkland in northern England are about £30 000 and are mainly met by visitor fees, donations and returns on investments. There is no government support. No animals are sold from the herd and this figure includes the opportunity cost represented by the income that would be received if the herd were replaced by a commercial beef enterprise. Less expensive schemes can be designed for other breeds whose husbandry is not inextricably tied to a particular area. For example, individual animals of conservation significance can be purchased by concerned people and placed in a conservation herd or flock. After the animal has bred its replacement it, or the proceeds of its sale, can be returned to the original purchaser.

Source and allocation of conservation funds

Global biodiversity conservation could be achieved if a small proportion of the environmentally damaging subsidies currently given by governments were redirected into parks and protected areas (James *et al.*, 1999). In the EU, these subsidies have been at the rate of $82 500 per km^2 per year (US figure: $16 100). Reform of the subsidy regimes for livestock would serve to protect livestock biodiversity. It is commonly stated that it is impossible to 'save all breeds' and this can all too easily turn into a justification for not attempting to 'save' anything. All of livestock biodiversity could be protected if conservation-oriented subsidy regimes were adopted worldwide.

Changes in policy of this kind, once agreed, would take at least 30 years to come into effect (James *et al.*, 1999) and in the meantime specific breeds will need to receive special protection. In May 1990 FAO proposed a list of breeds deemed particularly worthy of development and conservation (FAO, 1992). Criteria for inclusion in the list were essentially utilitarian – candidate breeds needed to have at least one highly desirable productive or adaptive trait, to be endangered or not be currently utilized efficiently and to have the potential to influence large animal populations. The list did not obviously take account of future market requirements being unpredictable. Hall and Bradley (1995) suggested that one way of conserving traits that are currently irrelevant but which might be of future value would be to prioritize breeds for conservation on the basis of their genetic distinctiveness. This was an extension of the proposal of May (1990) that evolutionary history is the commodity that conservation should concentrate upon.

Much genetic distancing work has now been done which has led to a new appreciation of the scientific interest of domestication and breed differentiation. However, it has been argued that the resources could have been better applied to direct phenotypic characterization and simple conservation measures (Ruane, 1999). A list of breeds ranked in order of priority for conservation would have to take account of several other criteria in addition to genetic distance from other breeds, if the list is to stand any chance of acceptance by the scientific, conservation and farming communities. Ruane (2000) suggests the following parameters for prioritization: degree of endangerment, traits of current economic value, special landscape value, traits of current scientific value, cultural/historical value and genetic uniqueness. There is a clear need for more discussion on this subject.

Scientific input into conservation of livestock biodiversity

Regarding livestock biodiversity, governments will tend to look to their animal science specialists for advice on policy. Governments will formulate policy whether or not the scientific community wishes or is able, to provide its opinions. At present livestock biodiversity does not have a large or coordinated scientific community behind it. Partly this is due to its multidisciplinary nature, partly to the general lack of government support for holistic animal science research, which means that animal science is fragmented and over specialized. In many countries undergraduate courses in agriculture are unpopular, reflecting decline in number of farmers (Hindmarch & Pienkowksi, 1998) and this is a major factor in the insecurity of the animal science community. Declining career opportunities in animal science have discouraged new entrants in many countries. In some countries sectors of the public are suspicious of biotechnology and new developments in animal production (Curry Report, 2002, Chapter 5). This situation can be contrasted with that in conservation biology, which is a relatively popular and media-friendly science.

Should livestock biodiversity become the property of conservation biology or should it remain within animal science? Do the paradigms of one of these sciences promise it more than those of the other? Perhaps animal science is now so molecular that no 'whole animal' researcher can feel at home there. Does it matter?

Conservation biology was originally applied ecology, it has incorporated elements of many other sciences, within a framework that is based on the simple axiom that the diversity of living things is good (Primack, 2002). These other sciences add to their popular appeal by referring to their contribution towards conservation.

There are parallels in animal welfare science (Appleby, 1999). Originally seen as part of veterinary medicine, this has taken elements of ethology, physiology and more recently of psychology and has synthesized them into a new experimental science. Methods and approaches have been adopted from other disciplines and animal welfare science now has a distinct identity and is contributing not just to important practical issues but also, through its contribution to bioethics, to questions of humanity's relationship to the living world.

Conservation biology has expressed its interest in livestock biodiversity (Groombridge, 1992; Heywood & Watson, 1995). Much good will come from the growing acceptance that animal genetic resources are part of world biodiversity, through access to new sources of funding, and (very importantly) from the recruiting of young biologists to their study. There are probably lessons here for animal science.

Conclusions

National governments have usually been heavily involved in agriculture but very few have ever given much support for conservation of genetic resources. The emergence of the concept of livestock biodiversity as something that is owned by the country, arising from the Convention on Biological Diversity, may lead to a change in this attitude. The general position is that conservation of livestock biodiversity has mainly been advocated by international organizations, by farmers' groups (mainly breed societies) and scientists and the lay (especially rural) public concerned with conservation and sustainable development.

Livestock biodiversity is difficult to value, because the characteristics that will be useful in the future are not predictable. Costs of setting up a new livestock project are high but cost of the animals themselves is a very small component. There is a serious lack of valuation methodologies. However, costs of maintenance of conservation populations are easily deduced. In principle a replacement cost can be calculated for a breed that has been lost. Comparative economic valuation of breeds could add to the criteria available for determining the allocation of conservation expenditure.

Umbrella organizations are often highly effective. Against a background of falling public support for agricultural research, funding for livestock biodiversity may be more likely from agencies concerned with social development and environmental protection, rather than with agricultural production.

Chapter 9
Conserving specific breeds

Introduction

Conservation can be *in situ* or *ex situ*. The first requirement for *in situ* conservation is that enough animals are kept and that the mating system acts to maintain heterozygosity. In livestock, *ex situ* conservation means cryoconservation of semen, embryos, oocytes or other germplasm.

How can effective *in situ* conservation be planned? Clearly it has two components, the maintenance of numbers and the maintenance of within-breed genetic variation. In principle this should be easy if the breed is under single ownership, but what if there are several or many private owners involved?

What are the prospects for cryoconservation? Can it be organized so it supports *in situ* conservation or is it fated to be accused of promoting an 'out of sight, out of mind' approach?

Conservation of within-breed variation

Within-breed variation needs to be conserved for two reasons. First, it means the population retains the capacity to respond to selection. Second, it means that individuals are less likely to become homozygous for harmful or sub-optimal alleles. This variation can only be measured directly in a few traits but patterns of change in it can be deduced from the rate of change in inbreeding. Genetic management of breeds is thus essentially a matter of controlling the rate of inbreeding.

Livestock breeds are usually under private ownership with each herd or flock being individually managed, a collective planned approach to genetic conservation is unusual. A conservation authority could relatively easily monitor genetic changes in a breed, by surveillance and review of annual registrations and it would also be helpful to have suggestions of matings that would rectify trends in registrations that could lead to loss of genetic variation.

At the other extreme a breed might be under fully centralized control and breeding strategies could be planned to purge harmful genes or to select for particular alleles. It may also be proposed to restore lost breeds, by mating survivors and upgrading to something close to the original phenotype. However one of the main challenges to genetic management is the conservation of within-breed variation of non-pedigreed populations, especially when there are genetic improvement schemes in progress.

Heterozygosity and effective population size

The most effective way of retaining genetic variation in a population is to maintain heterozygosity. As explained in Chapter 7, in a closed population heterozygosity will decline with time following the formula

$$H_t = H_o(1 - 1/2N_e)^t$$

with the average heterozygosity decreasing by a fraction $1/(2N_e)$ per generation from the original value of H_o.

Thus retention of heterozygosity is maximized by:

- Maximizing initial heterozygosity H_o
- Minimizing number of generations t
- Maximizing N_e

N_e is influenced by sex ratio, variation in family size, fluctuations in population size over generations and the overlap of generations (Frankham *et al.*, 2002, Chapter 10). It can be measured for a population by a set of demographic equations that account for these factors or it can be inferred from the rate of loss of genetic diversity from the population. Because it is inversely proportional to change of inbreeding ($\Delta F = 1/2\,N_e$) it can also be deduced from the observed rate of inbreeding and this is the method most commonly used in livestock breeds.

N_e can be deduced (Nicholas, 1987, p.455) from the numbers of male parents (m) and female parents (f) in the population following the formula:

$$N_e = 4\,mf/(m + f).$$

Variance in family size affects N_e thus:

$$N_e = 8\,N/(V_{km} + V_{kf} + 4)$$

where N is the census population size, V_{km} and V_{kf} are the variances of family sizes of males and females respectively (Falconer & Mackay, 1996, p.68). Thus, zero variance in family size would make N_e equal to double the census population size.

In livestock, long experience (FAO, 1999; Wright, 1977) has suggested N_e of 50, i.e. ΔF of 1% per generation, is compatible with long-term survival of breeds. This is an empirical finding. It emerged from the studies which began in the late 1940s on breed structure and quickly showed that in European, North American, Australian and New Zealand livestock generally, top groups of breeders bred the stud males which were used in the rest of the breed. This structure, coupled with high variance in family size meant that the effective number of males used per generation was usually around 30 (Barker, 1991). It appears that breeders were taking account of the need to restrict inbreeding in their choices of mate for their stock, as had been shown by Wright (1923) in relation to Shorthorn cattle.

In wild populations of animals and plants, N_e is usually about 11% of the total census population, N (Frankham *et al.*, 2002, Chapter 10), but the range is enormous, from 10^{-6} to 99% in the populations studied. In livestock N_e ranges from 15–258 and the N_e/N ratio ranges from 0.001%–29% (Table 9.1). Today, livestock values for N_e are similar to what they were in the early twentieth century. Summarizing 17 earlier livestock studies, Lush (1946) found inbreeding rates of between 0.2%–1.2% per generation implying N_e of 42–250.

Table 9.1 Effective population sizes in relation to census population sizes.

		Census	N_e inbrdg	N_e inbrdg/census (%)	Reference
Pigs	Gloucester Old Spot	474	50	10.6	1,9
	British Saddleback	462	50	10.8	1,9
	Large Black	316	50	15.8	1,9
	Berkshire	264	50	18.9	1,9
	Middle White	258	50	19.4	1,9
	British Lop	196	50	25.5	1,9
	Tamworth	173	50	28.9	1,9
Sheep	Manx Loghtan	1048	23–46	2.2–4.4	8
	Hebridean	1027	27–54	2.6–5.2	8
	Portland	369	14–27	3.7–7.4	8
Equine	Standardbred – Pacer	120 000	51	0.04	3,7,15
	Standardbred – Trotter	30 000	51	0.17	3,7,15
	Norwegian Trotter	10 000	45	0.45	2,4
	Døle	3000	158	5.3	4,11
	Nordland	2300	62	2.7	4,11
Cattle	Holstein (France)	3 100 000	46	0.001	5
	Austrian Simmental	1 891 030	258	0.01	13
	Normande	1 200 000	47	0.004	5
	Montbéliard	710 000	125	0.02	5
	Japanese Black	530 000	17	0.003	10
	Finnish Ayrshire	428 228	34	0.01	14
	Austrian Brown	231 346	109	0.05	13
	Abondance	150 000	106	0.07	5
	French Simmental	100 000	208	0.21	5
	Pinzgauer	53 875	232	0.43	13
	French Brown	40 000	56	0.14	5
	Tarentaise	30 000	27	0.09	5
	Grey Alpine	16 651	73	0.44	13
	Breton Black Pied	1000	28	2.8	6
	Kerry	227	38	16.7	12

References:
1: Anon., 1989; 2: Arnason, 2001; 3: Bixby *et al.*, 1994; 4: Bjørnstad, 2001; 5: Boichard *et al.*, 1996; 6: Colleau *et al.*, 2002; 7: Cothran *et al.*, 1986; 8: Hall, 1986; 9: Hall, 1989a; 10: Nomura *et al.*, 2001; 11: Olson & Klemetsdal, 2002; 12: O'Huigin & Cunningham, 1990; 13: Sölkner *et al.*, 1998; 14: Woolliams & Mäantysaari, 1995; 15: www.ustrotting.com/services/breed/harness.cfm.

Breeding plans

To maximize N_e breeding plans need to be developed, which restrict inbreeding and minimize variance in family size. A circular breeding plan can achieve this. Such a scheme starts with unrelated founder animals (equal numbers of males and females) that are mated following a predetermined scheme; each male is replaced by one of his sons and each female by one of her daughters. The scheme is applied to them and to subsequent generations (Frankham *et al.*, 2002, p.432) resulting in maximum avoidance of inbreeding. For Danish Shorthorn cattle, which currently number 96 breeding females, a modified circular mating scheme has been running for 50 years (Trinderup *et al.*, 1999a) but these schemes are difficult to organize especially when numbers in the breed rise above 1000 (Giraudeau *et al.*, 1991; Vu Tien *et al.*, 1994). A circular scheme was applied to the main British herd of Gloucester Old Spot pigs, but inbreeding in the breed as a whole was no less than in the other rare British pig breeds (Hall, 1989a).

Genetic variation can be conserved by subdividing a population into lines and then breeding within the lines. The lines will then become inbred and more homozygous (this is called the Wahlund effect), but because genetic differentiation will develop between the lines genetic variation will be conserved at the population level. In faunal and floral conservation, population fragmentation has harmful effects in the long term (Frankham *et al.*, 2002, Chapter 13) but the situation is different in livestock where selection is practised.

De Rochambeau and Chevalet (1982) compared, by simulation, the effects on mean inbreeding coefficient of cyclical and fixed, line-breeding schemes – over the first five to ten generations a cyclical scheme was more effective, provided the number of groups was large enough (at least ten). If the population could only be split into a few groups, say five, then the difference was not great and theory predicts that fixed schemes would be better in the long run (Kimura & Crow, 1963).

In France, several breeds have been managed following conservation breeding plans, some for over 20 years (Danchin-Burge & Avon, 2000). In Breton Black Pied cattle, for example, a cyclical breeding plan was in operation at first but this has now been replaced by a programme based on minimizing kinship (Colleau *et al.*, 2002; Frankham *et al.*, 2002, Chapter 17). Two rare French sheep breeds, the Solognot and the Précoce sheep, have followed a circular mating plan, but numbers are still low and the populations are ageing (Djellali *et al.*, 1994).

Various designs exist for controlling inbreeding in populations under selection (Hill, 2000), including optimization procedures that maximize rate of genetic progress for a given rate of inbreeding, by allocating specific mates. Aimed specifically towards genetic conservation, new computer programs are becoming available which use breed society registration data to calculate the inbreeding of actual individuals and of the offspring of hypothetical matings. Zoo animal population management software has not been found to be suitable for livestock breed management.

Effective population size is increased by equalizing family size. In practical terms this could be done by encouraging or subsidizing the use of more breeding males than in normal practice. Nomura *et al.* (2001) recommended placing an upper limit on the use in AI of popular sires of the Japanese Black cattle breed, because of the sharp recent decrease, through inbreeding, of N_e from about 30 in the 1980s to around 17 in the 1990s. However, imposition of a mating plan can require breeders to use animals for breeding that they would prefer to discard; equalizing family size implies relaxed selection and could reduce population viability.

Founder representation

Work with zoo animal breeding has led to new methods of intensive population management with optimal matings being chosen for each generation. These methods have been based on the genetic representation of each founder in the descendant population. The relatedness between each descendant and the founders can be calculated and summed for each founder. Founder contributions to the current generations can be calculated from computer simulations of the transmission of genes through the population.

Often, populations have undergone many generations of breeding which has led certain lineages or founders to be very over-represented in the pedigrees of the current generation. It has often been proposed that the genetic variation would be best conserved if the genetic contributions of the founders were made more uniform.

The process of gene dropping assumes selective neutrality and calculates the probability distribution for the number of extant copies of each founder allele. Each founder at the base of the pedigrees of the population is assigned two hypothetical distinct founder alleles and the chances are calculated of one or both of these alleles disappearing as a result of Mendelian segregation at each mating in the pedigree. From this the proportion remaining of the original heterozygosity of the founders can be calculated. In livestock, gene dropping has been applied to a number of breeds. In American Standardbred horses (MacCluer *et al.*, 1986), which have highly complex pedigrees, specific animals have been found to contribute heavily to different segments of the breed. Gandini *et al.* (1997) used gene dropping to work out how important in present-day pedigrees of the Haflinger breed of horse were genes derived from the Arab stallions (six in number) that have contributed to the breed during the last one hundred years. Gene dropping enabled the extinction probability of the Arab gene pool to be calculated at different periods during this time, and the conclusion was that the Arab contribution has been stabilized for several years and is now homogeneously distributed in the breed. The idea of linking performance characteristics with particular founder representation was taken further here – relationship with Arab founders had a small but significant effect on 'head expression, back and loin strength, movement and docility'. The historical function of this breed, as a packhorse, is being superseded by use for riding and selection towards a more Arab phenotype would be predicted.

Gene dropping also makes it possible to identify individuals which are especially likely to carry alleles from specific founders, so in principle breeding plans could be devised to increase the numbers of descendants of poorly represented founders. Trinderup *et al.* (1999a,b) described this process in Danish Shorthorn cattle, where animals of present-day generations that carry alleles of the traditional type have been identified.

Problems with breeding plans based on conservation of specific alleles have been discussed by Miller (1995) who showed that this could lead to over-representation of the genes of particular founders in the pedigrees of the conserved population and thus a loss of diversity in the rest of the genome.

Effective numbers of founders and of founder genomes

When each founder has the same expected contribution the effective number of founders is equal to the actual number. When contributions are not equal, the effective number, f_e, of founders (also called the number of founder equivalents) of a population will be lower. Where f_e is the number of equally contributing founders that would give the same genetic diversity as that observed in the actual population and is a statistic that can be compared between breeds (Lacy, 1989). It is defined thus:

$$f_e = 1/\Sigma(p_I^2)$$

where p_I is the proportion of the genes of the living descendant population contributed by founder i. In other words it is the number of equally contributing founders expected to produce the same genetic diversity as the population under study if none of its founder genes were lost by genetic drift (Toro *et al.*, 2000).

In the context of wild animals, f_e can be seen as the number of unrelated wild-caught individuals that would present the same amount of genetic diversity as the captive pedigreed population.

A proportion of the genome of each founder will have been lost by genetic drift. This value can be calculated directly for very simple pedigrees or found approximately by computer simulation; it is known as the equivalent number of founder genomes and is similar to f_e but the contribution of each founder is reduced by the proportion r_I of its genome that is deduced to have been lost by drift:

$$f_g = 1/\Sigma(p_I^2/r_I).$$

These values have been calculated in a number of livestock pedigree studies. Sölkner *et al.* (1998) in a new development derived, from founder genome equivalents, the distances between and within different subgroups, which fitted well with knowledge about the breed structure.

Migration between breeds

A major factor determining the genetic composition of a breed is migration. Genes can be introduced from outside the breed, for example the French dairy breeds (Boichard *et al.*, 1996). This happens when a breed society agrees that introgression from some other related breed can be permitted to help the breed to remain competitive, without losing its identity. In many cases, for example the Southdown sheep

(Hall, 1989b) and Lincoln Red cattle (Stennett, 2000) the traditional type can be almost lost. Blott *et al.* (1998) showed how the introduction of Canadian Hereford genes into the British Hereford population in the 1970s resulted in an increase of heterozygosity (measured from red cell antigens and blood proteins) but could ultimately result in a loss of the genetic variation characteristic of the original British type. This may well be an issue in many minority breeds (Figure 9.1).

A rare Polish breed of pig, the Zlotnicka Spotted, seems to have a valuable disease resistance gene at high frequency (0.63 for an allele associated with resistance to post-weaning diarrhoea). It was suggested that in the related Zlotnicka White, the allele had been reduced to low frequency through crossing with the Swedish Landrace in the 1950s (Klukowska *et al.*, 1999).

There have been several studies on the genetic differences between national populations of international breeds. For example Hanslik *et al.* (2000) describe how North American Holstein populations have significantly different microsatellite genotypes from German and Danish Friesians even though there has been much gene flow from the former. Similarly, Brenneman *et al.* (2001) were able to distinguish the Colombian and Costa Rican varieties of the Romosinuano cattle (a Criollo breed), on the basis of microsatellites, even though they look exactly the same. The Costa Rican animals sampled were known to be descended from Herefords graded up with Romosinuano semen while the Colombian cattle were registered and purebred.

Figure 9.1 South Country Cheviot sheep. Even though this breed is strong in numbers in southern Scotland and northern England, there must be some concern for its purity, in view of the considerable though unquantified use of North Country Cheviot and other rams.

Structuring within breeds

Pedigree studies can show how within a country a breed can be quite extensively differentiated. For example, in Austria, there are clearly distinct local populations of Simmental cattle. In contrast the Austrian Brown is losing this regional distinctiveness, apparently because of the widespread use of Brown Swiss semen from the USA (Sölkner *et al.*, 1998). In the Czech republic (Horín *et al.*, 1998) the grey and black subpopulations of the Kladruby horse are distinct in MHC and microsatellite variation, though similar as measured by blood groups and biochemical polymorphisms.

Another horse breed, the Lipitsa, which originated in 1580, is structured into maternal lineages and today's animals can each be traced back to a named female founder (Kavar *et al.*, 1999). This pedigree distinctiveness is mirrored in the mtDNA (Figure 9.2).

The French dairy breeds illustrate how narrow the genetic base of modern breeds can be (Boichard *et al.*, 1996). The number of ancestors whose contributions accounted for 50% of the genes of females born between 1988 and 1991 ranged between 29 for the French Holstein (2.3 million females born in that period) and seven for the Tarentaise (5817 females). The genetic variabilities of these breeds correspond to those of populations of 15 to 30 completely unrelated animals.

Putting theory into practice

If there is a supervisory authority overseeing genetic conservation in a breed, there are various policies that could be adopted. The FAO DAD-IS system is being developed to provide a framework for these activities (FAO, 1998). The work discussed in this chapter shows how the animals important to use for breeding can be identified – but how this information can be translated into breeding plans or mating systems is a much more difficult question (Caballero & Toro, 2000). There will be breed and species differences in the time scale, in the extent of genetic drift and migration, in age structure and in response to inbreeding.

One key point (Caballero & Toro, 2000) is that the genetic parameters relating to founder representation are all directly related to the average pairwise coancestry of a group of individuals, whose calculation is 'the most straightforward way of maintaining genetic variability' and which can lead to practical suggestions for choice of individuals for breeding. Pairwise coancestry is the inbreeding coefficient of the progeny of a hypothetical mating between the pair.

An urgent requirement (Woolliams & Meuwissen, 1999) is for a decision support system, which will enable the consequences at a population level of specific mating decisions to be predicted. Such a system must have regard to the realities of livestock breeding. The launch of the GENIPED software (Townsend, 2003b) is an important step. This program uses pedigree data to calculate individual and population inbreeding coefficients and kinships, founder animal genetic contributions and census and actual population sizes.

In most livestock populations, commercial considerations will play a part in mating decisions. Even if a mating is proposed and agreed, it may not result in

Haplotype \ Position	15494	15495	15496	15510	15534	15538	15542	15585	15596	15597	15602	15603	15604	15615	15616	15617	15635	15649	15650	15659	15666	15667	15703	15709	15720	15770	15771	15775	15776	15777	15806	15807	15809	15810	15826	15827		16361	16371	16407	16439	16476	16540	16543	16546	16556	16559	16563A	16605	16629
X79547	**T**	**T**	**A**	**T**	**C**	**A**	**C**	**G**	**A**	**A**	**C**	**T**	**G**	**A**	**A**	**T**	**C**	**A**	**A**	**T**	**G**	**A**	**T**	**C**	**G**	**C**	**C**	**C**	**T**	**A**	**C**	**C**	**A**	**A**	**A**	**A**	**.**	**A**	**T**	**C**	**C**	**C**	**C**	**T**	**T**	**T**	**C**	**~**	**A**	**A**
Batosta	.	C	.	.	.	.	.	A	.	.	T	.	.	.	.	.	.	.	G	.	.	.	.	.	A	.	.	.	.	.	.	.	.	.	G	.	.	.	C	.	.	.	.	.	.	.	.	C	.	.
Capriola	.	C	.	.	.	.	.	A	.	.	.	.	.	.	.	.	.	.	G	.	A	.	.	.	A	.	.	.	.	.	.	.	.	G	G	.	.	.	C	.	.	.	.	.	.	.	.	.	.	.
Slavina	.	C	.	.	.	.	T	A	.	.	T	.	.	.	.	.	T	.	G	.	A	.	C	.	A	.	.	.	.	.	.	.	.	.	.	.	.	.	C	.	.	.	.	.	.	.	.	.	.	.
Dubovina	.	C	.	.	.	.	T		.	G	T	.	.	.	.	.	T	.	G	.	A	.	C	.	A	.	.	.	.	.	.	.	.	.	.	.	.	.	C	.	.	.	.	.	.	.	.	.	.	.
Trompeta	.	C	.	.	.	G	.	.	G	.	T	.	.	.	.	.	.	.	G	.	.	.	.	T	A	.	T	.	.	.	.	.	.	.	G	.	.	.	C	.	T	.	.	.	.	.	.	.	.	.
Gaetana	.	C	.	.	.	.	.	.	.	G	T	.	A	.	.	.	.	.	.	.	.	G	C	.	A	.	T	.	.	G	.	.	G	.	.	.	.	.	C	.	.	.	.	A	.	.	.	C	.	G
Gratiosa	.	C	.	C	.	.	.	.	.	.	T	.	.	.	.	C	.	.	.	.	.	.	.	.	A	.	T	.	.	.	T	.	.	.	.	G	.	.	C	.	.	.	.	A	C	.	T	.	G	G
Strana	.	C	.	.	.	.	.	A	.	.	T	.	.	.	.	C	.	.	.	C	.	.	.	.	A	.	T	T	.	.	T	.	.	.	.	G	.	.	C	.	.	.	.	A	C	.	T	.	.	G
Thais	.	C	.	.	.	.	.	.	.	.	T	.	.	G	G	.	.	.	.	C	.	.	C	.	A	.	.	T	C	.	T	.	.	.	.	G	.	.	C	.	.	T	T	A	.	.	.	.	.	G
Allegra	C	C	G	.	T	.	.	.	.	.	T	C	.	.	.	.	.	G	.	.	.	.	.	.	A	.	T	.	.	.	.	.	.	.	.	.	.	.	C	~	.	.	.	.	.	.	.	C	.	G
Wera	C	C	G	.	T	.	.	A	.	.	.	C	.	.	.	.	.	G	.	.	.	.	.	.	A	.	T	.	.	.	.	T	.	.	.	.	.	.	C	~	.	.	.	.	.	.	.	C	.	G
Monteaura	C	C	G	.	T	.	.	A	.	.	T	C	.	.	.	.	.	G	.	.	.	.	.	.	A	.	T	.	.	.	.	.	.	.	.	.	.	G	C	~	.	.	.	.	.	C	.	C	.	G
Betalka	C	C	G	.	T	.	.	A	.	.	T	C	.	.	.	.	.	G	.	.	.	.	.	.	A	.	T	.	.	.	.	.	.	.	.	.	.	G	C	~	.	.	.	.	.	C	.	C	.	G

Figure 9.2 Mitochondrial DNA haplotypes in the matrilines of Lipitsa horses. Base pair substitutions within the mitochondrial D-loop sequence in each matriline of Lipitsa horses. The haplotypes are compared with a reference sequence from an unrelated horse (Kavar *et al.*, 1999).

viable offspring. For example, AI only succeeds in a proportion of cases. Management plans will need to operate in four stages:

- Recommend matings on the basis of genetic conservation
- Deduce the implications for genetic conservation of the matings proposed by the breeder
- Propose compromise matings
- Monitor the genetic changes that actually take place.

As indicated above, minimizing of coancestry would be an effective methodology for the conservation of within-breed variation. Parsons *et al.* (1996) describe a pedigree Merino flock that had been closed for 11 generations, following a mating plan whereby no matings were made between relatives closer than four generations apart. Molecular genetics showed a satisfactorily high degree of heterozygosity. However, the concept of retaining representation, in pedigrees of the current generation, of founder animals, should not be discarded. This is mainly because of the importance of the lineage concept to livestock breeders (Hall, 1989a). A decision support system must be able to signal the impending loss of such representation. One could imagine circumstances where minimization of kinship requires matings that would lead to the loss of such representation and a decision rule is needed as to how much reduction of kinship would justify how much loss of founder representation. This problem has been observed in Przewalski's horse where the representation of some founders occurs only in association with one specific founder that is over-represented in the pedigrees of the current population (Thompson, 1995). It was originally proposed that matings should be planned so that her genes would eventually disappear from the population. This idea was not adopted because of the high probability that other ancestral genes would also be lost as a result.

Non-registered livestock

Livestock which are evidently of a particular breed, but which are not pedigree or registered, are more difficult to document. Breed societies and government departments can usually give general information about their numbers and distribution and these are important data though expensive to obtain in any detail.

Scottish figures on cattle breed distributions are summarized by Wright *et al.* (2002). Certain regions are very important for certain breeds and regional agricultural, biosecurity and environmental policies should take account of this. Adaptations and histories of breeds can be better understood when these data are available (Figure 9.3a,b,c,d,e,f,g).

Even 200 years after these breeds began to be documented (Hall & Clutton-Brock, 1988) they are still firmly associated with the areas where they were developed – the Ayrshire, Galloway and Belted Galloway in the south-west grasslands, the Beef Shorthorn and Aberdeen Angus in the arable-dominated north-east. The Highland and its recent offshoot, the Luing, are concentrated in the central and west

Figure 9.3 Distribution of numbers (registered and unregistered combined) of Scottish native cattle breeds according to counties. Counties with over 20%, 10–19% and 5–10% of the national population of each breed are identified (black, dark grey, light grey respectively. The county of Shetland and the Shetland cattle breed are excluded because of low numbers (Wright *et al.*, 2002).

Highlands. Reflecting its wide use as a terminal sire and as a crossing sire for dairy cattle, the Aberdeen Angus is more evenly spread over the country than any of the others.

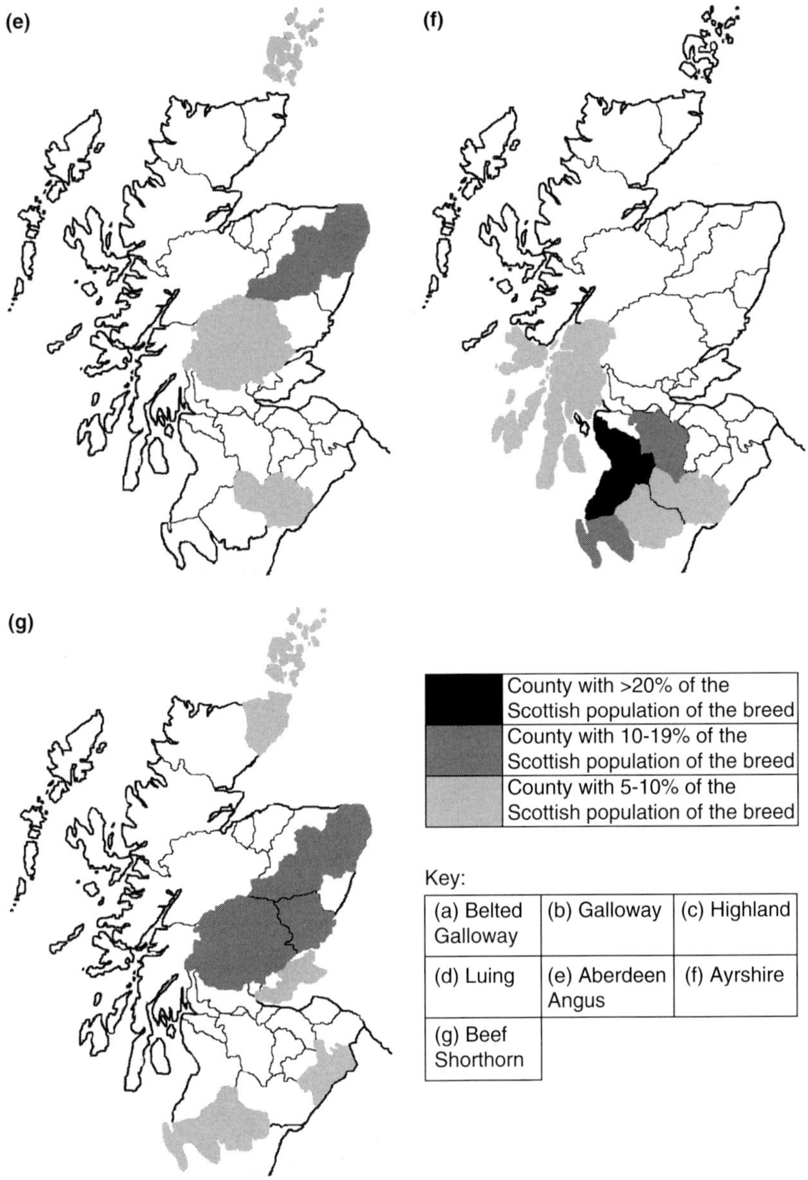

Figure 9.3 (*Continued*)

Management of traditional breeding system

The traditional breeding system could continue to operate and the authority could monitor changes in the genetic parameters mentioned. If cause for concern arises, a more active conservation policy could be initiated – some traditional breeding systems, especially in goats (Barker *et al.*, 2001) have led to highly inbred populations.

However, as awareness of the issues increases there is likely to be a demand for guidance regarding mating plans and this should be anticipated.

The most economical intervention would be at the level of promoting changes in population dynamics that would serve to maintain genetic variability. At its simplest, this would be increase in population size and in number of breeding groups. Another action, especially relevant when rare breeds are under private ownership, would be to promote interchange of breeding animals, discourage the emergence of breeding goals that would reduce genetic variation, and act against the formation of a hierarchical breed structure. These are all actions taken by, for example, the RBST, and they seem to have been effective because the rates of inbreeding observed in rare sheep and pigs are not excessive (Hall, 1986, 1989a).

Management as single conservation population

A breed can be managed as a single population. There are many instances where a conservation herd or flock might be established where the intention is to manage in some degree of isolation from the rest of the breed (for example traditional Lincoln Red cattle; Stennett, 2000). A group of such herds could be run independently, with occasional transfers of breeding stock between them, and perhaps with occasional use of cryopreserved semen (see Chapter 9).

When conservation programmes are started for endangered breeds, there must be some verification of provenance of the animals that are accepted as foundation stock. Microsatellites can enable individuals to be assigned to breeds with a high degree of confidence (Table 9.2). Detection of zebu introgression in taurine cattle in West Africa has been previously described (see Chapter 4) but the only example of this kind of verification is from rare British sheep, using blood groups and biochemical polymorphisms (Clarke *et al.*, 1989). Three breeds had similar degrees of heterozygosity, rare alleles were attributable to particular founders, and it was inferred that acceptance procedures for registration were equally rigorous in all the breeds.

Table 9.2 Assignment of simulated individuals to breed on the basis of microsatellites. This is a measure (under restrictive conditions) of the reliability with which an individual can be correctly assigned to breed.

Species	Number of breeds	Number of loci	Range of success (%)	Reference
Cattle	7	10	94–100	MacHugh *et al.* (1998)
Cattle[1]	5	20	95–100	Ciampolini *et al.* (2000)
Cattle	18	16	94–99	Cañon *et al.* (2001)
Horse	5	13	93–97	Cañon *et al.* (2000)
Horse	10	15	95	Vilà *et al.* (2001)
Sheep	10	10	90–100	Farid *et al.* (2000)
Camel[2] (Kenya)	7	14	39–48	Mburu *et al.* (2003)

[1] Double-blind procedure

[2] Low success of assignment prompts re-examination of the accepted classification of Kenyan camel breeds.

If even greater precision is desired the kinship of a group of candidate animals can in principle be done by genetic markers but very large numbers of markers (around 200) would be needed (Eding & Meuwissen, 2001). Confirming this, Toro *et al.* (2002) found coancestries calculated from pedigrees of Iberian pigs were not strongly correlated with those deduced from study of 49 microsatellite loci.

Purging of harmful alleles

It has long been thought that a history of occasional or gentle, inbreeding may equip populations to survive further genetic bottlenecks: '...more drastic declines [of fitness on inbreeding]...for example rabbits, livestock, man, Japanese quail, may be traceable to lesser exposure to previous bottlenecks of inbreeding during which deleterious genes could have been eliminated' (Wright, 1977, p.95).

The response of numerically small populations to inbreeding is difficult to predict. If it is caused by only a few major harmful alleles, then inbreeding could lead to the population being purged of these alleles (Frankham *et al.*, 2002). Selection reduces their frequency and they are then lost by genetic drift. Simulations suggest that mild inbreeding such as half-sib mating can be as effective at purging as more intense inbreeding, and the former carries less risk of extinction (Fu *et al.*, 1998). Some populations have been claimed to be purged of harmful alleles through slow inbreeding (Lacy & Ballou, 1998; Templeton & Read, 1994; Visscher *et al.*, 2001).

In a theoretical population, an optimal population size can be deduced so there is maximal purging at a given rate of inbreeding. However, in an apparently purged population, even though the harmful major alleles may be disappearing, minor detrimentals will still be accumulating. According to mathematical simulations, this could lead to extinction (Wang, J. L. *et al.*, 1999). There have been many studies (Christensen 1990; Berg & Christensen 1990) showing how inbreeding that fluctuates in its severity can result in purging of harmful recessive genes without accompanying loss of neutral variation. Purging may have been important in the histories of domestication and of breed differentiation but it is unlikely to be useful in management of breeds today and in the future, mainly because of the economic loss to the breeder.

Management of non-pedigreed populations

On the global scale, the main problem in livestock conservation is probably that of protecting numerically small breeds which do not have breed societies or a pedigree structure. Here, mating plans could simply be based on exchange of animals between groups. Much work needs to be done in this area, because interchange of breeding stock could lead to a breakdown of local adaptation, if it is not carefully managed.

Molecular genetics provides information of this kind. Golijow *et al.* (1996) deduced the breed structure of the 300 000-strong Argentine Criollo cattle population from molecular data. This breed appears to be geographically subdivided, with κ-casein alleles being at Hardy-Weinberg equilibrium and with different gene frequencies in different herds.

Selection programmes for traditional breeds

Experience is accumulating in the management of breeding programmes for traditional breeds in low-input systems (Galal *et al.*, 2000). Most of the documented examples relate to sheep and cattle while other species, of considerable importance to households, have been almost totally ignored (Figure 9.4). Some programmes were started explicitly to conserve, develop and disseminate adapted local breeds (for example N'Daṃa cattle in Gambia, D'Man sheep in Morocco). Others were started with a view to wider development issues, for example poultry in Bangladesh as a means of economic advancement of women.

While some systems are based on crossbreeding using exotic genotypes, often utilizing AI services, others are founded on pure breeding. All have systems for the dissemination of improved stocks, which are bred in centralized facilities. There is a wide variation in the systems used to choose members of these central flocks and herds. When there is a performance recording system, as is more easily established for milking animals than for meat producers, offspring of high performers might be selected for the central unit and if there is an AI system as well, their merit can be confirmed by progeny testing and these superior genotypes can be rapidly disseminated.

Programmes vary greatly in the degree of involvement of the producers themselves. Some programmes make no reference to the farmers or to the purchasers of

Figure 9.4 Kanuri woman with donkeys, north Nigeria. Donkeys of two distinct phenotypes are visible, against a backdrop of a large herd of Red Bororo cattle. Recently more attention has been paid to the so-called minor livestock species, but characterizations and information on their breed differentiation are still markedly lacking.

the animals or their products in determining the breeding goals or the methods of obtaining them. Some programmes are funded by governments or the international community while others are claimed to be self financing. In more recent years, programmes have tended to start with the establishment of farmers' cooperatives or breeders' associations. Some have used direct or indirect subsidy to encourage farmers to take part.

Few if any seem to have been based on a rigorous economic study or on a competitive evaluation of different breeds. Most have had to alter their goals, plans or organization during the course of the project. Some programmes can demonstrate genetic improvement while others can claim they have served as a vehicle for more general agricultural, economic and social development.

Some of the projects reviewed by Galal *et al.* (2000) have a long history, going back to institutional herds and flocks founded in the 1930s and 1940s. Purebreeding has become more favoured in more recent years for cattle and sheep. Numbers of reported goat, water buffalo, poultry and pig projects are too small to generalize but taking these species together, use of exotic genotypes for crossbreeding is fairly prominent. Only in more economically advanced areas is there evidence of consumer demand or niche marketing influencing breeding programmes.

Animal breeding programmes are generally a good investment provided they are properly designed, capitalized and carried out, though testing facilities, infrastructure and research are expensive (Simm, 1998). Nucleus breeding schemes (Nicholas, 1987) are theoretically attractive. These schemes are based on farmers contributing their best females to a central breeding unit and being allowed to purchase improved males there. Practical problems may limit their applicability. First, few people who have carried out fieldwork on livestock in the developing world would expect livestock owners to part with their best female stock, a basic requirement of such schemes. Second, there may be health problems. van Wyk *et al.* (1999) consider that because of the intensive husbandry practised in nucleus flocks and herds these schemes are an excellent way of spreading anthelmintic-resistant parasites ('stud worms') around the environment. In New Zealand, disease risks arising from élite animals being transported from one flock to another have been cited as a cause for open nucleus breeding schemes losing favour; modified forms of group breeding schemes can be developed (Roden, 1996).

In addition to these programmes which are designed following the principles of scientific animal breeding, it is important to consider traditional breeding systems. Conservation is more likely to succeed if farmers' knowledge is used in forming policy (Jabbar *et al.*, 1999). Khombe *et al.* (1994) reported how, in Zimbabwe, Mashona cattle were selected for beef performance on an experimental station; performance (birth weight and weaning weight) of calves of bulls from the selected herd and from commercial ranches was compared with that of calves sired by bulls purchased from the traditional smallholder sector. When the bulls were ranked according to the performance of their offspring it became clear that some of the best bulls were smallholder bulls. This implied that the selection programme had not led to greatly superior bulls. The conclusion drawn by Khombe *et al.* was that there are valuable genetic resources in the smallholder sector that should be utilized in commercial improvement programmes and that these make the importation of exotic breeds unnecessary. Another possibility exists that the authors of this paper did not mention: that selection had in fact been operating effectively in the smallholder population which may be far from being an unselected reference population against which genetic progress in the station and commercial herds can be measured.

Genetic studies can give insight into what selection programmes have been operating in traditional systems. For example the Spanish fighting bull shows lower microsatellite heterogeneity and more deviations from Hardy-Weinberg equilibrium than other Spanish cattle (Martin-Burriel *et al.*, 1999) apparently because herds are very self contained and tend to operate breeding programmes based on assortative mating (mating like with like). The same process seems to operate in the Hérens breed which is customarily used in cow fights in Switzerland (Schmid *et al.*, 1999).

A challenge is posed by the need to reconcile breed development through selective breeding, with maintenance of genetic variation. None of the breeding schemes reviewed by Galal *et al.* (2000) raise the issue of designing the selection programme to conserve variation though the desirability of restricting inbreeding is clearly acknowledged. Genetic gain and inbreeding can be expressed in the same framework so optimal solutions can be obtained with rate of response being maximized for a given rate of inbreeding (Hill, 2000), but practical breeding plans or decision support systems following this approach are not yet available.

Recovery of lost breeds

If a breed dies out, but stored semen (or perhaps a single surviving individual) remains, by progressive backcrossing it is possible to 'regenerate' the lost breed. Four generations of backcrossing will give a 95% chance that 90% of the genome of the individuals produced will come from the conserved breed (FAO, 1999, p.105). Often, the last few generations of existence of such a breed have featured attempts to restore vitality, which may not be well documented and the revived breed may be rather different from the original. This is well illustrated with the Norfolk Horn sheep, an ancestor of the Suffolk (Wade-Martins, 1993).

Microsatellites can show how rational crossbreeding decisions might be approached, if it is desired to combine breeds that are no longer demographically viable. As an example, the very scarce Pustertaler Sprinzen cattle are clearly genetically closer to Pinzgauer than to two other breeds (Figure 9.5). They are described as a variety of the Pinzgauer by Porter (2002). This study (Edwards *et al.*, 2000) also showed that the Pustertaler Sprinzen did not show reduced variability compared to other breeds; maybe restoration of numbers is a better approach than conservation crossbreeding in this case.

Another example of revival is the tarpan. Herds of wild horses, some probably with influence from feral animals, survived into the eighteenth and nineteenth centuries in Poland and eastern Europe. These were known as the tarpan and were quite thoroughly described by zoologists at the time. In the Ukraine, the last surviving tarpans died out altogether in 1918–19 (Hemmer, 1990). Around 1780, the last

also possible, but ejaculates are of poorer quality and there are welfare concerns for the animals (FAO, 1999). Whether semen is to be used fresh or frozen depends on the species and on market requirements.

In contrast, techniques for cryopreservation of poultry semen are not yet fully established (Tselutin *et al.*, 1995; Pisenti *et al.*, 1999, p.27). The avian embryo is large (40–50 000 cells) and complex when the egg is laid, so mammalian embryo storage techniques cannot be used. It may be possible to store primordial germ cells from the blood of early-stage chick embryos. These are the precursors of adult gametes and can be isolated, cultured and cryopreserved (Pisenti *et al.*, 1999, p.29).

Even in species whose semen is amenable to freezing, the process kills about 50% of sperm and many of those that survive may lose the ability to reach and fertilize the egg (Holt, 2000). Individual males can differ in the response their semen exhibit to freezing. Boars vary greatly in the fertility of their semen after storage in the frozen state (Crabo, 1991). The same is true of stallions. Twenty years ago only 20% of stallions showed good 'freezability' of sperm. Today the proportion is rather higher (freezing protocols are often modified empirically to suit individual stallions) but there is much frustration when AI fails; one problem is the difficulty of assessing the viability of sperm *in vitro* (Katila *et al.*, 2001). In sheep there might be breed differences in this respect, semen from Cambridge rams may be more sensitive to freezing and thawing (Owen, 1996).

Embryos

The technique of MOET is widely used in cattle breeding (Simm, 1998). It comprises the hormonal pretreatment of a cow so that at her next (induced) ovulation she yields large numbers of ova. Insemination *in vivo* or *in vitro* leads to the production of embryos which after being flushed out can then be placed in the uterus of other, hormonally primed cows (surrogate dams) or frozen. In cattle and in water buffalo this can be done non-surgically, but in sheep, goats and pigs it is a surgical procedure. In the horse, surgery is not needed for embryo recovery but fertile superovulation has not yet been achieved, so the overall embryo recovery rate is less than one embryo per recovery attempt (Allen, 2001). Attempts to freeze horse embryos have given disappointing results; it is more successful with small embryos but these are difficult to recover non-surgically. There are also problems with embryo collection from rabbits. In one study, rabbit donors were slaughtered and embryos retrieved. One in four donors gave no viable embryos and the numbers obtained varied widely (Joly *et al.*, 1994).

As well as species differences there are individual differences. Protocols now permit an average of 30–35 embryos per cow per year (van der Lende *et al.*, 1998, p.91) but within this global figure cows differ in their response to superovulation – embryos per recovery range from 0–40 with a mean of 5. This has encouraged the development of new techniques for the recovery of oocytes (i.e. presumptive ova, before their release from the ovarian follicle).

It used not to be possible to freeze pig embryos, but now it is found that removal of cytoplasmic lipid from 2- to 4-cell stage embryos abolishes their sensitivity to low temperature (Nagashima *et al.*, 1995). Another method is applying cytochalasin before freezing which depolymerizes microfibrils (Kleiner, 1998).

Cryopreservation of poultry embryos is considered impossible, but in an ingenious new approach Kino *et al.* (1997) stored chicken blastodermal cells in the frozen state and injected them, after thawing, into recipient eggs to yield chimeric offspring.

Embryos are generally frozen at the morula or blastocyst stage, when they have 30–120 cells, which is usually 4–10 days after fertilization depending on the species (FAO, 1999). It is critical to avoid formation of intracellular ice crystals and this can be done by reversible dehydration of the cells or by vitrification, which relies on the water forming a glassy solid phase without ice crystallization. Vitrified embryos are reportedly 10% less fertile than frozen.

Oocytes

Production of embryos for cryoconservation requires animals to be superovulated which, as they differ in their response to superovulation, could lead to certain animals being over-represented in embryo banks (Nicholas, 1996b). This variation has commercial disadvantages as well, meaning that superior females do not necessarily respond well to superovulation and this has encouraged the development of new techniques (van der Lende *et al.*, 1998) for the recovery of oocytes for *in vitro* fertilization (IVF).

Immature oocytes can be obtained from slaughterhouse ovaries or by ovum pickup (OPU or technically, repeated follicular aspiration). OPU is ultrasound-guided in water buffalo and horses, but endoscopic in sheep, goats and pigs and can be either in cattle. OPU does not require hormonal priming so there can be up to 80 recoveries per year in cattle (FAO, 1999). IVF is then performed. However, only about 30% of oocytes develop after fertilization and about 40% of transferred embryos result in a live offspring. Also, these cultured embryos survive less well (20%) than embryos recovered *in vivo* (40%). OPU has been successful in captive Burchell's zebra (6.6 per female; Meintjes *et al.*, 1997) but has not advanced far in horses. Only two foals have been born so far by IVF (Squires *et al.*, 2003).

Frozen oocytes have been thawed and fertilized *in vitro* to yield viable calves (Vajta *et al.*, 1998). Success is anticipated with goats (Begin *et al.*, 2003) but considerable work is needed before these techniques become routine. In one study of sheep, 24 oocytes from one slaughtered ewe yielded 11 embryos but only 2 live lambs (Wang *et al.*, 1998). Frozen oocytes have been thawed and implanted in mares which have been inseminated naturally, yielding foals (Maclellan *et al.*, 2002).

Material for cloning

Wells *et al.* (1998) obtained mural granulosa cells from ovaries of 'Lady', the last cow of a population of Shorthorns that had been isolated on a sub-Antarctic island since 1890. Her oocytes were of 'poor quality' and IVF initially failed. Their nuclei were transferred into enucleated oocytes from slaughterhouse ovaries and two calves were born from 22 embryos transferred with one calf surviving. Solti *et al.* (2000)

summarize similar projects on Korean and Japanese cattle. The first cloned equines, a mule and then a Haflinger horse, were foaled in May 2003 (Woods *et al.*, 2003; Galli *et al.*, 2003).

Cryoconservation programmes

Cryopreserved stocks could be used to re-establish an extinct breed, to help with development of new breeds, to support breeds that are being conserved *in situ* and to provide material for further research (FAO, 1999). The optimal amount of material to be stored can be planned if its likely use can be predicted. Ownership of the material must also be defined (Cardelinno, 2003).

Semen was originally stored at −78°C in dry ice but nowadays the medium is liquid nitrogen (LN_2) at −196°C. Thus any cryoconservation programme must have reliable access to LN_2. The health status of donor animals needs to be certified. Philpott (1993) reported that semen presented a greater hazard of transmitting disease than embryos. For example the foot-and-mouth virus is readily transmitted in bovine semen but the situation is very different with properly handled and processed embryos, in which the protective layer, the zona pellucida, has been kept intact. In one trial, even cows that exhibited foot-and-mouth virus in the reproductive tract did not transmit infectivity to the recipients of the embryos or to the resulting calves. The Colombian Romosinuano cattle mentioned above had been imported from Venezuela to Florida as embryos, the first bovine embryo importation permitted into the USA from a country with foot and mouth disease, following the development of embryo-washing techniques (Brennemann *et al.*, 2001).

Guidance is available on how to decide which individuals to include in a cryo-conserved stock (FAO, 1999). The overall costings of a programme will need to be assessed. Although frozen semen is the cheapest option, it would be very expensive to reestablish the breed by progressive upgrading. Lömker and Simon (1995) found a combination of live animal conservation and cryoconservation of semen to be the most cost-effective approach.

Data for the stores of semen and embryos that exist for some breeds is being accumulated in the FAO database. A survey in the early 1990s showed that in Europe, about a third of the breeds classified as endangered were represented in the form of frozen germplasm (Table 9.4).

Cryoconservation in relation to *in situ* conservation

Ideally, cryoconservation programmes should be integrated with *in situ* operations. The RBST operates a semen bank, which in 1999 had 79 644 straws of semen from 583 bulls of 30 breeds (Holt, 2003) and in 1997 it cost £2000 per year to run (Alderson, 1997). This bank serves a dual purpose, as a long-term store and as a source of supply for routine AI. For pigs the arrangement has been for rare breed boars to be kept at stud at commercial breeding facilities under partnership agreements but there are only 279 stored straws from a total of 28 collections from

Table 9.4 Numbers of endangered European breeds of livestock for which semen and embryo stores existed in the early 1990s (Simon & Buchenauer, 1993).

	Semen	Embryos and semen	Number of endangered breeds
Cattle	64	29	149
Sheep	8	2	134
Pigs	14	1	71
Horses	11	1	74
Goats	3	0	37
Totals	100	33	465

six rare breeds in the UK (Holt, 2003). During the 2001 foot-and-mouth outbreak, emergency collections of sheep semen were made. This was not straightforward because this is the season of reproductive quiescence for sheep (Figure 3.7). The newly-formed Sheep Trust (Bowles *et al.*, 2003) concentrated on collection from local minority breeds notably the Herdwick while for the RBST (Mansbridge, 2001) 1649 straws were collected during April 2001 from five breeds (range 80–908 per breed). Funding is being sought (Anon., 2002) for a major extension of this scheme to include all breeds on the RBST lists. Semen would be partitioned three ways for (a) storage in a national archive, (b) return to the owner of the animal and (c) use in conservation breeding schemes.

A general view of assisted reproduction in conservation is given by Solti *et al.* (2000) and the current state of breed cryoconservation in Europe is described by Planchenault (2003) and Simm *et al.* (2004).

The costs of conservation are easily calculated (a combination of live animals plus semen is the most economical; Lömker & Simon, 1995) and they are not excessively high, though the dependence on reliable supplies of liquid nitrogen is total. The problems of reliable methods for use in field conditions still remain (Nicholas, 1996b). A major constraint is likely to be the question of health. There are risks of disease transmission in the course of artificial insemination and embryo transfer (Philpott, 1993). Not all pathogens are fully detectable and much depends on good practice being followed in the collection, screening and processing of semen and embryos. It is easy to visualize semen or embryos of conservation significance, perhaps from an extinct breed or strain, being condemned because the standards that applied at its collection fell short of those in force at the time of its use (Henson, 1992). In addition, one might speculate that it is in the intensively farmed species, where genetic change and loss of variation is most rapid, that cryoconservation is most likely to be justified; but these are the species where health constraints are most likely to apply and thus where frozen germplasm is least likely to be used.

Some animals produce sperm that survive the processes of cryoconservation well, while other individuals do not. Consequently, a semen bank will not necessarily represent the totality of the breed. It would be worth establishing guidelines on how frozen semen could be used for a proportion of matings in a conservation population.

One way to overcome these difficulties would be a system whereby semen stores are rotated, semen being used while the health regulations in force continue to permit it and the stores are replenished by harvesting semen from the resulting progeny, under current health protocols.

Conclusions

To conserve variation within breeds N_e should be maximized and this can be achieved by equalizing family size and minimizing inbreeding. While pedigree data make it possible to determine whether the original foundation animals of a breed continue to be represented in current generations, planning matings so as to minimize inbreeding is equally efficient at conserving ancestral variation.

In reality, breeders will impose their own breeding plans and there is a great need for decision support programmes to help breed societies to maintain variation at breed level, while selection is carried out to adapt the breed to current conditions.

Collection and maintenance of material for cryoconservation is relatively cheap and easy but the regeneration of live animals from it is very expensive. Reproductive technology is far more advanced in some species than others. Generally, the cryoconservation approach has been rather separate from the *in situ* approach, and there is a need for the two to be more closely coordinated.

Part 5
Conclusions

Overview

Livestock are of world significance as providers of commodities that are in growing demand and as a key part of human cultural heritages. Domestication took place in many different places and the genetic base of domesticated livestock is correspondingly wide.

After domestication, animals showed substantial qualitative and quantitative changes in their phenotype, which related to the conditions in which they were kept. Mutation rates at gene and chromosome levels probably increased and so did recombination rates. Artificial selection led to rapid evolution of specific traits and population structure had effects on the nature and extent of heritable variation. Natural selection continued to act on domesticated animals, but probably did not contribute to the differentiation of breeds.

There are very many definitions of the term 'breed'. For conservation purposes, the definition of a breed as 'whatever a government says it is' is likely to be used generally. Breeds arose to fill cultural or economic niches and their diversity reflects human diversity. This is more obvious in the Old World, where the time depth of livestock husbandry is at least 5 000 years, than in the New World, where the time depth is only 500 years or less (the ancient livestock systems of pre-Hispanic America were almost totally eradicated).

Breeds show marked ecological adaptation. Within-breed differences in parasite and disease resistance are usually as great as between-breed differences, but there are some conditions (such as trypanotolerance and tick resistance), which differ markedly between breeds.

Quantitative genetics has been applied with great success to increasing the productive performance of farm animals. Many genetic mechanisms that could be important in determining traits not directly related to production have received much less attention. Such themes as the interactions between loci need much more theoretical and experimental investigation.

Crossbreeding, for livestock breeds, is both a threat and an opportunity. If genes are introduced into one breed from another, the former might become better adapted to current conditions but it could lose its identity in the process. If sustainable crossbreeding systems are set up and correctly managed, breed conservation can be assured and production will be able to exploit heterosis.

Livestock development projects can contribute to reducing poverty in the developing world if correctly designed, and publicly funded projects should normally be based on indigenous breeds. Crossbreeding definitely has its place. For both publicly and privately funded projects, genetic impact assessment should be conducted. Methods for such assessments need to be developed. In the wealthiest countries livestock biodiversity can produce high value-added and specialized niche products and can be used to manage landscapes and as part of cultural heritages. Indeed without these uses local breeds are unlikely to have a sustainable future in the globalized economy.

Behavioural differences among breeds need to be studied, especially with reference to welfare. Selection for reduced fearfulness is a likely route, but care must be taken not to select merely for reduced reactivity. Selection for production characteristics in intensive husbandry does not lead to correlated improvements in adaptation, at the cognitive level, to the systems.

FAO DAD-IS is a continually updated database of livestock breeds and also provides the basis for national conservation programmes. It is a repository of breed information. It has the advantages and disadvantages of being based on returns from national coordinators appointed by governments. Worldwide, very large amounts of data exist on breed performance and attributes, most of it on the most commercially significant breeds while some breeds of rarity and interest are hardly documented at all. Recently most of the internationally funded work on livestock biodiversity has been on deducing the genetic distances among breeds. A debate is beginning on the relative urgency of genetic distance work, and of characterizations of performance and other attributes. Recent genetic work, which has mainly used microsatellites, has provided important insights into the evolutionary history of breeds. Feral livestock, especially goats and pigs, thrive in many countries, but generally have not reverted to a wild-type physical appearance. In some places they damage the environment and cause disease risks, in others they add to local biodiversity. With growing interest in the use of free-ranging livestock for habitat management, ethical issues relating to their release into the wild need to be considered.

Within a livestock species roughly 50% of utilizable quantitative variation is between individuals within a breed, and 50% between breeds. Thus the loss of a livestock breed means a substantial loss to the overall accessible genetic diversity of the species. Several factors cause the loss of heterozygosity and allelic diversity from a population.

Economic valuations can be made of animal genetic resources, and in principle the cost of conservation is far lower than the benefits that could come from utilizing a conserved breed, and the costs of replacing a breed that has been lost. While the costs and benefits of animal breeding programmes are well understood, much needs to be done in the valuation of specific animal genetic resources, if their present values are to be properly understood.

For this genetic variation to be conserved, international and national policies are needed, and the FAO 'State of the World's Animal Genetic Resources' project will provide a framework for these. The genetics of conservation of within-breed variation are well understood and are, effectively, based on the minimizing of inbreeding through the use of pedigree data. The techniques are difficult to apply in a given breed with the resources that are typically available, and computer packages and expert systems are urgently needed.

Prospects for livestock biodiversity

In the developed world, rare breeds are on the whole well protected. The finance mainly comes from the private sector, probably from the same people who support conservation of flora and fauna. The place of livestock breeds in the cultural landscape is generally accepted.

In the commercial breeds, loss of within-breed diversity could be a serious issue in certain sectors, notably dairy cattle and turkeys and to a lesser extent other poultry, because of the structures of the industries and the reproductive technologies in use.

Globalization of food production demands that mass-market animal products must be produced at the lowest possible cost and this implies the intensive husbandry of a small number of highly selected breeds. Increasing sophistication of the market means there are opportunities for the minor and rare breeds to supply niche markets.

In the developing world, livestock should be seen as an instrument contributing to social development as well as a source of food and agricultural services. As urbanization spreads, the demand will be for cheap food and globalization will mean that world prices will apply. Livestock projects that seek only to increase food production are for the private sector to fund; public money should only go to projects that contribute to social equity and food security.

All livestock projects whether in the developed or the developing world should be subject to a genetic impact assessment, which would quantify the effects of the project on local or traditional breeds. Economic appraisal of animal breeding programmes is well understood and the techniques should be easy to develop. If a project is predicted to erode genetic variation or reduce numbers of the local breed, measures must be included to rectify this, perhaps by storing semen or embryos or subsidizing keepers of the traditional breed.

Scientific contribution

Scientifically, livestock biodiversity is a fertile field for theoretical, experimental, and descriptive studies. Certain issues in quantitative genetics need to be studied, notably questions about coadaptation (epistasis). Ideas on conservation prioritization need to be developed and rapid methods for characterizing livestock biodiversity are needed. Systems for the conservation management of pedigreed or non-pedigreed populations need to be developed which reconcile genetic improvement with maintenance of genetic variation which can be used by practical breeders. The interface between livestock genetic diversity, and human genetic diversity, seems worthy of exploration.

Livestock systems especially in the developing world, and most especially at the village level, need to be studied to identify constraints and to understand breeding and husbandry practices as they relate to the use of local breeds. Relationships between livestock and wildlife need to be understood and methods found for their coexistence.

Policy issues: funding

Spending on agricultural research is not increasing as fast as the rate of increase of human population. Certain areas are receiving less funding, and there is a general shift to private funding of R&D. Against this background at first sight it is difficult to see much prospect of generous funding for the conservation and sustainable development of livestock biodiversity.

Perhaps the falling-off of interest in public funding of agricultural R&D reflects, at the global level, acceptance of the proposition that the main impediment to people achieving food security is poverty rather than the global production of food. Poverty will be reduced by the fostering of sustainable livelihoods, and development funding is now being targeted more explicitly in that direction. One might caricature global research aims as being

- agricultural R&D: geared towards increased production
- development R&D: geared towards better distribution.

Traditional breeds can contribute towards increased production, but more in the sense of making it sustainable than of increasing gross output. This is probably of medium-term interest. They can also contribute to the improvement of rural and peri-urban livelihoods, thus increasing food security, and this is clearly of immediate interest.

The future

Market requirements for livestock are changing at a faster rate than ever before. Maintaining viable populations of the widest possible range of breeds is essential because these changes can only be responded to by capitalizing on between-breed differences. Within-breed development would be too slow when the new requirement is very novel.

It may be claimed that the livestock breeds of the world cannot all be conserved – and, indeed, this is so, if the only conservation mechanisms available were the organization of formal programmes. They can all be conserved if the appropriate policy environment is established; that is, if each country decides that in all livestock policies, local pre-existing breeds will be favoured. This principle is only impossible in agricultural systems where production regardless of environmental constraints, social health and welfare concerns and overall biological efficiency is the priority, and many commentators argue that over much of the world, such systems have had their day.

Appendix

Exchange rates at the time of writing were £1 = US$1.59 = 1.43 euro.

Over 300 breeds are mentioned in this book. For many of them some further background information will be helpful. Further details on breeds are in CAB International (2002), Hall and Clutton-Brock (1988), Porter (1993, 2002) and www.fao.org/dad-is.

Aberdeen Angus	Cattle	Scottish beef breed very influential worldwide. Black colour and lack of horns are dominant so animals with Aberdeen Angus genotype are easily recognized. Often known as Angus outside the UK.
Afrikander	Cattle	South African beef breed. Sanga type.
Angus	Cattle	US version of Aberdeen Angus. This and other non-UK populations of the breed have been selected for larger size and more rapid growth than the original UK populations.
Arab	Horse	Part origin of Thoroughbred racehorse.
Awassi	Sheep	Fat-tailed sheep, with dairy qualities. Originated in middle-east and now quite widespread in semi-arid lands.
Ayrshire	Cattle	Scottish dairy breed, influential worldwide, rather smaller than Friesian.
Bali	Cattle	Domesticated form of banteng *Bos javanicus* with some genetic contribution from zebu cattle.
Bedouin	Goat	Part of poorly defined breed group, the 'Syrian type'.
Belgian Blue and White	Cattle	Double-muscled meat sire breed.
Beltex	Sheep	Belgian sire breed for production of meat lambs, developed from Texel, often double-muscled.
Beulah Speckled Face	Sheep	Welsh hill sheep, apparently intermediate between Welsh Mountain and local Down breeds
Blonde d'Aquitaine	Cattle	Worldwide beef breed, French origin.
Blue Albion	Cattle	Blue roan cattle, apparently derived from cross of Shorthorn and Friesian. Documented history in late nineteenth–early twentieth century, but links to the few present day herds have not been strong enough for the breed to be recognized by the Rare Breeds Survival Trust.
Bluefaced Leicester	Sheep	See Border Leicester.

Booroola Merino	Sheep	Prolific strain of Merino.
Border Leicester	Sheep	British breed with long history. Mated with hill/mountain ewes to produce lowland breeding sheep. Has also been influential abroad. Gave rise to Bluefaced Leicester which is more important in the UK now.
Boreray	Sheep	Found on Boreray, small island of St. Kilda group, off Scotland. A relict of the early (late 1700s) crosses of Scottish Blackface with local sheep. Rare breed; a few on mainland Britain.
Brahman	Cattle	Zebu breed developed in southern USA, important beef breed in many semi-arid areas.
Brangus	Cattle	Beef cattle developed in southern USA from Brahman and Aberdeen Angus.
Braunvieh	Cattle	Correctly known as Brown Mountain – breed group including Swiss Brown, American Brown Swiss.
British White	Cattle	Hornless, white coat and black ears, associated with parkland. Rare English breed. Contributed to American White Park.
Brown Swiss	Cattle	Correctly known either as Swiss Brown (Switzerland), or as American Brown Swiss (USA). Semen from the latter has been extensively used in Europe.
Cambridge	Sheep	Highly prolific UK breed developed in late twentieth century by selection from a group of ewes chosen for their large litter sizes.
Charmoise	Sheep	French sire breed for production of meat lambs.
Charolais	Cattle	Worldwide beef breed, French origin.
Charollais	Sheep	Meat sire breed, French origin.
Cheviot	Sheep	British hill sheep breed, associated with relatively benign hill environments. Divisible into distinct local breeds.
Chianina	Cattle	Italian beef breed, one of the tallest cattle breeds in the world.
Chillingham	Cattle	Single herd usually numbering 54 (May 2004), associated for centuries with Chillingham Park in NE England. Reserve herd in NE Scotland. Long history, remarkably inbred yet still viable.
Criollo	Cattle and sheep	Central and South America, Caribbean: breeds of Spanish origin. The Texas Longhorn is a Criollo breed.
Danish Shorthorn	Cattle	Rare, developed from crosses of UK Shorthorn with local cattle.
Dexter	Cattle	Dwarf breed, originated from Kerry, mainly found in UK.
Dorper	Sheep	Developed in mid twentieth century in South Africa from crosses between Dorset Horn and Blackhead Persian.
Down breeds	Sheep	Group of UK meat sheep developed by crossing local breeds with Southdown from late 1700s onwards. Generally used for mating with ewes that have been selected for maternal qualities, to sire meat lambs. Examples: Shropshire, Dorset Down, Hampshire Down.
Duroc	Pig	American breed noted for high meat quality and ability to thrive outdoors. Rival to Hampshire.

Finn Dorset	Sheep	Crossbred of Finnish Landrace and Dorset Horn (or Polled Dorset): combining the prolificacy of the former and the extended breeding season of the latter.
Finnish Landrace	Sheep	Highly prolific northern short-tailed breed, widely used in breeding programmes to increase prolificacy.
Friesian	Cattle	Black and white dairy cattle, originated from Netherlands. Less extreme dairy type than the north American Holstein which was developed from it.
Galway	Sheep	Irish longwoolled breed.
Guernsey	Cattle	Dairy breed, milk fat content intermediate between Friesian and Jersey, found worldwide.
Hampshire	Pig	American breed, rival to Duroc.
Herdwick	Sheep	Native to Lake District of north-west England, very coarse-woolled breed, generally lacks commercial attributes but is of great historic interest and very strong local associations.
Hereford	Cattle	English beef breed very influential worldwide. White face is dominant so animals with Hereford genotype are easily recognized. Herefords outside UK have been selected for larger size and more rapid growth than the original UK populations.
Holstein	Cattle	North American dairy breed, usually black and white though red and white Holsteins are also popular. The world's most numerous dairy breed, developed from the Friesian of the Netherlands. Very high milk yield, poor beef conformation. In Canada, milk-recorded cows average 9 152 kg per lactation (www.dairyinfo.agr.ca). Name bears no relation to origin or history.
Holstein-Friesian	Cattle	Name for Friesian cattle which have received genes from North American Holsteins.
Hungarian Grey	Cattle	Long horned breed, strong cultural/historical associations.
Japanese Black	Cattle	Taurine breed noted for high meat quality, relatively uncommon.
Jersey	Cattle	Dairy breed, high milk fat content, found worldwide.
Kerry	Cattle	Irish rare breed, origin of the Dexter.
Kuri	Cattle	West African taurine breed, enormous horns, native to Lake Chad area. Not trypanotolerant.
Landrace	Pig	White breeds developed from Danish Landrace which traces back to Large White but had been intensively selected from about 1900, for production of lean meat.
Large Black	Pig	UK breed locally influential elsewhere, not suited to modern mass pigmeat market.
Large White	Pig	Developed in UK, world-wide breed, also known as Yorkshire in some countries.
Limousin	Cattle	Worldwide beef breed, French origin.
Lipizzaner	Horse	Light horse, very long history, associated with former Austro-Hungarian empire, also called Lipizzaner.

Lleyn	Sheep	Derived from Welsh Mountain, evidently rare in 1970s but today's high numbers presumably reflect upgrading and other breed development programmes. Selected for prolificacy.
Longhorn	Cattle	English breed, improved by Robert Bakewell from about 1760. Historically significant but far less influential than the Shorthorn. Distinct from Texas Longhorn which is of Spanish origin.
Maine-Anjou	Cattle	Worldwide beef breed, French origin, clearly influenced by Shorthorn.
Meishan	Pig	Chinese breed, very prolific though too fat, being crossbred with Large White, Landrace and others to produce new commercial breeds.
Merino	Sheep	Group of breeds of Spanish origin with heavy fleece of very fine wool.
Mule		(a) Hybrid of a donkey stallion and a horse mare; (b) popular British crossbred sheep, sired by Bluefaced Leicester out of hill ewe.
N'Dama	Cattle	West African taurine breed (Guinea, Gambia, Senegal), trypanotolerant, exported to many African countries.
Navajo-Churro	Sheep	Developed from Spanish-origin sheep in western USA since about 1600.
Nguni	Cattle	Swaziland and South Africa, sanga breed (i.e. has both taurine and zebu appearance, though affinities stronger with taurine). Is being used in semi-arid beef production outside Africa.
Norfolk Horn	Sheep	Rare English breed, was crossed with Southdown to give rise to Suffolk.
North Ronaldsay	Sheep	Primitive northern short-tailed breed, essentially a relict of the early introductions of Roman fine-woolled sheep into the native type whose relict today is the Soay.
Oxford Sandy and Black	Pig	English breed. The last boar licensed under this designation died in 1964. Claims that the breed survived were not accepted by the Rare Breeds Survival Trust. The breed has been reconstructed with a revived breed society since 1985.
Piétrain	Pig	Very lean and muscular Belgian breed, extensively used in intensive pig production.
Poitou	Donkey	Very tall French donkey breed formerly kept to breed army mules.
Poll Dorset	Sheep	Genetically hornless development of Dorset Horn.
Portland	Sheep	Rare English breed, may be relict of original stock that gave rise to Dorset Horn.
Quarter Horse	Horse	American breed, with Thoroughbred influence, traditionally used for races over a quarter mile.
Red Bororo	Cattle	Long-horned, usually deep red, zebu breed from Sahel of west Africa. Commonly known as Rahaji. Not trypanotolerant.
Red Poll	Cattle	English breed, also quite widespread and fairly successful in warm climates.

Rhiw	Sheep	Variant of Welsh Mountain, sought in 1970s but found to have gone extinct.
Romanov	Sheep	Northern short-tailed breed, highly prolific, from Russia, used quite widely in developing new breeds.
Sahiwal	Cattle	Pakistan, though main purebred population is now in Kenya. One of the very few zebu breeds with a reputation for dairy merit.
Scottish Blackface	Sheep	Most numerous sheep breed in Britain. Hardy. Associated with the more rigorous hill/mountain environments.
Shetland	Sheep	Two stocks recognized: the numerous commercial sheep kept on the Shetland Islands and the traditional type with much variation in fleece colour which are almost all on the UK mainland.
Shorthorn	Cattle	UK breed used worldwide in nineteenth and early twentieth centuries to upgrade local stocks, now much less common but still widespread.
Shropshire	Sheep	See Down breeds
Simmental	Cattle	Group of breeds, developed essentially by crosses of Swiss Simmental with local populations. Dairy and beef aptitudes.
Soay	Sheep	Primitive northern short-tailed breed; famous feral populations on islands of Soay and Hirta in the St. Kilda group, off Scotland; some husbanded populations elsewhere. A relict of the first sheep farmed in Britain, 5 000 years ago.
St. Kilda	Sheep	The old name, now disused, for the Hebridean sheep. Not to be applied to Soay sheep.
Southdown	Sheep	The original Down breed, developed in England from late eighteenth century.
Standardbred	Horse	Same as American Trotter, developed from Thoroughbred and others in early nineteenth century.
Suffolk	Sheep	Widely used British meat sire breed, very important in many countries.
Teeswater	Sheep	English breed for production of breeding ewes from hill/mountain ewes. Similar function to Wensleydale. Rare.
Texel	Sheep	Dutch meat sire breed, noted for lean carcase. Very popular worldwide.
Thoroughbred	Horse	World-wide breed with long history and pedigrees back to early 1700s; race horse, often crossed with heavier breeds to produce riding horses known as 'warmbloods'.
Tuli	Cattle	Zimbabwe, sanga breed (i.e. has both taurine and zebu appearance, though affinities stronger with taurine). Is being used in semi-arid beef production outside Africa.
Welsh Mountain	Sheep	Second most numerous sheep breed in Britain. Hardy.
Wensleydale	Sheep	Similar function to Teeswater. Rare British breed.
West African Dwarf goat	Goat	Dwarf breed, trypanotolerant.

West African Shorthorn	Cattle	Dwarf taurine breed, associated with humid forest areas. Trypanotolerant. Populations also in savannah areas. No connection with the Shorthorn breed.
White Fulani	Cattle	Medium-horned zebu breed from rather more humid savannah areas of west Africa. Commonly known as Bunaji. Not trypanotolerant.
White Park	Cattle	Rare horned British breed, a composite of the surviving parkland herds, with the exception of Chillingham cattle which are totally separate (though Chillingham bulls were used in one of these herds). Gave rise to American White Park.
White bred shorthorn	Cattle	Rare variety of Shorthorn, crossing sire for production of Blue-Grey crossbreds from Galloway cows. Local to Cumbria, NW England.

References

Abbott, A. (2003) Chicken flu races through Dutch poultry farms. *Nature*, **422**, 247.

Abplanalp, H. (1992) Inbred lines as genetic resources of chickens. *Poultry Science Review*, **4**, 29–39.

Achukwi, M.D., Tanya, V.N., Hill, E.W., *et al.* (1997) Susceptibility of the Namchi and Kapsiki cattle of Cameroon to trypanosome infection. *Tropical Animal Health and Production*, **29**, 219–26.

Ageeb, A.G. (1991) Performance and heterosis from crossing local cattle in the Sudan. *Tropical Animal Health and Production*, **23**, 251–7.

Ajmone-Marsan, P., Negrini, R., Crepaldi, P., *et al.* (2001) Assessing genetic diversity in Italian goat populations using AFLP(R) markers. *Animal Genetics*, **32**, 281–8.

Ajmone-Marsan, P., Valentini, A., Cassandro, M., *et al.* (1997) AFLP™ markers for DNA fingerprinting in cattle. *Animal Genetics*, **28**, 418–26.

Alaku, O. & Moruppa, S.M. (1983) Body and carcass weight losses in goats during the advanced periods of the west African sahelian dry seasons. *World Review of Animal Production*, **19**, 49–54.

Alderson, G.L.H. (1981) The conservation of animal genetic resources in the United Kingdom. In: *Animal genetic resources. Conservation and management. FAO Animal Production and Health Paper 24* (Ed. by FAO/UNEP), pp. 53–76. Food and Agriculture Organization of the United Nations, Rome, Italy.

Alderson, G.L.H. (1997) Trust activities. *Ark*, **25**, 26–7.

Allen, W.R. (2001) Equine embryo transfer: a brief update on techniques and progress. *Ark*, **29**, 146–7.

Aller, J.F., Rebuffi, G.E., Cancino, A.K., *et al.* (2002) Successful transfer of vitrified llama (*Llama glama*) embryos. *Animal Reproduction Science*, **73**, 121–7.

Allies, P., Cuff, J. & Mills, J. (1999) *Working together. Communities, conservation and rural economies*. Royal Society for the Protection of Birds, Sandy, Bedfordshire, UK.

Amakiri, S.F. (1974) Sweat gland measurements in some tropical and temperate breeds of cattle in Nigeria. *Animal Production*, **18**, 285–91

Amakiri, S.F. & Mordi, R. (1975) The rate of cutaneous evaporation in some tropical and temperate breeds of cattle in Nigeria. *Animal Production*, **20**, 63–8.

Amos, W. (1999) Two problems with the measurement of genetic diversity and genetic distance. In: *Genetics and the extinction of species* (Ed. by L.F. Landweber & A.P. Dobson), pp.75–100. Princeton University Press, Princeton, NJ, USA.

Andelt, W.F. & Hopper, S.N. (2000) Livestock guard dogs reduce predation on domestic sheep in Colorado. *Journal of Range Management*, **53**, 259–67.

Anderson, D.E. (1991) Genetic study of eye cancer in cattle. *Journal of Heredity*, **82**, 21–6.

Anderson, D.M. & Urquhart, N.S. (1986) Using digital pedometers to monitor travel of cows grazing arid rangeland. *Applied Animal Behaviour Science*, **16**, 11–23.

Andersson, M., Nordin, E. & Jensen, P. (2001) Domestication effects on foraging strategies in fowl. *Applied Animal Behaviour Science*, **72**, 51–62.

Anon. (1989) Rare breed pig line survey. *Ark*, **16**, 53–60.

Anon. (2002) Re*gene*ration project update – equine, ovine, caprine and porcine semen collection. *Ark*, **30**, 131.

Apanius, V., Penn, D., Slev, P.R, *et al.* (1997) The nature of selection on the major histocompatibility complex. *Critical Reviews in Immunology*, **17**, 179–224.

Appleby, M.C. (1999) *What should we do about animal welfare?* Blackwell Science, Oxford, UK.

Aranguren-Méndez, J.A., Jordana, J. & Gómez, M. (2001) Genetic relationships between Spanish donkey breeds from analysis of microsatellite markers. *ITEA*, **22**, 164–6.

Arapovic, M. (1997) The contemporary features of the reconstruction of cattle breeding in the Croatian republic of Bosnia and Herzegovina. *Agronomski Glasnik*, **59**, 95–107.

Archetti, E.P. (1997) *Guinea-pigs. Food, symbol and conflict of knowledge in Ecuador*. Berg, Oxford and New York, USA.

Armitage, P.L. (1982) Developments in British cattle husbandry from the Romano-British period to early modern times. *Ark*, **9**, 50–4.

Arnason, T. (2001) Inbreeding – threat to trotting horses? *Proceedings of the 2001 World Trotting Conference*. Helsinki and Mikkeli, Finland. (www.hippos.fi/wtc2001/conference %20papers/Thorvaldur%20Arnason.doc)

Arranz, J.J., Bayon, Y. & San Primitivo, F. (1996) Comparison of protein markers and microsatellites in differentiation of cattle populations. *Animal Genetics*, **27**, 415–9.

Arthur, P.F., Hearnshaw, H. & Stephenson, P.D. (1999) Direct and maternal additive and heterosis effects from crossing *Bos indicus and Bos taurus* cattle: cow and calf performance in two environments. *Livestock Production Science*, **57**, 231–41.

Ashton, C. & Ashton, M. (2001) *The domestic duck*. Crowood Press, Marlborough, Wiltshire, UK.

Astolfi, P., Pagnacco, G. & Guglielmino-Matessi, C.R. (1983) Phylogenetic analysis of native Italian cattle breeds. *Zeitschrift für Tierzuchtung und Zuchtungsbiologie*, **100**, 87–100.

Audiot, A., Bougler, J. & Verrier, E. (1993) [Is it possible to have conservation without exploitation]? *Ethnozootechnie*, **52**, 19–24.

Ayalew, W., Rischkowsky, B., King, J. M. & Bruns, E. (2003) Crossbreds did not generate more net benefits than indigenous goats in Ethiopian small holdings. *Agricultural Systems*, 76, 137–56.

Badyaev, A.V. (2002) Growing apart: an ontogenetic perspective on the evolution of sexual size dimorphism. *Trends in Ecology and Evolution*, **17**, 369–78.

Bailey, J.F., Richards, M.B., Macaulay, V.A., *et al.* (1996) Ancient DNA suggests a recent expansion of European cattle from a diverse wild progenitor species. *Proceedings of the Royal Society of London Series B*, **263**, 1467–73.

Baker, R.L. (1991) Breeding for disease resistance – some historical perspectives, problems and prospects. *Proceedings of the New Zealand Society of Animal Production*, **51**, 1–13.

Baker, R.L., Mwamachi, D.M., Audho, J.O., *et al.* (1999) Genetic resistance to gastrointestinal nematode parasites in Red Maasai, Dorper and Red Maasai x Dorper ewes in the sub-humid tropics. *Animal Science*, **69**, 335–44.

Baldock, D., Bignal, E.M., Jones, A.B., *et al.* (2000) New national proposals for the Less Favoured Areas – EFNCP seminar, Brussels, March 2000. *La Cañada. Newsletter of the European Forum on Nature Conservation and Pastoralism*, no. 12, 1–4.

Baldwin, J.A. (1983) Pre-Cookian pigs in Australia? *Journal of Cultural Geography*, **4**, 17–27.

Bansback, B. (1995) Towards a broader understanding of meat demand – presidential address. *Journal of Agricultural Economics*, **46**, 287–308.

Barker, J.S.F. (1991) Population structure. In: *Evolution and animal breeding. Reviews on molecular and quantitative approaches in honour of Alan Robertson* (Ed. by W.G. Hill & T.F.C. Mackay), pp. 75–80. CAB International, Wallingford, UK.

Barker, J.S.F. (1994) Animal breeding and conservation genetics. In: *Conservation genetics* (Ed. by V. Loeschcke, J. Tomiuk & S.K. Jain), pp. 381–95. Birkhäuser Verlag, Basel, Switzerland.

Barker, J.S.F. (2000) Conservation and management of genetic diversity – a domestic animal perspective. *Canadian Journal of Forest Research*, **31**, 588–95.

Barker, J.S.F., Bradley, D.G., Fries, R., *et al.* (1993) *An integrated global programme to establish the genetic relationships among the breeds of each domestic animal species.* Animal Production and Health Division, Food and Agriculture Organization of the United Nations, Rome, Italy.

Barker, J.S.F., Moore, S.S., Hetzel, D.J.S., *et al.* (1997a) Genetic diversity of Asian water buffalo (*Bubalus bubalis*): microsatellite variation and a comparison with protein-coding loci. *Animal Genetics*, **28**, 103–15.

Barker, J.S.F., Tan, S.G., Moore, S.S., *et al.* (2001) Genetic variation within and relationships among populations of Asian goats (*Capra hircus*). *Journal of Animal Breeding and Genetics*, **118**, 213–33.

Barker, J.S.F., Tan, S.G., Selvaraj, O.S., *et al.* (1997b) Genetic variation within and relationships among populations of Asian water buffalo *(Bubalus bubalis). Animal Genetics*, **28**, 1–13.

Barlow, R. (1981) Experimental evidence for interaction between heterosis and environment in animals. *Animal Breeding Abstracts*, **49**, 715–7.

Barton, R.A. (1984) The role of breed societies: past, present and probable. *Proceedings of the 2nd World Congress on Sheep and Beef Cattle Breeding, Pretoria, South Africa*, pp. 177–85. South African Stud Book and Livestock Improvement Association, Pretoria, South Africa.

Bastian, C.T., Van Tassell, L.W., Cotton, A.C., *et al.* (1999) Opportunity costs related to feral horses: a Wyoming case study. *Journal of Range Management*, **52**, 104–12.

Bayer, W. & Waters-Bayer, A. (1998) *The tropical agriculturalist. Forage husbandry*. Macmillan and Technical Centre for Agricultural and Rural Cooperation, London, UK and Wageningen, The Netherlands.

Beaufoy, G. (1998) The proposed Rural Development Regulation – a view from Spain. *La Cañada. Newsletter of the European Forum on Nature Conservation and Pastoralism*, no. 10, 4–6.

Becerril, C.M., Wilcox, C.J., Lawlor, T.J., *et al.* (1993) Effects of percentage of white coat color on Holstein production and reproduction in a subtropical environment. *Journal of Dairy Science*, **76**, 2286–91.

Begin, I., Bhatia, B., Baldassarre, H., *et al.* (2003) Cryopreservation of goat oocytes and *in vivo* derived 2- to 4-cell embryos using the cryoloop (CLV) and solid-surface vitrification (SSV) methods. *Theriogenology*, **59**, 1839–50.

Berg, P. & Christensen, K. (1990) Cyclical inbreeding in every third generation and its effect on gene fixation in relation to fitness. *Journal of Animal Breeding and Genetics*, **107**, 254–60.

Bignal, E.M. (1998) Using an ecological understanding of farmland to reconcile nature conservation requirements, EU agriculture policy and world trade agreements. *Journal of Applied Ecology*, **35**, 949–54.

Bignal, E.M. (2003) Editorial: the mid-term review of the CAP and high nature value farming systems. *La Cañada. Newsletter of the European Forum on Nature Conservation and Pastoralism*, no. 16, 1–3.

Bijlsma, R., Bundgaard, J. & van Putten, W.F. (1999) Environmental dependence of inbreeding depression and purging in *Drosophila melanogaster. Journal of Evolutionary Biology*, **12**, 1125–37.

Bishop, S.C. & Gettinby, G. (2000) Modelling farm animal diseases. In: *Breeding for disease resistance in farm animals*. 2nd edn. (Ed. by R.F.E. Axford, S.C. Bishop, F.W. Nicholas & J.B. Owen), pp. 27–45. CABI Publishing, Wallingford, UK.

Bixby, D.E., Christman, C.J., Ehrman, C.J., *et al.* (1994) *Taking stock. The North American livestock census*. McDonald and Woodward, Blacksburg, Virginia, USA.

Bjørnstad, G. (2001) *Genetic diversity of Norwegian horses with emphasis on native breeds*. D.Sci. thesis, Norwegian School of Veterinary Science.

Blanc, C.P. & Ennesser, Y. (1989) [Zoogeographical approach to the subspecific differentiation of the one-humped camel]. *Revue d'Elevage et de Médecine vétérinaire des Pays tropicaux*, **42**, 573–87.

Blench, R. (1995) A history of domestic animals in Northeastern Nigeria. *Cahiers des Sciences humaines*, **31**, 181–237.

Blench, R. (1999) Agriculture and the environment in northeastern Ghana: a comparison of high and medium population density areas. In: *Natural resource management in Ghana and its socio-economic context* (Ed. by R.M. Blench), pp. 21–43. Overseas Development Institute, London.

Blench, R.M. (1994) The expansion and adaptation of Fulbe pastoralism to subhumid and humid conditions in Nigeria. *Cahiers d'Etudes africaines*, **133–135**, 197–212.

Blench, R.M. (2001) ' "til the cows come home". Why conserve livestock biodiversity?' In: *Living off biodiversity. Exploring livelihoods and biodiversity issues in natural resource management* (Ed. by I. Koziell & J. Saunders), pp. 113–47. International Institute for Environment and Development, London.

Blench, R.M. & MacDonald, K.C. (Eds.) (2000) *The origins and development of African livestock. Archaeology, genetics, linguistics and ethnography*. UCL Press, London.

Blesbois, E. & Labbé, C. (2003) Main improvements in semen and embryo cryopreservation for fish and fowl. In: *Workshop on cryopreservation of animal genetic resources in Europe. Paris, February 2003*. (Ed. by D. Planchenault), pp. 55–65. Salon International de l'Agriculture, Paris, France.

Blott, S.C., Williams, J.L. & Haley, C.S. (1998) Genetic variation within the Hereford breed of cattle. *Animal Genetics*, **29**, 202–11.

Blows, M.W. & Hoffmann, A.A. (1996) Evidence for an association between nonadditive genetic variation and extreme expression of a trait. *American Naturalist*, **148**, 576–87.

Bohlken, H. (1964) [Comparative studies on the skulls of wild and domesticated cattle]. *Zeitschrift für wissenschaftliche Zoologie*, **170**, 323–418.

Boichard, D., Maignel, L. & Verrier, E. (1996) [Pedigree analysis of French dairy breeds]. *INRA Productions Animales*, **9**, 323–35.

Boland, M., MacGibbon, A. & Hill, J. (2001) Designer milks for the new millennium. *Livestock Production Science*, **72**, 99–109.

Bourn, D.M. & Blench, R.M. (1999) *Can wildlife and livestock co-exist? An interdisciplinary approach.* Overseas Development Institute/Environmental Research Group Oxford, London and Oxford, UK.

Bowles, D., Gilmartin, P., Holt, W.V., *et al.* (2003) The emergency of 2001: cryopreservation of sheep germplasm during the UK crisis of foot and mouth disease. In: *Workshop on cryopreservation of animal genetic resources in Europe. Paris, February 2003.* (Ed. by D. Planchenault), pp. 117–22. Salon International de l'Agriculture, Paris, France.

Bowman, J.C. & Aindow, C.T. (1973) *Genetic conservation and the less common breeds of British cattle, pigs and sheep. Study no. 13*, University of Reading Department of Agriculture and Horticulture, Reading, UK.

Bradley, D.G., MacHugh, D.E., Cunningham, E.P., *et al.* (1996) Mitochondrial diversity and the origins of African and European cattle. *Proceedings of the National Academy of Sciences of the USA*, **93**, 5131–5.

Bradley, D.G., MacHugh, D.E., Loftus, R.T., *et al.* (1994) Zebu-taurine variation in Y chromosomal DNA: a sensitive assay for genetic introgression in West African trypanotolerant cattle populations. *Animal Genetics*, **25**, 7–12.

Brant, A. W. (1998) A brief history of the turkey. *World's Poultry Science Journal*, **54**, 365–73.

Brenneman, R.A., Chase, C.C. & Coleman, S.W. (2001) Detection of origin between two bloodlines of Romosinuano cattle using microsatellite screening. *Tropical Agriculture*, **78**, 255–62.

Brisbin, I.L. (1990) A consideration of feral swine (*Sus scrofa*) as a component of conservation concerns and research priorities for the Suidae. *Bongo*, **18**, 283–93.

Broom, D.M. & Johnson, K.G. (1993) *Stress and animal welfare*, Chapman and Hall, London, UK.

Brossard, M. (1998) The use of vaccines and genetically resistant animals in tick control. *Revue scientifique et technique de l'Office Internationale des Epizooties*, **17**, 188–99.

Browman, D.L. (1989) Origins and development of Andean pastoralism: an overview of the past 6000 years. In: *The walking larder. Patterns of domestication, pastoralism, and predation* (Ed. by J. Clutton-Brock), pp. 256–68. Unwin Hyman, London, UK.

Brown, E. (1892) *Poultry keeping as an industry for farmers and cottagers.* The Fanciers' Gazette, Limited, London, UK.

Bruford, M.W. (2004) Conservation genetics of UK livestock: from molecules to management. In: *Farm animal genetic resources. BSAS Publication no. 30.* (Ed. by G. Simm, B. Villanueva, K.D. Sinclair *et al.*), pp. 151–69. Nottingham University Press/British Society of Animal Science, Nottingham, UK.

Bruns, E. (1992) Synthesis of research methodology. In: *African animal genetic resources: their characterization, conservation and utilisation. Proceedings of the research planning workshop held at ILCA, Addis Ababa, Ethiopia, 19–21 February 1992.* (Ed. by J.E.O. Rege & M.E. Lipner), pp. 125–34. International Livestock Centre for Africa, Addis Ababa, Ethiopia.

Buchanan, F.C., Adams, L.J., Littlejohn, R.P., *et al.* (1994) Determination of evolutionary relationships among sheep breeds using microsatellites. *Genomics*, **22**, 397–403.

Bulfield, G. (1998) No rustling at Roslin. *Nature*, **396**, 108.

Bunch, T.D., Mitchell, R.M. & Maciulis, A. (1990) G-banded chromosomes of the Gansu argali (*Ovis ammon jubata*) and their implications in the evolution of the *Ovis* karyotype. *Journal of Heredity*, **81**, 227–30.

Bunzel-Drüke, M. (2001) Ecological substitute for wild horse (*Equus ferus*, Boddaert 1785 = *E. przewalskii*, Poljakov 1881) and aurochs (*Bos primigenius*, Bojanus 1827). *Natur- und Kulturlandschaft*, **4**, 240–52.

Burfening, P.J. & Carpio, M. (1995) Improving Criollo sheep in Peru through crossbreeding. *Small Ruminant Research*, **17**, 31–5.

Burrow, H.M. (1993) The effects of inbreeding in beef cattle. *Animal Breeding Abstracts*, **61**, 737–51.

Burt, A. & Bell, G. (1987) Mammalian chiasma frequencies as a test of two theories of recombination. *Nature*, **326**, 803–5.

Bush, G.L. (1996) Stasipatric speciation and rapid evolution in animals. In: *Evolution and speciation, Essays in honor of M.J.D. White* (Ed. by W.R. Atchley & D.S. Woodruff), pp. 201–18. Cambridge University Press, Cambridge, UK.

Butler, D. (2001) A wolf in sheep's clothing. *Nature*, **414**, 576–7.

Butler-Hogg, B.W. (1984) The growth of Clun and Southdown sheep: body composition and the partitioning of total body fat. *Animal Production*, **39**, 405–11.

Butler-Hogg, B.W. & Wood, J.D. (1982) The partition of body fat in British Friesian and Jersey steers. *Animal Production*, **35**, 253–62.

CAB International (2002) *Animal health and production compendium.* CAB International, Wallingford, UK.

Caballero, A. & Toro, M.A. (2000) Interrelations between effective population size and other pedigree tools for the management of conserved populations. *Genetical Research*, **75**, 331–43.

Caballero, A. & Toro, M.A. (2002) Analysis of genetic diversity for the management of conserved subdivided populations. *Conservation Genetics*, **3**, 289–99.

Campbell, B.M., Doré, D., Luckert, M., *et al.* (2000) Economic comparisons of livestock production in communal grazing lands in Zimbabwe. *Ecological Economics*, **33**, 413–38.

Camac, R. (1997) The Poitou: a breed threatened and saved. In: *The professional handbook of the donkey. 3rd edn.* (Ed. by E.D. Svendsen), pp. 155–65. Whittet Books, London, UK.

Camus, E., Landais, E. & Poivey, J.P. (1981) [Genetic structure of sedentary cattle in the north of Côte d'Ivoire. Future prospects following the spread of zebu crossing]. *Revue d'Elevage et de Médecine vétérinaire des Pays tropicaux*, **34**, 187–98.

Cañon, J., Alexandrino, P., Bessa, I., *et al.* (2001) Genetic diversity measures of local European beef cattle breeds for conservation purposes. *Genetics Selection Evolution*, **33**, 311–32.

Cañon, J., Checa, M.L., Carleos, C., *et al.* (2000) The genetic structure of Spanish Celtic horse breeds inferred from microsatellite data. *Animal Genetics*, **31**, 39–48.

Cardelinno, R.A. (2003) Ownership and use of biological material from a cryobank. In: *Workshop on cryopreservation of animal genetic resources in Europe. Paris, February 2003.* (Ed. by D. Planchenault), pp. 123–32. Salon International de l'Agriculture, Paris, France.

Cassens, R.G. (2000) Historical perspectives and current aspects of pork meat quality in the USA. *Food Chemistry*, **69**, 357–63.

Cavalli-Sforza, L.L., Menozzi, P. & Piazza, A. (1994) *The history and geography of human genes.* Princeton University Press, Princeton, NJ, USA.

Cepica, S., Wolf, J., Hojny, J., *et al.* (1995) Relations between genetic distance of parental pig breeds and heterozygosity of their F1 crosses measured by genetic markers. *Animal Genetics*, **26**, 135–40.

Chardon, P., Renard, C., Gaillard, C.R., *et al.* (2000) The porcine major histocompatibility complex and related paralogous regions: a review. *Genetics Selection Evolution*, **32**, 109–28.

Cheng, P. (1984) *Livestock breeds of China. FAO Animal Production and Health Paper 46.* Food and Agricultural Organization of the United Nations, Rome, Italy.

Chenoweth, P.J. (1981) Libido and mating behaviour in bulls, boars and rams. A review. *Theriogenology*, **16**, 155–77.

Cheverud, J.M., Vaughn, T.T., Pletscher, L.S., *et al.* (1999) Epistasis and the evolution of additive genetic variance in populations that pass through a bottleneck. *Evolution*, **53**, 1009–18.

Chicoteau, P. (1991) [Reproduction in tropical bovines]. *Receuil de Médecine Vétérinaire*, **167**, 241–247.

Chilliard, Y., Ferlay, A., Faulconnier, Y., *et al.* (2000) Adipose tissue metabolism and its role in adaptations to undernutrition in ruminants. *Proceedings of the Nutrition Society*, **59**, 127–34.

Choi, N. J., Enser, M., Wood, J. D., *et al.* (2000) Effect of breed on the deposition in beef muscle and adipose tissue of dietary n-3 polyunsaturated fatty acids. *Animal Science*, **71**, 509–19.

Choquenot, D. (1993) Growth, body condition and demography of wild banteng (*Bos javanicus*) on Cobourg Peninsula, northern Australia. *Journal of Zoology, London*, **231**, 533–42.

Choquenot, D., Lukins, B. & Curran, G. (1997) Assessing lamb predation by feral pigs in Australia's semi-arid rangelands. *Journal of Applied Ecology*, **34**, 1445–54.

Christensen, K. (1990) Simulation study of cyclical in- and out-breeding: effect on gene fixation in relation to fitness and induction of tolerance to inbreeding. *Hereditas*, **112**, 2.

Christman, C.J., Sponenberg, D.P. & Bixby, D.E. (1997) *A rare breeds album of American livestock.* American Livestock Breeds Conservancy, Pittsboro, North Carolina, USA.

Ciampolini, R., Leveziel, H., Mazzanti, E. *et al.* (2001) Genomic identification of the breed of an individual or its tissue. *Meat Science*, **54**, 35–40.

Ciobanu, D.C., Day, A.E., Nagy, A., *et al.* (2001) Genetic variation in two conserved local Romanian pig breeds using type 1 DNA markers. *Genetics Selection Evolution*, **33**, 417–32.

Clarke, C.M.H. & Dzieciolowski, R.M. (1991a) Feral pigs in the northern South Island New Zealand: 1. Origin, distribution and density. *Journal of the Royal Society of New Zealand*, **21**, 237–47.

Clarke, C.M.H. & Dzieciolowski, R.M. (1991b) Feral pigs in the northern South Island New Zealand: II. Breed composition of present population. *Journal of the Royal Society of New Zealand*, **21**, 249–60.

Clarke, J.N. (1995) New sheep – a catalyst for change. *Proceedings of the New Zealand Society of Animal Production*, **55**, 318–20.

Clarke, S.W., Tucker, E.M. & Hall, S.J.G. (1989) Genetic polymorphisms and their relationships with inbreeding and breed structure in rare British sheep: the Portland, Manx Loghtan, and Hebridean. *Conservation Biology*, **3**, 381–8.

Clayton, G.A. (1972) Effects of selection on reproduction in avian species. *Journal of Reproduction and Fertility, Supplement*, **15**, 1–21.

Clutton-Brock, J. (1992a) The process of domestication. *Mammal Review*, **22**, 79–85.

Clutton-Brock, J. (1992b) *Horse power. A history of the horse and the donkey in human societies*. Natural History Museum Publications, London, UK.

Clutton-Brock, J. (1999) *A natural history of domesticated mammals*. 2nd edn. Cambridge University Press, Cambridge, UK.

Clutton-Brock, T.H. & Pemberton, J.M. (Eds.) (2004) *Soay sheep. Dynamics and selection in an island population*. Cambridge University Press, Cambridge, UK.

Cobey, S. (2001) Beekeeping on the only mite-free continent – Australia. *American Bee Journal*, **141**, 495–9.

Cockett, N.E., Jackson, P.S., Snowder, G.D., *et al.* (1999). The callipyge phenomenon: evidence for unusual genetic inheritance. *Journal of Animal Science*, **77**, supplement 2, 221–7.

Coghlan, A. (1996) Early warning for sore udders. *New Scientist*, **150**, no. 2036, p. 19.

Cognié, Y., Baril, G., Paulin, N., *et al.* (2003) Current status of embroyo technologies in sheep and goat *Theriogenology*, **59**, 171–88.

Colleau, J.J., Quéméré, P., Lorroque, H., *et al.* (2002) [Genetic management of the Breton Black Pied cattle breed]. *INRA Productions Animales*, **15**, 221–30.

Coltman, D.W., Pilkington, J., Kruuk, L.E.B., *et al.* (2001) Positive genetic correlation between parasite resistance and body size in a free-living ungulate population. *Evolution*, **55**, 2116–25.

Coltman, D.W., Pilkington, J.G., Smith, J.A., *et al.* (1999) Parasite-mediated selection against inbred Soay sheep in a free-living, island population. *Evolution*, **53**, 1259–67.

Conelly, W.T. (1998) Colonial era livestock development policy: introduction of improved dairy cattle in high-potential farming areas of Kenya. *World Development*, **26**, 1733–48.

Côrte-Real, H.B.S.M., Macaulay, V.A. & Richards, M.B. (1996) Genetic diversity in the Iberian Peninsula determined from mtDNA sequence analysis. *Annals of Human Genetics*, **60**, 331–50.

Cothran, E.G., MacCluer, J.W., Weitkamp, L.R., *et al.* (1986) Genetic variability, inbreeding & reproductive performance in Standardbred horses. *Zoo Biology*, **5**, 191–201.

Cothran, E.G. & van Dyk, E. (1998) Genetic analysis of three South African horse breeds. *Journal of the South African Veterinary Association*, **69**, 120–5.

Cothran, E.G., van Dyk, E. & van der Merwe, F.J. (2001) Genetic variation in the feral horses of the Namib Desert, Namibia. *Journal of the South African Veterinary Association*, **72**, 18–22.

Coulter, G.H. & Foote, R.H. (1979) Bovine testicular measurements as indicators of reproductive performance and their relationship to productive traits in cattle: a review. *Theriogenology*, **11**, 297–311.

Cousins, D.V. & Roberts, J.L. (2001) Australia's campaign to eradicate bovine tuberculosis: the battle for freedom and beyond. *Tuberculosis*, **81**, 5–15.

Crabbe, J.C., Wahlsten, D. & Dudek, B.C. (1999) Genetics of mouse behavior: interactions with laboratory environment. *Science*, **284**, 1670–2.

Crabo, B.G. (1991) Preservation of boar semen: a worldwide perspective. In: *Boar semen preservation II. Proceedings of the 2nd international conference on boar semen preservation held at Beltsville, Maryland, USA, August 1990*. pp. 3–9. Verlag Paul Parey, Berlin, Germany.

Crawford, A.M., Knappes, S.M., Paterson, K.A., *et al.* (1998) Microsatellite evolution: testing the ascertainment bias hypothesis. *Journal of Molecular Evolution*, **46**, 256–60.

Crepaldi, P., Negrini, R., Milanesi, E., Gorni, C., *et al.* (2001) Diversity in five goat populations of the Lombardy Alps: comparison of estimates obtained from morphometric traits and molecular markers. *Journal of Animal Breeding and Genetics*, **118**, 173–80.

Cronje, P.B. (1992) Glucose metabolism and adrenal function in goats bred for fibre production (Angora goat) or meat production (Boer goat). *South African Journal of Animal Science*, **22**, 149–53.

Croston, D., Kempster, A.J., Guy, D.R., *et al.* (1987) Carcass composition of crossbred lambs by ten sire breeds compared at the same carcass subcutaneous fat proportion. *Animal Production*, **44**, 99–106.

Crystal, D. (Ed.) (1997) *The Cambridge encyclopedia. Third edition.* Cambridge University Press, Cambridge, UK.

Cuartas, P. & Garcia-Gonzalez, R. (1992) *Quercus ilex* browse utilization by Caprini in Sierra de Cazorla and Segura (Spain). *Vegetatio*, **99–100**, 317–30.

Cundiff, L.V. (1977) Foundations for animal breeding research. *Journal of Animal Science*, **44**, 311–9.

Cunningham, E.P. (1983) European Friesians – the Canadian and American invasion. *Animal Genetic Resources Information*, **1**, 21–3.

Cunningham, E.P. (1999) The application of biotechnologies to enhance animal production in different farming systems. *Livestock Production Science*, **58**, 1–24.

Cunningham, E.P., Dooley, J.J., Splan, R.K., *et al.* (2001) Microsatellite diversity, pedigree relatedness and the contributions of founder lineages to thoroughbred horses. *Animal Genetics*, **32**, 360–4.

Cunningham, E.P. & Syrstad, O. (1987) *Crossbreeding Bos taurus and Bos indicus for milk production in the tropics. FAO Animal Production and Health Paper 68.* Food and Agricultural Organization of the United Nations, Rome, Italy.

Curry Report (2002) *Farming and food: a sustainable future.* Policy Commission on the Future of Farming and Food. Cabinet Office, HM Government of the UK. (www.cabinet-office.-gov.uk/farming)

Cymbron, T., Loftus, R.T., Malheiro, M.I., *et al.* (1999) Mitochondrial sequence variation suggests an African influence in Portuguese cattle. *Proceedings of the Royal Society of London Series B*, **266**, 597–603.

d'Ieteren, G.D.M., Authié, E., Wissocq, N., *et al.* (2000) Exploitation of resistance to trypanosomes and trypanosomosis. In: *Breeding for disease resistance in farm animals.* 2nd edn. (Ed. by R.F.E. Axford, S.C. Bishop, F.W. Nicholas & J.B. Owen), pp. 195–216. CABI Publishing, Wallingford, UK.

Daly, J.J. (1981) Beef cattle breeds. *Queensland Agricultural Journal*, **107**, 28–30.

Danchin-Burge, C. & Avon, L. (2000) [Genetic variability study in French rare cattle breeds, after twenty years of conservation]. *Renc. Rech. Ruminants*, **7**, 145–8.

Darwin, C. (1868) *The variation of animals and plants under domestication. Vols. I and II. Popular edition, 1905.* John Murray, London, UK.

Davidson, H.R. & Coey, W.E. (1966) *The production and marketing of pigs.* 3rd edn. Longmans, London, UK.

Davis, S. & Payne, S. (1993) A barrow full of cattle skulls. *Antiquity*, **67**, 12–22.

Davis, S.J.M. (1987) *The archaeology of animals.* B.T. Batsford, London, UK.

Dawson, M., Hoinville, L., Hosie, B.D., *et al.* (1998) Developments in PrP genotyping. *Proceedings of the Sheep Veterinary Society*, **22**, 141–3.

Day, P.R., Strauss, M.S. & Barton, J.H. (1994) Domestic animal genetic resources and the Convention on Biological Diversity. *Proceedings of the 5th World Congress on Genetics applied to Livestock Production. Guelph, Ontario, Canada*, **21**, 470–6.

de Haan, C., van Veen, T.S., Brandenburg, B., *et al.* (2001). *Livestock development. Implications for rural poverty, the environment, and global food security. Report 23241, November 2001.* The World Bank, Washington, DC, USA.

de Rochambeau, H. & Chevalet, C. (1982) Some aspects of the genetic management of small breeds. *Proceedings of the 2nd World Congress on Genetics applied to Livestock Production, Madrid*, 282–7.

de Vaccaro, L.P. (1990) Survival of European dairy breeds and their crosses with zebus in the tropics. *Animal Breeding Abstracts*, **58**, 475–94.

DEFRA (2002) UK country report on farm animal genetic resources. Department for Environment, Food and Rural Affairs, London, UK.

Delany, M.E. & Pisenti, J.M. (1998) Conservation of poultry genetic research resources: consideration of the past, present and future. *Poultry and Avian Biology Review*, **9**, 25–42.

DeMartini, J.C., de la Concha-Bermejillo, A., Carlson, J.O., *et al.* (2000) Diseases caused by maedi-visna and other ovine lentiviruses. In: *Breeding for disease resistance in farm animals*. 2nd edn. (Ed. by R.F.E. Axford, S.C. Bishop, F.W. Nicholas & J.B. Owen), pp. 301–24. CABI Publishing, Wallingford, UK.

Desforges, M.F. & Wood-Gush, D.G.M. (1976) Behavioural comparison of Aylesbury and mallard ducks: sexual behaviour. *Animal Behaviour*, **24**, 391–7.

Devendra, C. & Burns, M. (1983) *Goat production in the tropics*. CAB, Wallingford, UK.

Devlin, R.H. (1994) Extraordinary salmon growth. *Nature*, **371**, 209–10.

Devlin, R.H., Biagi, C.A., Yesaki, T.Y., *et al.* (2001) Growth of domesticated transgenic fish. *Nature*, **409**, 781–2.

Dickerson, G.E. (1978) Animal size and efficiency: basic concepts. *Animal Production*, **27**, 367–79.

Djellali, A., Vu Tien, J., de Rochambeau, H., *et al.* (1994) Evaluation of the genetic conservation programmes for the sheep breeds Solognote and Mérinos-Précoce. *Genetics Selection Evolution*, **26(S1)**, 255–65.

Dmi'el, R., Robertshaw, D. & Choshniak, I. (1979) Sweat gland secretion in the black bedouin goat. *Physiological Zoology*, **52**, 558–64.

Doebley, J., Stec, A. & Gustus, C. (1995) *Teosinte branched1* and the origin of maize: evidence for epistasis and the evolution of dominance. *Genetics*, **141**, 333–46.

Drucker, A.G., Gomez, V. & Anderson, S. (2001) The economic valuation of farm animal genetic resources: a survey of available methods. *Ecological Economics*, **36**, 1–18.

Eckhardt, R. B., Eckhardt, D. A. & Eckhardt, J. T. (1988) Are racehorses becoming faster? *Nature*, **335**, 773.

Eding, J.H. & Laval, G. (1999) Measuring genetic uniqueness in livestock. In: *Genebanks and the conservation of farm animal genetic resources* (Ed. by J.K. Oldenbroek), pp. 33–58. DLO Institute for Animal Science and Health, Lelystad, The Netherlands.

Eding, J.H. & Meuwissen, T.H.E. (2001) Marker-based estimates of between and within population kinships for the conservation of genetic diversity. *Journal of Animal Breeding and Genetics*, **118**, 141–59.

Edwards, C.J., Dolf, G., Looft, C., *et al.* (2000) Relationships between the endangered Pustertaler-Sprinzen and three related European cattle breeds as analysed with 20 microsatellite loci. *Animal Genetics*, **31**, 329–32.

Edwards, S.V. & Hedrick, P.W. (1998) Evolution and ecology of MHC molecules: from genomics to sexual selection. *Trends in Ecology and Evolution*, **13**, 305–11.

Edwards, S.V. & Potts, W.K. (1996) Polymorphism of genes in the major histocompatibility complex (MHC): implications for conservation genetics of vertebrates. In: *Molecular genetic approaches in conservation* (Ed. by T.B. Smith & R.K. Wayne), pp. 214–37. Oxford University Press, Oxford, UK.

Ehling, C. & Niemann, H. (2000) Establishment of embryo banks to maintain endangered sheep breeds. *Zuchtungskunde*, **72**, 140–52.

Ellis, N.E., Heal, O.W., Dent, J.B., *et al.* (1999) Pluriactivity, farm household socio-economics and the botanical characteristics of grass fields in the Grampian region of Scotland. *Agriculture Ecosystems and Environment*, **76**, 121–34.

Ellis, S.A. & Ballingall, K.T. (1999) Cattle MHC: evolution in action? *Immunological Reviews*, **167**, 159–68.

Emmerson, D.A., Anthony, N.B. & Nestor, K.E. (1991) Genetics of growth and reproduction in the turkey. II. Evidence of nonadditive genetic variation. *Poultry Science*, **70**, 1084–91.

Ercanbrack, S.K. & Knight, A.D. (1981) Weaning trait comparisons among inbred lines and selected noninbred and randomly bred control groups of Rambouillet, Targhee and Columbia sheep. *Journal of Animal Science*, **52**, 977–88.

Evans, N. & Yarwood, R. (2000) The politicisation of livestock: rare breeds and countryside conservation. *Sociologia ruralis*, **40**, 228–48.

Fabbrocini, A., Del Sorbo, C., Fasano, G., *et al.* (2000) Effect of differential addition of glycerol and pyruvate to extender on cryopreservation of Mediterranean buffalo (*B. bubalis*) spermatozoa. *Theriogenology*, **54**, 193–207.

Falconer, D.S. & Mackay, T.F.C. (1996) *Introduction to quantitative genetics*. 4th edn. Longman, Harlow, UK.

FAO (1992) *The management of global animal genetic resources. FAO Animal Production and Health Paper 104*. Food and Agriculture Organization of the United Nations, Rome, Italy.

FAO (1998) *User's manual for national coordinators for the management of farm animal genetic resources*. Food and Agriculture Organization of the United Nations, Rome, Italy.

FAO (1999) *Secondary guidelines for development of national farm animal genetic resources management plans. Management of small populations at risk*. Food and Agricultural Organization of the United Nations, Rome, Italy.

FAO/UNEP (1981) *Animal genetic resources. Conservation and management. Proceedings of the FAO/UNEP technical consultation. FAO Animal Production and Health Paper 24*. Food and Agriculture Organization of the United Nations, Rome, Italy.

Farid, A., O'reily, E., Dollard C., *et al.* (2000) Genetic analysis of ten sheep breeds using microsatellite markers. *Canadian Journal of Animal Science*, **80**, 9–17.

FARM-Africa (1996) *Goat types of Ethiopia and Eritrea. Physical description and management systems*. FARM-Africa, International Livestock Research Institute, London and Nairobi.

Fell, L.R. (1991) Behavioural and physiological effects in sheep of a chronic stressor and a parasite challenge. *Australian Journal of Agricultural Research*, **42**, 1335–46.

Ferguson, N.M., Ghani, A.C., Donnelly, C.A., *et al.* (2002) Estimating the human health risk from possible BSE infection of the British sheep flock. *Nature*, **415**, 420–4.

Ferrante, V., Verga, M., Mangiagalli, M.G., *et al.* (2001) Behavioural reactions, semen quality and testosterone levels in cocks: genetic implications. *Animal Welfare*, **10**, 269–79.

Ferraz, J.B.S. & Eler, J.P. (2002) The South American perspective on the use of animal resources. *Landbauforschung Völkenrode*, Special Issue **228**, 75–84.

Festing, M.F.W. (1997) Fat rats and carcinogenesis screening. *Nature*, **388**, 321–2.

Fiems, L.O., Moermans, R.J., Boucqué, C.V., *et al.* (1999) Voluntary feed intake in dual-purpose and double-muscled Belgian White-Blue bulls. In: *Regulation of feed intake* (Eds. D. van der Heide, E.A. Huisman, E. Kanis, *et al.*), pp. 21–5. CABI Publishing, Wallingford, UK.

Finch, V.A. (1981) Black cattle in the cool pastoral highlands of Kenya. *Journal of Agricultural Science, Cambridge*, **97**, 739–42.

Finch, V.A., Bennett, I.L. & Holmes, C.R. (1982) Sweating response in cattle and its relation to rectal temperature, tolerance of sun and metabolic rate. *Journal of Agricultural Science, Cambridge*, **99**, 479–87.

Finch, V.A., Dmi'el, R., Boxman, R., *et al.* (1980) Why black goats in hot deserts? Effects of coat color on heat exchanges of wild and domestic goats. *Physiological Zoology*, **53,** 19–25.

Finch, V.A. & Western, D. (1977) Cattle colors in pastoral herds: natural selection or social preference? *Ecology*, **58,** 1384–92.

Fisher, A.V., Enser, M., Richardson, R.I., *et al.* (2000) Fatty acid composition and eating quality of lamb types derived from four diverse breed x production systems. *Meat Science*, **55,** 141–7.

Fisher, R.A. & Holt, S.B. (1944) The experimental modification of dominance in Danforth's short-tailed mutant mice. *Annals of Eugenics*, **12,** 102–20.

Flamand, J.R.B., Vankan, D., Gairhe, K.P., *et al.* (2003) Genetic identification of wild Asian water buffalo in Nepal. *Animal Conservation*, **6,** 265–70.

Fleming, I.A. & Einum, S. (1997) Experimental tests of genetic divergence of farmed from wild Atlantic salmon due to domestication. *ICES Journal of Marine Science*, **54,** 1051–63.

Fleming, I.A., Jonsson, B., Gross, M.R., *et al.* (1996) An experimental study of the reproductive behaviour and success of farmed and wild Atlantic salmon (*Salmo salar*). *Journal of Applied Ecology*, **33,** 893–905.

Folch, P. & Jordana, J. (1997) Characterization, reference ranges and the influence of gender on morphological parameters of the endangered Catalonian donkey breed. *Journal of Equine Veterinary Science*, **17,** 102–11.

Forbes, S.H., Hogg, J.T., Buchanan, F.C., *et al.* (1995) Microsatellite evolution in congeneric mammals: domestic and bighorn sheep. *Molecular Biology and Evolution*, **12,** 1106–13.

Frankel, W.N. & Schork, N.J. (1996) Who's afraid of epistasis? *Nature Genetics*, **14,** 371–3.

Frankham, R., Ballou, J.D. & Briscoe, D.A. (Eds.) (2002) *Introduction to conservation genetics*. Cambridge University Press, Cambridge, UK.

Fraser, A.F. & Broom, D.M. (1990) *Farm animal behaviour and welfare*. 3rd edn. Baillière Tindall, London, UK.

Fraser, D., Weary, D.M., Pajor, E.A., *et al.* (1997) A scientific conception of animal welfare that reflects ethical concerns. *Animal Welfare*, **6,** 187–205.

Fraser, K.W., Cone, J.M. & Whitford, E.J. (2000) A revision of the established ranges and new populations of 11 introduced ungulate species in New Zealand. *Journal of the Royal Society of New Zealand*, **30,** 419–37.

Freeland, W.J. & Boulton, W.J. (1990) Feral water buffalo (*Bubalus bubalis*) in the major floodplains of the 'Top End', Northern Territory, Australia: population growth and the brucellosis and tuberculosis eradication campaign. *Australian Wildlife Research*, **17,** 411–20.

Friend, A. (1999) SAC trustee reports. *Send a Cow Lifeline*, no. 29.

Frisch, J.E. (1987) Physiological reasons for heterosis in growth of *Bos indicus* x *Bos taurus*. *Journal of Agricultural Science, Cambridge*, **109,** 213–30.

Frisch, J.E. (1990) More meat in the northern sandwich. *Agricultural Science (Special issue): International Agriculture*, 71–7.

Frisch, J.E., Drinkwater, R., Harrison, B., *et al.* (1997) Classification of the southern African sanga and east African shorthorned zebu. *Animal Genetics*, **28,** 77–83.

Frisch, J.E. & O'Neill, C.J. (1998a) Comparative evaluation of beef cattle breeds of African, European and Indian origins. 1. Live weights and heterosis at birth, weaning and 18 months. *Animal Science*, **67,** 27–38.

Frisch, J.E. & O'Neill, C.J. (1998b) Comparative evaluation of beef cattle breeds of African, European and Indian origins. 2. Resistance to cattle ticks and gastrointestinal nematodes. *Animal Science*, **67,** 39–48.

Frisch, J.E., O'Neill, C.J. & Kelly, M.J. (2000) Using genetics to control cattle parasites – the Rockhampton experience. *International Journal of Parasitology*, **30,** 253–64.

Fu, Y.B., Namkoong, G. & Carlson, J.E. (1998) Comparison of breeding strategies for purging inbreeding depression via simulation. *Conservation Biology*, **12,** 856–64.

Fuller, Z., Cox, J.E., Argo, C.McG. (2001) Photoperiodic entrainment of seasonal changes in the appetite, feeding behaviour, growth rate and pelage of pony colts. *Animal Science*, **72,** 65–74.

Fumihito, A., Miyake, T., Sumi, S.I., *et al.* (1994) One subspecies of the red jungle fowl (*Gallus gallus gallus*) suffices as the matriarchic ancestor of all domestic breeds. *Proceedings of the National Academy of Sciences of the USA*, **91,** 12505–9.

Gaffney, B. & Cunningham, E.P. (1988) Estimation of genetic trend in racing performance of thoroughbred horses. *Nature*, **332,** 722–4.

Galal, S., Boyazoglu, J. & Hammond, K. (2000) *Developing breeding strategies for lower input animal production environments. ICAR Technical Series no. 3*. International Committee for Animal Recording, Rome, Italy.

Gall, C.F. (1991) Breed differences in adaptation of goats. In: *World animal science, B8. Genetic resources of pig, sheep and goat* (Ed. by K. Maijala), pp. 413–29. Elsevier, Amsterdam, The Netherlands.

Galli, C., Lagutina, I., Crotti, G. *et al.* (2003) A cloned horse born to its dam twin. *Nature*, **424,** 635.

Galtier, N., Depaulis, F. & Barton, N.H. (2000) Detecting bottlenecks and selective sweeps from DNA sequence polymorphisms. *Genetics*, **155,** 981–7.

Gandini, G.C. & Oldenbroek, J.K. (1999) Choosing the conservation strategy. In: *Genebanks and the conservation of farm animal genetic resources* (Ed. by J.K. Oldenbroek), pp. 11–31. DLO Institute for Animal Science and Health, Lelystad, The Netherlands.

Gandini, G.C., Samore, A. & Pagnacco, G. (1997) Genetic contribution of the Arabian to the Italian Haflinger horse. *Journal of Animal Breeding and Genetics*, **114,** 457–64.

Gandini, G.C. & Villa, E. (2003) Analysis of the cultural value of local livestock breeds: a methodology. *Journal of Animal Breeding and Genetics*, **120,** 1–11.

Gaughan, J.B., Mader, T.L., Holt, S.M., *et al.* (1999) Heat tolerance of Boran and Tuli crossbred steers. *Journal of Animal Science*, **77,** 2398–405.

Gazzola, C., O'Neill, C.J. & Frisch, J.E. (1999) Comparative evaluation of the meat quality of beef cattle breeds of Indian, African and European origins. *Animal Science*, **69,** 135–42.

Georges, M. (1999) Towards marker assisted selection in livestock. *Reproduction Nutrition Development*, **39,** 555–61.

Georges, M., Dietz, A.B., Mishra, A., *et al.* (1993) Microsatellite mapping of the gene causing weaver disease in cattle will allow the study of an associated QTL. *Proceedings of the National Academy of Sciences of the USA*, **90,** 1058–62.

Gibson, J.P. (1986) Efficiency and performance of genetically high and low milk-producing British Friesian and Jersey cattle. *Animal Production*, **42,** 161–82.

Gingerich, P.D. (1983) Rates of evolution: effects of time and temporal scaling. *Science*, **222,** 159–61.

Giraudeau, L., Casañe, D., Verrier, E., *et al.* (1991) [Analysis of the genetic structure of a breed of limited size – the case of the Parthenaise cattle breed.] *Elevage et Insémination*, **242,** 3–14.

Giuffra, E., Evans, G., Tornsten, A. *et al.* (1999) The Belt mutation in pigs is an allele at the dominant white *(I/KIT)* locus. *Mammalian Genome*, **10,** 1132–6.

Giuffra, E., Kijas, J.M.H., Amarger, V., *et al.* (2000) The origin of the domestic pig: independent domestication and subsequent introgression. *Genetics*, **154,** 1785–91.

Givens, D.I. & Moss, A.R. (1994) Effect of breed, age and bodyweight of sheep on the measurement of apparent digestibility of dried grass. *Animal Feed Science and Technology*, **46,** 155–62.

Goddard, M.E. (2003) Animal breeding in the (post-) genomic era. *Animal Science*, **76,** 353–365.

Goddard, P.J., Fawcett, A.R., Macdonald, A.J., *et al.* (2000) The behavioural, physiological and immunological responses of lambs from two rearing systems and two genotypes to exposure to humans. *Applied Animal Behaviour Science*, **66,** 305–21.

Godfrey, R.W., Smith, S.D., Guthrie, M.J., *et al.* (1991) Physiological responses of newborn *Bos indicus* and *Bos indicus* x *Bos taurus* calves after exposure to cold. *Journal of Animal Science*, **69,** 258–63.

Goldmann, W. (1998) PrP protein genotypes in sheep and their association with scrapie. *Proceedings of the Sheep Veterinary Society*, **28,** 139–40.

Golijow, C.D., Giovambattista, G., Poli, M., *et al.* (1996) Kappa-casein gene frequencies support subdivision and historical origin of Argentine Creole cattle. *Brazilian Journal of Genetics*, **19,** 583–6.

Goodnight, C.J. (2000a) Modeling gene interaction in structured populations. In: *Epistasis and the evolutionary process* (Ed. by J.B. Wolf, E.D. Brodie & M.J. Wade), pp.129–45. Oxford University Press, Oxford, UK.

Goodnight, C.J. (2000b) Quantitative trait loci and gene interaction: the quantitative genetics of metapopulations. *Heredity*, **84,** 587–98.

Gosler, A.G., Greenwood, J.J.D. & Perrins, C. (1995) Predation risk and the cost of being fat. *Nature*, **377,** 621–3.

Graczyk, R. (1989) An attempt at a reintroduction of forest tarpan (*Equus gmelini silvatica*) in north-western Poland. *Abstracts of papers and posters, 5th International Theriological Congress, Rome*, **1,** 249–50.

Grayson, F.W. (2000) The Grazing Animals Project – a collaborative approach to improving conservation grazing in Britain. *La Cañada. Newsletter of the European Forum on Nature Conservation and Pastoralism*, no. 12, 8–9.

Green, R.E. (1997) The influence of numbers released on the outcome of attempts to introduce exotic bird species to New Zealand. *Journal of Animal Ecology*, **66,** 25–35.

Greger, P.D. & Romney, E.M. (1999) High foal mortality limits growth of a desert feral horse population in Nevada. *Great Basin Naturalist*, **59,** 374–9.

Gregory, K.E. & Cundiff, L.V. (1980) Crossbreeding in beef cattle: evaluation of systems. *Journal of Animal Science*, **51,** 1224–42.

Grenot, C.J. (1992) Ecophysiological characteristics of large herbivorous mammals in arid Africa and the Middle East. *Journal of Arid Environments*, **23,** 125–55.

Grigson, C. (1982) Cattle in prehistoric Britain. *Ark*, **9,** 47–9.

Grobet, L., Martin, L.J., Poncelet, D., *et al.* (1997) A deletion in the bovine myostatin gene causes the double-muscled phenotype in cattle. *Nature Genetics*, **17,** 71–4.

Groombridge, B. (1992) *Global biodiversity. Status of the Earth's living resources*. Chapman and Hall, London, and World Conservation Monitoring Centre, Cambridge, UK.

Grunenfelder, H.P. (1992) Protection of genetic resources in Central and Eastern Europe. *Animal Genetic Resources Information*, **9,** 9–15.

Grunenfelder, H.P., Gugic, G. & Punz, F. (1994) Saving the Turopolje pig in Croatia. *Animal Genetic Resources Information*, **14,** 27–32.

Guglielmino, C.R., Viganotti, C., Hewlett, B., *et al.* (1995) Cultural variation in Africa: role of mechanisms of transmission and adaptation. *Proceedings of the National Academy of Sciences of the USA*, **92,** 7585–9.

Gura, T. (2000) Bones, molecules . . . or both? *Nature*, **406,** 230–3.

Gustafsson, M., Jensen, P., deJonge, F.H., *et al.* (1999) Domestication effects on foraging strategies in pigs (*Sus scrofa*). *Applied Animal Behaviour Science*, **65,** 29–42.

Haley, C.S. (1995) Livestock QTLs – bringing home the bacon? *Trends in Genetics*, **11,** 488–92.

Hall, J.G. & Purser, A.F. (1979) Lamb production and haemoglobin type. *Animal Breeding Research Organisation Report*, 1979, 9–12.

Hall, S.J.G. (1986) Genetic conservation of rare British sheep: the Portland, Manx Loghtan and Hebridean breeds. *Journal of Agricultural Science, Cambridge*, **107,** 133–44.

Hall, S.J.G. (1989a) Breed structures of rare pigs: implications for conservation of the Berkshire, Tamworth, Middle White, Large Black, Gloucester Old Spot, British Saddleback, and British Lop. *Conservation Biology*, **3,** 30–8.

Hall, S.J.G. (1989b) Conserving and developing minority British breeds of sheep: the example of the Southdown. *Journal of Agricultural Science, Cambridge*, **112,** 39–45.

Hall, S.J.G. (1991) Body dimensions of Nigerian cattle, sheep and goats. *Animal Production*, **53,** 61–9.

Hall, S.J.G. (1996) Human ecology and the evolution of livestock. *Anthrozoös*, **9,** 81–4.

Hall, S.J.G. (1998) Traditional livestock in semi-arid north eastern Zimbabwe: Mashona cattle. *Tropical Animal Health and Production*, **30,** 351–60.

Hall, S.J.G. & Bradley, D.G. (1995) Conserving livestock breed biodiversity. *Trends in Ecology and Evolution*, **10,** 267–70.

Hall, S.J.G. & Clutton-Brock, J. (1988) *Two hundred years of British farm livestock*. British Museum (Natural History), London, UK.

Hall, S.J.G., Gnaho, L.K. & Meghen, C. (1995) [A study of Somba cattle in Benin]. *Revue d'Elevage et de Médecine vétérinaire des Pays tropicaux*, **48,** 77–83.

Hall, S.J.G. & Hall, J.G. (1988) Inbreeding and population dynamics of the Chillingham cattle *(Bos taurus). Journal of Zoology, London*, **216,** 479–493.

Hall, S.J.G. & Moore, G.F. (1986) Feral cattle of Swona, Orkney Islands. *Mammal Review*, **16,** 89–96.

Hall, S.J.G. & Ruane, J. (1993) Livestock breeds and their conservation: a global overview. *Conservation Biology*, **7,** 815–25.

Hall, S.J.G., Russel, A.J.F. & Redden, H. (1996) Coat fibres of Nigerian sheep and goats: A preliminary characterization. *Small Ruminant Research*, **22,** 169–75.

Hamann, O. (1993) On vegetation recovery, goats and giant tortoises on Pinta Island, Galápagos, Ecuador. *Biodiversity and Conservation*, **2,** 138–51.

Hammond, A.C. (1995) *Leucaena* toxicosis and its control in ruminants. *Journal of Animal Science*, **73,** 1487–92.

Hammond, J. (1963) *Animal breeding*. Edward Arnold, London, UK.

Hammond, K. & Leitch, H. (1995) Towards better management of animal genetic resources. *World Review of Animal Production*, **84,** 48–53.

Hanotte, O., Bradley, D.G., Ochieng, J., *et al.* (2002) African pastoralism: genetic imprints of origins and migrations. *Science*, **296,** 336–9.

Hanotte, O., Tawah, C.L., Bradley, D.G., *et al.* (2000) Geographic distribution and frequency of a taurine *Bos taurus* and an indicine *Bos indicus* Y specific allele amongst sub-Saharan African cattle breeds. *Molecular Ecology*, **9,** 387–96.

Hansen, C., Shrestha, J.N.B., Parker, R.J., *et al.* (2002) Genetic diversity among Canadienne, Brown Swiss, Holstein, and Jersey cattle of Canada based on 15 bovine microsatellite markers. *Genome*, **45,** 897–904.

Hansen, H.R., Hector, B.L. & Feldmann, J. (2003) A qualitative and quantitative evaluation of the seaweed diet of North Ronaldsay sheep. *Animal Feed Science and Technology*, **105,** 21–8.

Hansen, P.J. (1985) Seasonal modulation of puberty and the postpartum anestrus in cattle: a review. *Livestock Production Science*, **12,** 309–27.

Hanslik, S., Harr, B., Brem, G., *et al.* (2000) Microsatellite analysis reveals substantial genetic differentiation between contemporary new world and old world Holstein Friesian populations. *Animal Genetics*, **31,** 31–8.

Harisah, M., Azmi, T.I., Hilmi, M., *et al.* (1989) Identification of crossbred buffalo genotypes and their chromosome segregation patterns. *Genome*, **32,** 999–1002.

Harris, D.L. & Newman, S. (1994) Breeding for profit: synergism between genetic improvement and livestock production. *Journal of Animal Science*, **72,** 2178–200.

Hartmann, W. (1989) From Mendel to multi national in poultry breeding. *World's Poultry Science Journal*, **45,** 5–26.

Hastings, I.M. (1996) The genetics and physiology of size reduction in mice. *Symposium of the Zoological Society of London*, **69,** 129–42

Hemmer, H. (1990) *Domestication: the decline of environmental appreciation*. Cambridge University Press, Cambridge, UK.

Hendry, A.P. & Kinnison, M.T. (1999) Perspective: the pace of modern life: measuring rates of contemporary microevolution. *Evolution*, **53,** 1637–53.

Henryon, M., Berg, P., Jensen, J., *et al.* (2001) Genetic variation for resistance to clinical and subclinical diseases exists in growing pigs. *Animal Science*, **73,** 375–87.

Henson, E.L. (1992) In situ *conservation of livestock and poultry. FAO Animal Production and Health Paper 99*. Food and Agricultural Organization of the United Nations, Rome, Italy.

Herbergs, J., Siwek, M., Crooijmans, R.P.M.A., *et al.* (2000) Multicolour fluorescent detection and mapping of AFLP markers in chicken (*Gallus domesticus*). *Animal Genetics*, **30,** 274–85.

Heringstad, B., Klemetsdal, G. & Ruane, J. (2000) Selection for mastitis resistance in dairy cattle: a review with focus on the situation in the Nordic countries. *Livestock Production Science*, **64,** 95–106.

Hernández, L., Barral, H. & Anaya, E. (1996) [Revival of an 18th century type of husbandry in northern Mexico]. *Cahiers des Sciences Humaines*, **32,** 65–84.

Hernández, L., Barral, H., Halffter, G., *et al.* (1999) A note on the behavior of feral cattle in the Chihuahuan Desert of Mexico. *Applied Animal Behaviour Science*, **63,** 259–67.

Herpin, P., Damon, M., Ledividich, J. (2002) Development of thermoregulation and neonatal survival in pigs. *Livestock Production Science*, **78,** 25–45.

Hertrampf, J. (1990) Pigs in Papua New Guinea. *World Farmers' Times*, (March), 23–4.

Heywood, V.H. & Watson, R.T. (1995) *Global biodiversity assessment.* United Nations Environment Programme/Cambridge University Press, Cambridge, UK.

Hickman, G. (1999) Improving returns from common grazings in the New Forest. *Faunus. Newsletter of the LSIRD network*, no. 6, 4.

Hiendleder, S., Mainz, K., Plante, Y., *et al.* (1998) Analysis of mitochondrial DNA indicates that domestic sheep are derived from two different ancestral maternal sources: no evidence for contributions from urial and argali sheep. *Journal of Heredity*, **89,** 113–20.

Hill, W.G. (1982) Predictions of response to artificial selection from new mutations. *Genetical Research, Cambridge*, **40,** 255–78.

Hill, W.G. (1988) Why aren't horses faster? *Nature*, **332,** 678.

Hill, W.G. (2000) Maintenance of quantitative genetic variation in animal breeding programmes. *Livestock Production Science*, **63,** 99–109.

Hillel, J., Gibbins, A.M.V., Etches, R.J., *et al.* (1993) Strategies for the rapid introgression of a specific gene modification into a commercial poultry flock from a single carrier. *Poultry Science*, **72,** 1197–211.

Hinde, R.A. & Barden, L.A. (1985) The evolution of the teddy bear. *Animal Behaviour*, **33,** 1371–2.

Hindmarch, C. & Pienkowski, M. (1998) *Land management: the hidden costs. BES Ecological Issues Series*. British Ecological Society/Blackwell Science, London and Oxford.

Hoag, H. (2003) US fails to quantify threat of West Nile virus. *Nature*, **423,** 104.

Hodges, J. (2002) Conservation of farm animal biodiversity: history and prospects. *Animal Genetic Resources Information*, **32,** 1–12.

Hodgson, J. (2001) Grassland production and management – trends and perspectives for the 21st century. *Proceedings of the XIX International Grassland Congress*, 1–3. Brazilian Society of Animal Husbandry.

Hodgson, J. & Illius, A.W. (Eds) (1996) *The ecology and management of grazing systems.* CAB International, Wallingford, UK.

Hoekstra, J., van der Lugt, A.W., van der Werf, J.H.J., *et al.* (1994) Genetic and phenotypic parameters for milk production and fertility traits in upgraded dairy cattle. *Livestock Production Science*, **40,** 225–32.

Hogan, J. (2003) Dolly firm in trouble after transgenic milk fails to flow. *Nature*, **423,** 907.

Hohenboken, W.D. (1986) Inheritance of behavioural characteristics in livestock. A review. *Animal Breeding Abstracts*, **54,** 623–39.

Holm, L. & Mohl, M. (2000) The role of meat in everyday food culture: an analysis of an interview study in Copenhagen. *Appetite*, **34,** 277–83.

Holmes, E.C. & Ellis, S.A. (1999) Evolutionary history of MHC Class 1 genes in the mammalian order Perissodactyla. *Journal of Molecular Evolution*, **49,** 316–24.

Holmes, I.S. & Hastings, I.M. (1995) Behavioural changes as a correlated response to selection. *Genetical Research, Cambridge*, **66,** 27–33.

Holt, W.V. (2000) Fundamental aspects of sperm cryobiology: the importance of species and individual differences. *Theriogenology*, **53,** 47–58.

Holt, W.V. (2003) Germplasm cryobanking for sheep, pig and cattle breeds in the United Kingdom: the British experience. In: *Workshop on cryopreservation of animal genetic resources in Europe. Paris, February 2003* (Ed. by D. Planchenault), pp. 9–13. Salon International de l'Agriculture, Paris, France.

Homer, D.B., Cuthbertson, A., Homer, D.L.M., *et al.* (1997) Eating quality of beef from different sire breeds. *Animal Science*, **64,** 403–8.

Homewood, K. & Lewis, J. (1987) Impact of drought on pastoral livestock in Baringo, Kenya 1983–85. *Journal of Applied Ecology*, **24,** 615–31.

Hone, J. (1990) How many feral pigs in Australia? *Australian Wildlife Research*, **17,** 571–2.

Hooper, S. (1998) What do the CAP reform proposals mean for the UK uplands? *Faunus. Newsletter of the LSIRD network*, no. **5,** 10.

Horín, P., Cothran, E.G., Trtková, K., *et al.* (1998) Polymorphism of Old Kladruber horses, a surviving but endangered baroque breed. *European Journal of Immunogenetics*, **25,** 357–63.

Houston, F., Goldmann, W., Chong, A., *et al.* (2003) BSE in sheep bred for resistance to infection. *Nature*, **423,** 498.

Huang, Y.-F., Shi, X.-W. & Zhang, Y.-P. (1999) Mitochondrial genetic variation in Chinese pigs and wild boars. *Biochemical Genetics*, **37,** 335–43.

Huelsenbeck, J.P., Bull, J.J. & Cunningham, C.W. (1996) Combining data in phylogenetic analysis. *Trends in Ecology and Evolution*, **11,** 152–8.

Hughes, A.L. & Hughes, M.K. (1995) Small genomes for better flyers. *Nature*, **377,** 391.

Hunter, N., Cairns, D., Foster, J.D., *et al.* (1997) Is scrapie solely a genetic disease? *Nature*, **386,** 137.

Hunter, G.L. (1957) Maternal influence on size in sheep. *Journal of Agricultural Science.* Cambridge, **48**, 36–60.

Hunter, N., Goldmann, W., Smith, G., *et al.* (1994) Frequencies of PrP gene variants in healthy cattle and cattle with BSE in Scotland. *Veterinary Record*, **135,** 400–3.

Hunter, R.A. & Siebert, B.D. (1986) The effects of genotype, age, pregnancy, lactation and rumen characteristics on voluntary intake of roughage diets by cattle. *Australian Journal of Agricultural Research*, **37,** 549–60.

Hutton, S.A. & Giller, P.S. (2003) The effects of the intensification of agriculture on northern temperate dung beetle communities. *Journal of Applied Ecology*, **40,** 994–1007.

Iason, G.R., Mantecon, A.R., Milne, J.A., *et al.* (1992) The effect of pattern of food supply on performance, compensatory growth and carcass composition of Beulah and Welsh Mountain lambs. *Animal Production*, **54,** 235–41.

Iason, G.R., Sim, D.A. & Foreman, E. (1995) Seasonal changes in intake and digestion of chopped timothy hay by three breeds of sheep. *Journal of Agricultural Science, Cambridge*, **125,** 273–80.

Iason, G.R., Sim, D.A., Foreman, E., *et al.* (1994) Seasonal variation of voluntary food intake and metabolic rate in three contrasting breeds of sheep. *Animal Production*, **58,** 381–7.

Illius, A.W., Albon, S.D., Pemberton, J.M., *et al.* (1995) Selection for foraging efficiency during a population crash in Soay sheep. *Journal of Animal Ecology*, **64,** 481–92.

Illius, A.W. & Gordon, I.J. (1987) The allometry of food intake in grazing ruminants. *Journal of Animal Ecology*, **56,** 989–1000.

Insausti, K., Beriain, M.J., Purroy, A., *et al.* (1999) Colour stability of beef from different Spanish native cattle breeds stored under vacuum and modified atmosphere. *Meat Science*, **53,** 241–9.

Ishida, N., Oyunsuren, T., Mashima, S., *et al.* (1995) Mitochondrial DNA sequences of various species of the genus *Equus* with special reference to the phylogenetic relationship between Przewalskii's wild horse and the domestic horse. *Journal of Molecular Evolution*, **41,** 180–8.

Jabbar, M.A., Swallow, B.M., d'Ieteren, G.D.M., *et al.* (1997) *Farmer preferences and market values of cattle breeds of west and central Africa. Socioeconomic and policy research working paper no. 21*, pp. 1–2. Livestock Policy Analysis Programme, International Livestock Research Institute, Addis Ababa, Ethiopia.

Jabbar, M.A., Swallow, B.M. & Rege, J.E.O. (1999) Incorporation of farmer knowledge and preferences in designing breeding policy and conservation strategy for domestic animals. *Outlook on Agriculture*, **28,** 239–43.

James, A.N., Gaston, K.J. & Balmford, A. (1999) Balancing the Earth's accounts. *Nature*, **401,** 323–4.

Jansen, T., Forster, P., Levine, M.A., *et al.* (2002) Mitochondrial DNA and the origins of the domestic horse. *Proceedings of the National Academy of Sciences of the USA*, **99,** 1 0905–10.

Jensen, P. (Ed.) (2002) *The ethology of domestic animals. An introductory text.* CABI Publishing, Wallingford, UK.

Jialin, B., Mingqiang, W., Zhonglin, L., *et al.* (1998) Meat production from crossbred and domestic yaks in China. *Animal Science*, **66,** 465–9.

Joly, T., Theau-Clément, M., Drouet-Viard, F., *et al.* (1994) Cryopreservation of embryos applied to the protection of genetic resources in the rabbit. *Genetics Selection Evolution*, **26,** supplement 1, 267s–78s.

Jones, R.B. & Hocking, P.M. (1999) Genetic selection for poultry behaviour: big bad wolf or friend in need? *Animal Welfare*, **8,** 343–59.

Jones, R.J. & Lowry, J.B. (1984) Australian goats detoxify the goitrogen 3-hydroxy-4(1H) pyridone (DHP) after rumen infusion from an Indonesian goat. *Experientia*, **40,** 1435–6.

Jonsson, N.N., Mayer, D.G., Matschoss, A.L., *et al.* (1998) Production effects of cattle tick (*Boophilus microplus*) infestation of high yielding dairy cows. *Veterinary Parasitology*, **78,** 65–77.

Jordana, J., Piedrafita, J. & Sanchez, A. (1992) Gentic relationships in Spanish dog breeds: II. The analysis of biochemical polymorphism. *Genetics Selection Evolution*, **24**, 245–63.

Jorritsma, I.T.M., van Hees, A.F.M. & Mohren, G.M.J. (1999) Forest development in relation to ungulate grazing: a modeling approach. *Forest Ecology and Management*, **120,** 23–34.

Joss, G.E. & Mason, R. (1983) Operation Noah's Ark. *Veterinary Record*, **113,** 476–7.

Kaasschieter, G.A., de Jong, R., Schiere, J.B., *et al.* (1992) Towards a sustainable livestock production in developing countries and the importance of animal health strategy therein. *Veterinary Quarterly*, **14,** 66–75.

Kadwell, M., Fernandez, M., Stanley, H.F., *et al.* (2001). Genetic analysis reveals the wild ancestors of the llama and the alpaca. *Proceedings of the Royal Society of London Series B*, **268,** 2575–84.

Kampf, H. (2000) The role of large grazing animals in nature conservation – a Dutch perspective. *British Wildlife*, October, 37–46.

Kanis, E., van den Belt, H., Groen, A. F. *et al.* (2004) Breeding for improved welfare in pigs: a conceptual framework and its use in practice. *Animal Science*, **78,** 315–29.

Kantanen, J., Olsaker, I., Holm, L.-E., *et al.* (2000) Genetic diversity and population structure of 20 north European cattle breeds. *Journal of Heredity*, **91,** 446–57.

Karaca, M., Johnson, E. & Lamont, S.J. (1999) Genetic line and major histocompatibility complex effects on primary and secondary antibody responses to T-dependent and T-independent antigens. *Poultry Science*, **78,** 1518–25.

Kasa, I.W., Thwaites, C.J. & Hill, M.K. (1999) Effects of treadmill exercise on sweating in three breeds of goats. *Australian Veterinary Journal*, **77,** 660–1.

Katila, T. (2001) *In vitro* evaluation of frozen-thawed stallion semen: a review. *Acta veterinaria scandinavica*, **42,** 201–17.

Katunguka-Rwakishaya, E., Murray, M. & Holmes, P.H. (1997) Pathophysiology of *Trypanosoma congolense* infection in two breeds of sheep, Scottish Blackface and Finn Dorset. *Veterinary Parasitology*, **68,** 215–25.

Kaufman, J. & Wallny, H.J. (1998) Chicken MHC molecules, disease resistance and the evolutionary origin of birds. *Current Topics in Microbiology and Immunology*, **212,** 129–41.

Kavar, T., Habe, F., Brem, G. *et al.* (1999) Mitochondrial D-loop sequence variation among the 16 maternal lines of the Lipizzan horse breed. *Animal Genetics*, **30,** 423–30.

Kay, R.N.B., Milne, J.A. & Hamilton, W.J. (1984) Nutrition of red deer for meat production. *Proceedings of the Royal Society of Edinburgh*, **82B,** 231–42.

Kemp, S.J., Iraqi, F., Darvasi, A., *et al.* (1997) Localization of genes controlling resistance to trypanosomiasis in mice. *Nature Genetics*, **16,** 194–6.

Kemp, S.J. & Teale, A.J. (1998) The genetic basis of trypanotolerance in cattle and mice. *Parasitology Today*, **14,** 450–4.

Kempster, A.J. (1980) Fat partition and distribution in the carcasses of cattle sheep and pigs – a review. *Meat Science*, **5,** 83–98.

Kempster, A.J., Croston, D., Guy, D.R., *et al.* (1987) Growth and carcass characteristics of crossbred lambs by ten sire breeds, compared at the same estimated carcass subcutaneous fat proportion. *Animal Production*, **44,** 83–98.

Kennedy, P.M. (1982) Ruminal and intestinal digestion in Brahman crossbred and Hereford cattle fed alfalfa or tropical pasture hay. *Journal of Animal Science*, **55,** 1190–9.

Khombe, C.T., Hayes, J.F., Tawonezvi, H.P.R., *et al.* (1994) Estimation of genetic differences between Mashona bulls from different production systems using an animal model. *Journal of Agricultural Science, Cambridge*, **122,** 459–63.

Kijas, J.M.H. & Andersson, L. (2001) A phylogenetic study of the origin of the domestic pig estimated from the near-complete mtDNA genome. *Journal of Molecular Evolution*, **52,** 302–8.

Kijas, J.M.H., Moller, M., Plastow, G., *et al.* (2001) A frameshift mutation in *MC1R* and a high frequency of somatic reversions cause black spotting in pigs. *Genetics*, **158,** 779–85.

Kijas, J.M.H., Wales, R., Törnsten, A., *et al.* (1998) Melanocortin receptor 1 (*MC1R*) mutations and coat color in pigs. *Genetics*, **150,** 1177–85.

Kikkawa, Y., Amano, T. & Suzuki, H. (1995) Analysis of genetic diversity of domestic cattle in east and southeast Asia in terms of variations in restriction sites and sequences of mitochondrial DNA. *Biochemical Genetics*, **33,** 51–60.

Kimura, M. & Crow, J.F. (1963) On the maximum avoidance of inbreeding. *Genetical Research, Cambridge*, **4,** 399–415.

King, J.W.B. (1991) Pig breeds of the world: their distribution and adaptation. *World animal science, B8. Genetic resources of pig, sheep and goat* (Ed. by K. Maijala), pp. 51–63. Elsevier, Amsterdam, The Netherlands.

King, K.L. (1993) The potential for avermectins to affect the nutrient economy of grazed pastures. *Veterinary Parasitology*, **48,** 261–71.

Kingwell, R.S., Abadi Ghadim, A.K. & Robinson, S.D. (1995) Introducing Awassi sheep to Australia: an application of farming system models. *Agricultural Systems*, **47,** 451–71.

Kino, K., Pain, B., Leibo, S.P., *et al.* (1997) Production of chicken chimeras from injection of frozen-thawed blastodermal cells. *Poultry Science*, **76,** 753–60.

Kirby, J.P. (1997) White, red and black: colour classification and illness – management in northern Ghana. *Social Science and Medicine*, **20,** 1–22.

Kleiner, K. (1998) Frozen little piggies. *New Scientist*, **157,** no. 2127, 17.

Klukowska, J., Urbaniak, B. & Switonski, M. (1999) High frequency of *M307(A)* mutation at *FUT1* locus, causing resistance to oedema disease, in an autochthonous Polish pig breed, the Zlotnicka Spotted. *Journal of Animal Breeding and Genetics*, **116,** 519–24.

Klungland, H., Olsen, H.G., Hassanane, M.S., *et al.* (2000) Coat colour genes in diversity studies. *Journal of Animal Breeding and Genetics*, **117,** 217–24.

Knorr, C., Moser, G., Müller, E., *et al.* (1997) Associations of GH gene variants with performance traits in F2 generations of European wild boar, Piétrain and Meishan pigs. *Animal Genetics*, **28,** 124–8.

Kohane, M.J. & Parsons, P.A. (1988) Domestication: evolutionary change under stress. *Evolutionary biology vol. 23* (Ed. by M.K. Hecht & B. Wallace), pp. 31–48. Plenum Publishing Corporation, USA.

Kohler-Rollefson, I. (1993) About camel breeds: a reevaluation of current classification systems. *Journal of Animal Breeding and Genetics*, **110,** 66–73.

Kohler-Rollefson, I. (1995) The role of pastoralists in preserving livestock genetic diversity. *Overseas Development Institute, Pastoral Development Network Newsletter*, **36,** 9–11.

Kohler-Rollefson, I. (1997) Indigenous practices of animal genetic resource management and their relevance for the conservation of domestic animal diversity in developing countries. *Journal of Animal Breeding and Genetics*, **114,** 231–8.

Kohler-Rollefson, I. & Rathore, H.S. (1995) The Malvi camel breed: an important genetic resource. *Journal of Camel Practice and Research*, **2,** 13–6.

Kolmodin, R., Strandberg, E., Jorjani, H., *et al.* (2003) Selection in the presence of a genotype by environment interaction: response in environmental sensitivity. *Animal Science*, **76,** 375–85.

Koziell, I. & Saunders, J. (Eds.) (2001) *Living off biodiversity. Exploring livelihoods and biodiversity issues in natural resources management.* International Institute for Environment and Development, London, UK.

Kristjanson, P., Rowlands, J., Swallow, B. *et al.* (1999) *Using the economic surplus model to measure potential returns to international livestock research. The case of trypanosomosis vaccine research.* ILRI Impact Assessment Series 4. International Livestock Research Institute, Nairobi, Kenya.

Kyle, R. (1987) *A feast in the wild.* Kudu Publishing, Oxford, UK.

Lacy, R.C. (1989) Analysis of founder representation in pedigrees: founder equivalents and founder genome equivalents. *Zoo Biology*, **8,** 111–23.

Lacy, R.C. & Ballou, J.D. (1998) Effectiveness of selection in reducing the genetic load in populations of *Peromyscus polionotus* during generations of inbreeding. *Evolution*, **52,** 900–9.

Laiolo, P., Dondero, F., Ciliento, E. *et al.* (2004) Consequences of pastoral abandonment for the structure and diversity of the alpine avifauna. *Journal of Applied Ecology*, **41,** 294–304.

Land, R.B. (1973) The expression of female sex-limited characters in the male. *Nature*, **241,** 208.

Land, R.B., Atkins, K.D. & Roberts, R.C. (1983) Genetic improvement of reproductive performance. In: *Sheep production* (Ed. by W. Haresign), pp. 515–35. Butterworths, London, UK.

Landa, A., Gudvangen, K., Swenson, J.E., *et al.* (1999) Factors associated with wolverine *Gulo gulo* predation on domestic sheep. *Journal of Applied Ecology*, **36,** 963–73.

Lande, R. (1998) Anthropogenic, ecological and genetic factors in extinction and conservation. *Researches on Population Ecology*, **40,** 259–69.

Lasota-Moskalewska, A. & Kobryn, H. (1989) Description of intermediate forms in the evolution of *Bos primigenius taurus* on the basis of osteometric characteristics. *Acta Theriologica*, **34,** 625–42.

Lauvergne, J.J. (1993) Breed development and breed differentiation. In: *Proceedings of CEC Workshop on data collection, conservation and use of farm animal genetic resources, 8–9 December 1992* (Ed. by D. Simon & D. Buchenauer) pp. 53–64. Hannover, Germany.

Lauvergne, J.J., Bourzat, D., Souvenir Zafindrajoana, P., *et al.* (1993) [Indices of primarity in the goats of north Cameroon and of Chad]. *Revue d'Elevage et de Médecine vétérinaire des Pays tropicaux*, **46,** 651–65.

Laval, G., Iannuccelli, N., Legault, C., *et al.* (2000) Genetic diversity of eleven European pig breeds. *Genetics Selection Evolution*, **32,** 187–203.

Lawrence, A.B., Pryce, J.E. & Stott, A. (2000) Animal welfare and public perceptions as potential constraints to genetic change in animal production. *Occasional Publication, British Society of Animal Science*, **27,** 81–2.

Lazo, A. (1994) Social segregation and the maintenance of social stability in a feral cattle population. *Animal Behaviour*, **48,** 1133–41.

Le Neindre, P. & Garel, J.P. (1990) [Adaptation of cattle to husbandry systems]. *INRA Productions Animales*, **3,** 189–94.

Le Neindre, P., Trillat, G., Sapa, J., *et al.* (1995) Individual differences in docility in Limousin cattle. *Journal of Animal Science*, **73,** 2249–53.

Le Tissier, P., Stoye, J.P., Takeuchi, Y., *et al.* (1997) Two sets of human-tropic pig retrovirus. *Nature*, **389,** 681–2.

Ledividich, J., Mormède, P., Catheline, M., *et al.* (1991) Body composition and cold resistance of the neonatal pig from European (Large White) and Chinese (Meishan) breeds. *Biology of the Neonate*, **59,** 268–77.

Leibo, S.P., Semple, M.E. & Kroetsch, T.G. (1994) *In vitro* fertilization of oocytes by 37-year-old cryopreserved bovine spermatozoa. *Theriogenology*, **42,** 1257–62.

Lerner, I.M. & Donald, H.P. (1966) *Modern developments in animal breeding*. Academic Press, London and New York.

Levin, P.S., Ellis, J., Petrik, R. *et al.* (2002) Indirect effects of feral horses on estuarine communities. *Conservation Biology*, **16,** 1364–71.

Levin, P.S. & Williams, J.G. (2002) Indirect effects of artifically propagated fish: an additional conservation risk for salmon. *Conservation Biology*, **16,** 1581–7.

Lickliter, R. (1991) Context and animal behaviour II: The role of conspecifics in species – typical perceptual development. *Ecological Psychology*, **3,** 11–23.

Lidauer, M. & Mäantysaari, E. (1996) Genetic constitution of the Finnish black and white cattle population and the influence of Holsteinization on protein yield, days open and somatic cell count. *Acta Agriculturae Scandinavica*, **46,** 193–200.

Lightfoot, J. (1987) The Awassi fat tail sheep project. *Journal of Agriculture, Western Australian Department of Agriculture*, no. 4, 1–7.

Lincoln, G.A. (1989) Seasonal cycles in testicular activity in mouflon, Soay sheep and domesticated breeds of sheep: breeding seasons modified by domestication. *Zoological Journal of the Linnaean Society*, **95,** 137–47.

Linklater, W.L. (2000) Adaptive explanation in socio-ecology: lessons from the Equidae. *Biological Reviews*, **75,** 1–20.

Linthicum, K.J., Anyamba, A., Tucker, C.J., *et al.* (1999) Climate and satellite indicators to forecast Rift Valley fever epidemics in Kenya. *Science*, **285,** 397–400.

Lister, A.M. (1989) Rapid dwarfing of deer on Jersey in the Last Interglacial. *Nature*, **342,** 539–42.

Lively, C.M., Craddock, C. & Vrijenhoek, R.C. (1990) Red Queen hypothesis supported by parasitism in sexual and clonal fish. *Nature*, **344,** 864–6.

Loftus, R.T., Ertugrul, O., Harba, A.H., *et al.* (1999) A microsatellite survey of cattle from a centre of origin: the Near East. *Molecular Ecology*, **8,** 2015–22.

Long, J.L. (2003) *Introduced mammals of the world. Their history, distribution and influence.* CABI Publishing/CSIRO Publishing, Wallingford, UK and Collingwood, Australia.

Lopez-Bote, C.J. (1998) Sustained utilization of the Iberian pig breed. *Meat Science*, **49,** supplement 1, S17–S27.

López-Fanjul, C., Fernández, A. & Toro, M.A. (2002) The effect of epistasis on the excess of the additive and nonadditive variances after population bottlenecks. *Evolution*, **56,** 865–76.

López-Gatius, F. (2003) Is fertility declining in dairy cattle? A retrospective study in northern Spain. *Theriogenology*, **60,** 89–99.

Lopez-Villalobos, N., Garrick, D.J., Blair, H.T., *et al.* (2000a) Possible effects of 25 years of selection and crossbreeding on the genetic merit and productivity of New Zealand dairy cattle. *Journal of Dairy Science*, **83,** 154–63.

Lopez-Villalobos, N., Garrick, D.J., Holmes, C.W., *et al.* (2000b) Profitabilities of some mating systems for dairy herds in New Zealand. *Journal of Dairy Science*, **83,** 144–53.

Lopez-Villalobos, N., Garrick, D.J., Holmes, C.W, *et al.* (2000c) Effects of selection and crossbreeding strategies on industry profit in the New Zealand dairy industry. *Journal of Dairy Science*, **83,** 164–72.

Lord, G.M., Matarese, G., Howard, J.K., *et al.* (1998) Leptin modulates the T-cell immune response and reverses starvation-induced immunosuppression. *Nature*, **394,** 897–901.

Lömker, R. & Simon, D.L. (1995) Conservation strategies for endangered cattle breeds in view of costs and inbreeding. *Stocarstvo*, **49,** 159–66.

Luikart, G., Gielly, L., Excoffier, L., *et al.* (2001) Multiple maternal origins and weak phylogeographic structure in domestic goats. *Proceedings of the National Academy of Sciences of the USA*, **98,** 5927–32.

Lundén, A., Edfors-Lilja, I., Johansson, K., *et al.* (1993) Associations between major histocompatibility complex genes and production traits in White Leghorns. *Poultry Science*, **72,** 989–99.

Lundstrom, K. (1995) Production, carcass and meat quality traits of F2 crosses between European wild pigs and domestic pigs including halothane gene carriers. *Animal Science*, **61,** 325–31.

Lush, J.L. (1943) *Animal breeding plans.* 2nd edn. Iowa State College Press, Ames, Iowa.

Lush, J.L. (1946) Chance as a cause of changes in gene frequencies within pure breeds of livestock. *American Naturalist*, **80,** 318–42.

MacCluer, J.W., Vandeberg, J.L., Read, B., *et al.* (1986) Pedigree analysis by computer simulation. *Zoo Biology*, **5,** 147–60.

Machado, T.M.M., Chakir, M. & Lauvergne, J.J. (2000) Genetic distances and taxonomic trees between goats of Ceará State (Brazil) and goats of the Mediterranean region (Europe and Africa). *Genetics and Molecular Biology*, **23,** 121–5.

MacHugh, D.E. & Bradley, D. G. (2001) Livestock genetic origins: goats buck the trend. *Proceedings of the National Academy of Sciences of the USA*, **98,** 5382–4.

MacHugh, D.E., Loftus, R.T., Bradley, D.G., *et al.* (1994) Microsatellite DNA variation within and among European cattle breeds. *Proceedings of the Royal Society of London Series B*, **256,** 25–31.

MacHugh, D.E., Loftus, R.T., Cunningham, P., *et al.* (1998) Genetic structure of seven European cattle breeds assessed using 20 microsatellite markers. *Animal Genetics*, **29,** 333–40.

MacHugh, D.E., Shriver, M.D., Loftus, R.T., *et al.* (1997) Microsatellite DNA variation and the evolution, domestication and phylogeography of taurine and Zebu cattle (*Bos taurus* and *Bos indicus*). *Genetics*, **146,** 1071–86.

MacKenzie, D. (1993) Saving Haiti's bacon. *New Scientist*, **139,** no. 1882, pp.35–8.

Mackinnon, M.J., Hetzel, D.J.S., Corbet, N.J., *et al.* (1990) Correlated responses to selection for fertility in a tropical beef herd. *Animal Production*, **50,** 417–24.

Mackinnon, M.J., Thorpe, W. & Baker, R.L. (1996) Sources of genetic variation for milk production in a crossbred herd in the tropics. *Animal Science*, **62,** 5–16.

Maclellan, L.J., Carnevale, E.M., Coutinho da Silva, M.A., *et al.* (2002) Pregnancies from vitrified equine oocytes collected from super-stimulated and non-stimulated mares. *Theriogenology*, **58,** 911–9.

MAFF (1975) *Energy allowances and feeding systems for ruminants. Reference book 433*. Her Majesty's Stationery Office, London, UK.

Magee, D.A., Meghen, C., Harrison, S., *et al.* (2002) A partial African ancestry for the Creole cattle populations of the Caribbean. *Journal of Heredity*, **93,** 429–32.

Maijala K. (2000) Cow milk and human development and well-being. *Livestock Production Science*, **65,** 1–18.

Maijala, K., Cherekaev, A.V., Devillard, J.-M., *et al.* (1984) Conservation of animal genetic resources in Europe. Final report of an EAAP working party. *Livestock Production Science*, **11,** 3–22.

Maijala, K. & Terrill, C.E. (1991) Breed structure, dynamics, and new breed development in sheep. *World Animal Science, B8. Genetic resources of pig, sheep and goat* (Ed. by K. Maijala), pp. 305–26. Elsevier, Amsterdam, The Netherlands.

Maillard, J.C., Kemp, S.J., Naves, M., *et al.* (1993) [An attempt to correlate cattle breed origins and diseases associated with or transmitted by the tick *Amblyomma variegatum* in the French West Indies]. *Revue d'Elevage et de Médecine vétérinaire des Pays tropicaux*, **46,** 283–90.

Main, K.L.& Reynolds, B. (Eds) (1993) *Selective breeding of fishes in Asia and the United States. Proceedings of a workshop in Honolulu, Hawaii, May 3–7 1993*. Oceanic Institute, Honolulu, Hawaii, USA.

Mansbridge, R. (2001) RBST funded ovine semen collection. *Ark*, **29,** 105.

Manwell, C. & Baker, C.M.A. (1977) Genetic distance between the Australian Merino and the Poll Dorset sheep. *Genetical Research, Cambridge*, **29,** 239–53.

Manwell, C. & Baker, C.M.A. (1980) Chemical classification of cattle. 2. Phylogenetic tree and specific status of the zebu. *Animal Blood Groups and Biochemical Genetics*, **11,** 151–62.

Mariner, J.C. & Roeder, P.L. (2003) Use of participatory epidemiology in studies of the persistence of lineage 2 rinderpest virus in East Africa. *Veterinary Record*, **151,** 641–7.

Marklund, S., Kijas, J.M.H., Rodriguez-Martinez, H., *et al.* (1998) Molecular basis for the dominant white phenotype in the domestic pig. *Genome Research*, **8,** 826–33.

Markowitz, T.M., Dally, M.R., Gursky, K., *et al.* (1998) Early handling increases lamb affinity for humans. *Animal Behaviour*, **55,** 573–87.

Martin-Burriel, I., Garcia-Muro, E. & Zaragoza, P. (1999) Genetic diversity analysis of six Spanish native cattle breeds using microsatellites. *Animal Genetics*, **30,** 177–82.

Martin, G.B., Hotzel, M.J., Blache, D., *et al.* (2002) Determinants of the annual pattern of reproduction in mature male Merino and Suffolk sheep: modification of responses to photoperiod by an annual cycle in food supply. *Reproduction Fertility and Development*, **14,** 165–75.

Martin, G.S., Carstens, G.E., King, M.D., *et al.* (1999) Metabolism and morphology of brown adipose tissue from Brahman and Angus newborn calves. *Journal of Animal Science*, **77,** 388–99.

Mason, G.J., Cooper, J. & Clarebrough, C. (2001) Frustrations of fur-farmed mink. *Nature*, **410,** 35–6.

Mason, I.L. (1973) The role of natural and artificial selection in the origin of breeds of farm animals. A critique of Darwin's 'The variation of animals and plants under domestication'. *Zeitschrift für Tierzuchtung und Zuchtungsbiologie*, **90,** 229–44.

Mason, I.L. (1988) *A world dictionary of livestock breeds types and varieties*. 3rd edn. CAB International, Wallingford, UK.

Mason, I.L. (1996) *A world dictionary of livestock breeds, types and varieties*. 4th edn. CAB International, Wallingford, UK.

Matheron, G. & Planchenault, D. (1992) Breed characterization: the *IEMVT/CIRAD* experience. *African animal genetic resources: their characterization, conservation and utilisation. Proceedings of the research planning workshop held at ILCA, Addis Ababa, Ethiopia, 19–21 February 1992*. (Ed. by J.E.O. Rege & M.E. Lipner), pp. 31–4. International Livestock Centre for Africa, Addis Ababa, Ethiopia.

Matsuzaki, M., Sato, T., Morita, S., *et al.* (2001) Pulsatile growth hormone secretion, circulating insulin-like growth factor-1 concentration and cellular density of somatotrophs differ between Wagyu and Holstein steers. *Animal Science*, **73,** 425–32.

May, R.M. (1990) Taxonomy as destiny. *Nature*, **347,** 129–30.

Mayer, J.J., Brisbin, I.L. & Sweeney, J.M. (1989) Temporal dynamics of color phenotypes in an isolated population of feral swine. *Acta Theriologica*, **34,** 247–52.

Mburu, D.N., Ochieng, J.W., Kuria, S.G., *et al.* (2003) Genetic diversity and relationships of indigenous Kenyan camel (*Camelus dromedarius*) populations: implications for their classification. *Animal Genetics*, **34,** 26–32.

McDowell, R.E., Wilk, J.C. & Talbott, C.W. (1996) Economic viability of crosses of *Bos taurus* and *Bos indicus* for dairying in warm climates. *Journal of Dairy Science*, **79,** 1292–303.

McKay, J.C., Barton, N.F., Koerhuis, A.N.M., *et al.* (2000). The challenge of genetic change in the broiler chicken. *Occasional Publication, British Society of Animal Science*, **27,** 1–7.

McKenny Hughes, T. (1896) On the more important breeds of cattle which have been recognised in the British Isles in successive periods, and their relation to other archaeological and historical discoveries. *Archaeologia*, **55,** 125–58.

McMillan, I., Fairfull, R.W., Gowe, R.S., *et al.* (1990) Evidence for genetic improvement of layer stocks of chickens during 1950–1980. *World's Poultry Science Journal*, **46,** 235–45.

Meagher, S. (2000) Genetic diversity and *Capillaria hepatica* (Nematoda) prevalence in Michigan deer mouse populations. *Evolution*, **53,** 1318–24.

Medjugorac, I., Kustermann, W., Lazar, P., *et al.* (1994) Marker-derived phylogeny of European cattle supports demic expansion of agriculture. *Animal Genetics*, **25,** supplement 1, 19–27.

Meffert, L.M. (1999) How speciation experiments relate to conservation biology – the assumption of captive breeding strategies, that founder events reduce variation, may not always be correct. *Bioscience*, **49,** 701–11.

Meghen, C., MacHugh, D.E., Sauveroche, B., *et al.* (2000) Characterization of the Kuri cattle of Lake Chad using molecular genetic techniques. In: *The origins and development of African livestock. Archaeology, genetics, linguistics and ethnography* (Ed. by R.M. Blench & K.C. MacDonald), pp. 259–68. UCL Press, London, UK.

Meintjes, M., Bezuidenhout, C., Bartels, P., *et al.* (1997) *In vitro* maturation and fertilization of oocytes recovered from free-ranging Burchell's zebra (*Equus burchelli*) and Hartmann's zebra (*Equus zebra hartmannae*). *Journal of Zoo and Wildlife Medicine*, **28,** 251–9.

Menard, C., Duncan, P., Fleurance, G., *et al.* (2002) Comparative foraging and nutrition of horses and cattle in European wetlands. *Journal of Applied Ecology*, **39,** 120–33.

Mendizabal, J.A., Alberti, P., Eguinoa, P., *et al.* (1999) Adipocyte size and lipogenic enzyme activities in different adipose tissue depots in steers of local Spanish breeds. *Animal Science*, **69,** 115–21.

Mercer, J.T., Lewis, R.M. & Alderson, G.L.H. (1997) *The adaptation of rare breeds of British livestock to different environments: a review*. Scottish Agricultural Colleges, Penicuik, Midlothian, UK.

Merks, J.W.M. (2000) One century of genetic change in pigs and the future needs. *Occasional Publication, British Society of Animal Science*, **27,** 8–19.

Mezöszentgyörgyi, D., Husvéth, F., Lengyel, A., *et al.* (2001) Genotype-related variations in subcutaneous fat composition in sheep. *Animal Science*, **72,** 607–12.

Micol, T. & Jouventin, P. (1995) Restoration of Amsterdam Island, South Indian Ocean, following control of feral cattle. *Biological Conservation*, **72,** 199–206.

Miglior, F., Burnside, E.B. & Dekkers, J.C.M. (1995) Nonadditive genetic effects and inbreeding depression for somatic cell counts of Holstein cattle. *Journal of Dairy Science*, **78,** 1168–73.

Milan, D., Jeon, J.-T., Looft, C., *et al.* (2000) A mutation in *PRKAG3* associated with excess glycogen content in pig skeletal muscle. *Science*, **288,** 1248–51.

Miller, P.S. (1995) Selective breeding programs for rare alleles: examples from the Przewalski's horse and California condor pedigrees. *Conservation Biology*, **9,** 1262–73.

Mills, A.D., Beilharz, R.G. & Hocking, P.M. (1997) Genetic selection. In: *Animal welfare* (Ed. by M.C. Appleby & B.O. Hughes), pp. 219–31. CAB International, Wallingford, UK.

Mills, A.D., Wood-Gush, D.G.M. & Hughes, B.O. (1985) Genetic analysis of strain differences in pre-laying behaviour in battery cages. *British Poultry Science*, **26,** 187–97.

Milne, J.A. (1996) Environmental effects of low intensity systems of animal production in the hills and uplands of the UK. *Animal Science*, **63,** 363–71.

Milner, J.M., Albon, S.D., Illius, A.W., *et al.* (1999) Repeated selection of morphometric traits in the Soay sheep on St. Kilda. *Journal of Animal Ecology*, **68,** 472–88.

Mingay, G.E. (1982) *British Friesians: an epic of progress.* British Friesian Cattle Society, Rickmansworth, UK.

Moffat, A.S. (1998) Down on the animal pharm. *Science*, **282,** 2177.

Molénat, M., Luquet, M. & Giffard, C. (1993) [A research and development strategy for the Gascon pig]. *Ethnozootechnie*, **52,** 43–54.

Mommens, G., Peelman, L.J., Van Zeveren, A., *et al.* (1999) Microsatellite variation between an African and five European taurine breeds results in a geographical phylogenetic tree with a bison outgroup. *Journal of Animal Breeding and Genetics*, **116,** 325–30.

Monnerot, M., Vigne, J.D., Biju-Duval, C., *et al.* (1994) Rabbit and man: genetic and historical approach. *Genetics Selection Evolution*, **26,** supplement 1, S167–S182.

Mooring, M.S., Benjamin, J.E., Harte, C.R., *et al.* (2000) Testing the interspecific body size principle in ungulates: the smaller they come, the harder they groom. *Animal Behaviour*, **60,** 35–45.

Moriya, T. (1996) Cryopreservation of semen from Japanese white rabbits for use in teratological studies. *Experimental Animals*, **45,** 403–4.

Morris, C.A., Baker, R.L., Hickey, S.M., *et al.* (1993) Evidence of genotype by environment interaction for reproductive and maternal traits in beef cattle. *Animal Production*, **56,** 69–83.

Morris, P.H., Reddy, V. & Bunting, R.C. (1995) The survival of the cutest: who's responsible for the evolution of the teddy bear? *Animal Behaviour*, **50,** 1697–700.

Moulia, C., Lebrun, N., Loubes, C., *et al.* (1995) Hybrid vigor against parasites in interspecific crosses between 2 mice species. *Heredity*, **74,** 48–52.

Mualem, R., Choshniak, I. & Shkolnik, A. (1995) Enviromental heat load, bioenergetics and water economy in two breeds of goats: The Mamber goat versus the desert Bedouin goat. *World Review of Animal Production*, **25,** 91–5.

Mueller, U.G. & Wolfenbarger, L.L. (1999) AFLP genotyping and fingerprinting. *Trends in Ecology and Evolution*, **14,** 389–94.

Muir, W.M. (1996) Group selection for adaptation to multiple-hen cages: selection program and direct responses. *Poultry Science*, **75,** 447–58.

Mukai, F., Tsuji, S., Ohtagaki, S., *et al.* (1990) Body size and reproductive performance of a closed herd of Japanese Black cattle. *Journal of Animal Breeding and Genetics*, **107,** 282–90.

Muñoz, C.E. & Tejon, D.T. (1986) [*Catalogue of native Spanish breeds 1. Sheep and goats*] *Catalogo de razas autoctonas espanolas 1. – Especies ovina y caprina.* Ministerio de agricultura, pesca y alimentacion, Madrid, Spain.

Muzzolini, A. (1985) [The variability of size among European and African aurochs *Bos primigenius*]. *Bulletin de la Société Méridionale de Spéléologie et de Préhistoire*, **25,** 23–36.

Mysterud, A. (2000) The relationship between ecological segregation and sexual body size dimorphism in large herbivores. *Oecologia*, **124,** 40–54.

Nagashima, H., Kashiwazaki, N., Ashman, R.J., *et al.* (1995) Cryopreservation of porcine embryos. *Nature*, **374,** 416.

Naylor, R.L., Goldburg, R.J., Primavera, J.H., *et al.* (2000) Effect of aquaculture on world fish supplies. *Nature*, **405,** 1017–24.

Neel, J.V. & Rothman, E. (1981) Is there a difference among human populations in the rate with which mutation produces electrophoretic variants? *Proceedings of the National Academy of Sciences of the USA*, **78,** 3108–12.

Nei, M. (1987) *Molecular evolutionary genetics*. Columbia University Press, New York, USA.

Nevison, C.M., Hurst, J.L. & Barnard, C.J. (1999) Strain-specific effects of cage enrichment in male laboratory mice (*Mus musculus*). *Animal Welfare*, **8,** 361–79.

Nicholas, F.W. (1987) *Veterinary genetics*. Clarendon Press, Oxford, UK.

Nicholas, F.W. (1996a) *Introduction to veterinary genetics*. Oxford University Press, Oxford, UK.

Nicholas, F.W. (1996b) Genetic improvement through reproductive technology. *Animal Reproduction Science*, **42,** 205–14.

Nicholson, M.J. (1985) The water requirements of livestock in Africa. *Outlook on Agriculture*, **14,** 156–64.

Nicholson, M.J. (1987) Effects of night enclosure and extensive walking on the productivity of zebu cattle. *Journal of Agricultural Science, Cambridge*, **109,** 445–52.

Nielsen, D.I., Ebert, P.R., Hunt, G.J., *et al.* (1999) Identification of africanized honey bees (Hymenoptera: Apidae) incorporating morphometrics and an improved polymerase chain reaction mitotyping procedure. *Annals of the Entomological Society of America*, **92,** 167–74.

Nijman, I.J., Bradley, D.G., Hanotte, O., *et al.* (1999) Satellite DNA polymorphisms and AFLP correlate with *Bos indicus-taurus* hybridization. *Animal Genetics*, **30,** 265–73.

Nijman, I.J., Otsen, M., Verkaar, E.L.C., *et al.* (2003) Hybridization of banteng (*Bos javanicus*) and zebu (*Bos indicus*) revealed by mitochondrial DNA, satellite DNA, AFLP and microsatellites. *Heredity*, **90,** 10–6.

Nimon, A.J. & Broom, D.M. (2001) The welfare of farmed foxes *Vulpes vulpes* and *Alopex lagopus* in relation to housing and management: a review. *Animal Welfare*, **10,** 223–48.

Nix, J. & Hill, P. (2001) *Farm management pocketbook*. 32nd edn. (2002). Wye College, University of London, Ashford, Kent.

Nomura, T., Honda, T. & Mukai, F. (2001) Inbreeding and effective population size of Japanese Black cattle. *Journal of Animal Science*, **79,** 366–70.

Notter, D.R. (1999) The importance of genetic diversity in livestock populations of the future. *Journal of Animal Science*, **77,** 61–9.

Nurminsky, D.I., Nurminskaya, M.V., De Aguiar, D., *et al.* (1998) Selective sweep of a newly evolved sperm-specific gene in *Drosophila*. *Nature*, **396,** 572–5.

Ohba, Y., Kitagawa, H., Kitoh, K., *et al.* (2001) Clinical features of renal tubular dysplasia, a new hereditary disease in Japanese Black cattle. *Veterinary Record*, **149,** 115–8.

Oliver, W.L.R. & Brisbin, I.L. (1993) Introduced and feral pigs: problems, policy, and priorities. In: *Pigs, peccaries and hippos. Status survey and conservation action plan of the IUCN World Conservation Union* (Ed. by W.L.R. Oliver), pp. 179–91. IUCN, Gland, Switzerland.

Ollivier, L. (1995) Genetic differences in recombination frequency in the pig (*Sus scrofa*). *Genome*, **38,** 1048–51.

Olsen, H.F. & Klemetsdal, G. (2002) Genetic variation in the Døle horse and the Nordland horse. *Husdryforsøksmotet*, (2002), 561–4 (www.nordgen.org/ download/ prosjekt-hannefjerdigby.doc).

Ormerod, S.J., Marshall, E.J.P., Kerby, J. *et al.* (2003) Meeting the ecological challenges of agricultural change: editors' introduction. *Journal of Applied Ecology*, **40,** 939–46.

Osoro, K., Olivan, M., Celaya, R., *et al.* (1999) Effects of genotype on the performance and intake characteristics of sheep grazing contrasting hill vegetation communities. *Animal Science*, **69,** 419–26.

Otto, S.P. & Barton, N.H. (2001) Selection for recombination in small populations. *Evolution*, **55,** 1921–31.

Ovenden, G.N., Swash, A.R.H. & Smallshire, D. (1998) Agri-environment schemes and their contribution to the conservation of biodiversity in England. *Journal of Applied Ecology*, **35,** 955–60.

Ovilo, C., Cervera, M.T., Castellanos, C., *et al.* (2000) Characterization of Iberian pig genotypes using AFLP markers. *Animal Genetics*, **31,** 117–22.

Owen, E., Smith, T., Steele, M.A. *et al.* (Eds) (2004) *Responding to the livestock revolution. The role of globalisation and implications for poverty alleviation. BSAS Publication no. 33.* Nottingham University Press/British Society for Animal Science, Nottingham, UK.

Owen, J.B. (1996) The Cambridge breed. In: *Prolific sheep* (Ed. by M.H. Fahmy), pp. 161–73. CAB International, Wallingford, UK.

O'Huigin, C.O. & Cunningham, E.P. (1990) Conservation of the Kerry breed. *Farm and Food Research*, Jan–Mar, 25–7.

Ørskov, E.R. (1993) *Reality in rural development aid with emphasis on livestock.* Rowett Research Services Limited, Aberdeen, UK.

Ørskov, E.R. (1998) Feed evaluation with emphasis on fibrous roughages and fluctuating supply of nutrients: a review. *Small Ruminant Research*, **28,** 1–8.

Parer, J.T.K. (1963) Wool length and radiant heating effects in sheep. *Journal of Agricultural Science, Cambridge*, **60,** 141–4.

Parham, P. (1999) Soaring costs in defence. *Nature*, **401,** 870–1.

Parkes, J.P. (1993) Feral goats – designing solutions for a designer pest. *New Zealand Journal of Ecology*, **17,** 71–83.

Parsons, P.A. (1990) Fluctuating asymmetry: an epigenetic measure of stress. *Biological Reviews*, **65,** 131–45.

Parsons, Y.M., Cooper, D.W. & Piper, L.R. (1994) Evidence of linkage between high-glycine-tyrosine keratin gene loci and wool fibre diameter in a Merino half-sib family. *Animal Genetics*, **25,** 105–8.

Parsons, Y.M., Cooper, D.W. & Piper, L.R. (1996) Genetic variation in Australian Merino sheep. *Animal Genetics*, **27,** 223–8.

Paterson, S., Wilson, K. & Pemberton, J.M. (1998) Major histocompatibility complex variation associated with juvenile survival and parasite resistance in a large unmanaged ungulate population (*Ovis aries* L.). *Proceedings of the National Academy of Sciences of the USA*, **95,** 3714–9.

Pavlov, P.M., Crome, F.H.J. & Moore, L.A. (1992) Feral pigs, rainforest conservation and exotic disease in North Queensland. *Wildlife Research*, **19,** 179–93.

Pearce, F. (2004) Brazil's beef trade wrecks rainforest. *New Scientist*, **182,** (2442, 10 April 2004), 14–15.

Pelant, R.K., Chandra, B., Pu, J.B., *et al.* (1999) Small ruminants in development: the Heifer Project International experience in Asia. *Small Ruminant Research*, **34,** 249–57.

Penning, P.D., Rook, A.J. & Orr, R.J. (1991) Patterns of ingestive behaviour of sheep continuously stocked on monocultures of ryegrass or white clover. *Applied Animal Behaviour Science*, **31,** 237–50.

Perry, B.D., Carter, M.E., Hill, F.W.G., *et al.* (1987) Mastitis and milk production in cattle in a communal land of Zimbabwe. *British Veterinary Journal*, **143,** 44–50.

Petersson, E., Järvi, T., Steffner, N.G., *et al.* (1996) The effect of domestication on some life history traits of sea trout and Atlantic salmon. *Journal of Fish Biology*, **48,** 776–91.

Philippe, M.-A. (1995) [The Poitou donkey, a breed threatened by its own success]. *Ethnozootechnie*, **56,** 43–54.

Philipsson, J. (2000) Sustainability of dairy cattle breeding systems utilising artificial insemination in less developed countries – examples of problems and prospects. In: *Developing breeding strategies for lower input animal production environments. ICAR Technical Series no. 3* (Ed. by S. Galal, J. Boyazoglu & K. Hammond), pp. 551–62. International Committee for Animal Recording, Rome, Italy.

Phillips, C.J.C. (1997) Animal welfare considerations in future breeding programmes for farm livestock. *Animal Breeding Abstracts*, **65,** 645–54.

Phillips, R.W. (1981) The identification, conservation and effective use of valuable animal genetic resources. *Animal genetic resources. Conservation and management. FAO Animal Production and Health Paper 24* (Ed. by FAO/UNEP), pp. 1–5. Food and Agriculture Organization of the United Nations, Rome, Italy.

Philpott, M. (1993) The dangers of disease transmission by artificial insemination and embryo transfer. *British Veterinary Journal*, **149,** 339–69.

Piazza, A. (1998) Towards a genetic history of China. *Nature*, **395,** 636–9.

Piazza, A., Rendine, S., Minch, E., *et al.* (1995) Genetics and the origin of European languages. *Proceedings of the National Academy of Sciences of the USA*, **92,** 5836–40.

Pienkowski, M. (1998) Conservation of biodiversity by supporting high-nature-value farming systems. *Journal of Applied Ecology*, **35,** 948.

Pintado, B. & Gutierrez-Adan, A. (1999) Transgenesis in large domestic species: future development for milk modification. *Reproduction Nutrition Development*, **39,** 535–44.

Pisenti, J.M., Delany, M.E., Taylor, R.L., *et al.* (1999) *Avian genetic resources at risk: an assessment and proposal for conservation of genetic stocks in the USA and Canada. Report no. 20. Genetic resources conservation program.* Division of Agriculture and Natural Resources, University of California, Davis, California, USA.

Planchenault, D. (Ed.) (2003) *Workshop on cryopreservation of animal genetic resources in Europe. Paris, February 2003.* Salon International de l'Agriculture, Paris, France.

Platt, J.L. (2001) New risks, new gains. *Nature*, **407,** 27–30.

Plusquellec, P. & Bouissou, M.F. (2001) Behavioural characteristics of two dairy breeds of cows selected (Hérens) or not (Brune des Alpes) for fighting and dominance ability. *Applied Animal Behaviour Science*, **72,** 1–21.

Polziehn, R.O., Strobeck, C., Sheraton, J., *et al.* (1995) Bovine mtDNA discovered in North American bison populations. *Conservation Biology*, **9,** 1638–43.

Pond, W.G. (1983) Modern pork production. *Scientific American*, **248,** 78–87.

Popescu, C.P. & Tixier, M. (1984) The frequency of chromosomal abnormalities in domestic animals and their economic consequences. *Annales de Génétique*, **27,** 69–72.

Porter, V. (1993) *Pigs. A handbook to the breeds of the world.* Helm Information, Robertsbridge, UK.

Porter, V. (2002) *Mason's world dictionary of livestock breeds, types and varieties.* 5th edn. CABI Publishing, Wallingford, UK.

Pratchett, D. & Triglone, T. (1989) Prospects for *Leucaena* on the Ord. *New directions for Kimberley cattle. Bulletin no. 4164*. Western Australian Department of Agriculture and Australian Meat and Livestock Research and Development Corporation.

Prather, R.S., Hawley, R.J., Carter, D.B., *et al.* (2003) Transgenic swine for biomedicine and agriculture. *Theriogenology*, **59,** 115–123.

Preisinger, R. (2002) Breeding strategies for sustainable layer breeding. *Landbau Forschung Völkenrode*, Special Issue, **228,** 23–7.

Price, E.O. (2002) *Animal domestication and behavior*. CABI Publishing, Wallingford, UK.

Price, M.A. (1977) The effects of severe feed restriction on bulls and steers Part 1. Live weight loss, behaviour and non carcass body composition. *Australian Journal of Agricultural Research*, **28,** 521–8.

Primack, R.B. (2002) *Essentials of conservation biology*. 3rd edn. Sinauer Associates, Sunderland, Massachusetts, USA.

Pritz, R.K., Launchbaugh, K.L. & Taylor, C.A. (1997) Effects of breed and dietary experience on juniper consumption by goats. *Journal of Range Management*, **50,** 600–6.

Purohit, G.N. (1999) Biotechnologies in camelid reproduction: current status and future prospectives. *Journal of Camel Practice and Research*, **6,** 1–13.

Quéval, R., Petit, J.P., Tacher, G., *et al.* (1971) [The Kuri: a cattle breed from Lake Chad. 1. General introduction, with reference to zootechnical and biochemical studies, its origins and ecology]. *Revue d'Elevage et de Médecine vétérinaire des Pays tropicaux*, **24,** 667–87.

Rakyan, V.K., Preis, J., Morgan, H.D., *et al.* (2001) The marks, mechanisms and memory of epigenetic states in mammals. *Biochemical Journal*, **356,** 1–10.

Rangel-Figueiredo, T. & Iannuzzi, L. (1991) A cattle breed close to 58 diploid number due to high frequency of rob (1:29). *Hereditas*, **115,** 73–8.

Ranilla, M.J., Carro, M.D., Valdes, C., *et al.* (1997) A comparative study of ruminal activity in Churra and Merino sheep offered alfalfa hay. *Animal Science*, **65,** 121–8.

Ranilla, M.J., Lopez, S., Giraldez, F.J., Valdes, C., *et al.* (1998) Comparative digestibility and digesta flow kinetics in two breeds of sheep. *Animal Science*, **66,** 389–96.

Rauw, W.M., Kanis, E., Noordhuizen-Stassen, E.N., *et al.* (1998) Undesirable side effects of selection for high production efficiency in farm animals: a review. *Livestock Production Science*, **56,** 15–33.

Reinsch, N., Thomsen, H., Xu, N., *et al.* (1999) A QTL for the degree of spotting in cattle shows synteny with the *KIT* locus on chromosome 6. *Journal of Heredity*, **90,** 629–34.

Reist-Marti, S.B., Simianer, H., Gibson, J. *et al.* (2003) Weitzman's approach and conservation of breed diversity: an application to African cattle breeds. *Conservation Biology*, **17,** 1299–311.

Reynolds, J.D. & Harvey, P.H. (1994) Sexual selection and the evolution of sex differences. *The differences between the sexes* (Ed. by R.V. Short & E. Balaban), pp. 53–70. Cambridge University Press, Cambridge, UK.

Réale, D., Boussès, P., Pisanu, B., *et al.* (2000a) Biannual reproductive cycle in the Kerguelen feral sheep population. *Journal of Mammalogy*, **81,** 169–78.

Réale, D., Festa-Bianchet, M. & Jorgenson, J.T. (2000b) Heritability of body mass varies with age and season in wild bighorn sheep. *Heredity*, **83,** 526–32.

Richards, R.A. (1998) Darwin, domestic breeding and artificial selection. *Endeavour*, **22,** 106–9.

Riha, J., Cunat, L., McKelvey, W.A.C., *et al.* (1994) Transfer of imported frozen cashmere goat embryos. *Zivocisna Vyroba*, **39,** 881–8.

RIM (1994) *Nigerian livestock resources*. Resource Inventory and Management Ltd., Federal Department of Livestock and Pest Control Services, St. Helier, Jersey and Abuja, Nigeria.

Robelin, J. (1986) Growth of adipose tissues in cattle: partitioning between depots, chemical composition and cellularity. A review. *Livestock Production Science*, **14,** 349–64.

Roberts, D.F., Jorde, L.B. & Mitchell, R.J. (1981) Genetic structure in Cumbria. *Journal of Biosocial Science*, **13,** 317–36.

Roden, J.A. (1996) A comparison of alternative nucleus breeding systems and a sire referencing scheme for sheep improvement. *Animal Science*, **62,** 265–70.

Rodero, E. & Herrera, M. (2000) [The breed concept. An epistemological approach]. *Archivos de Zootecnia*, **49,** 5–16.

Rodero, E., Camacho, M.E., Delgado, J.V., *et al.* (1992) Study of the Andalusian minor breeds: evaluation of the priorities of conservation. *Animal Genetic Resources Information*, **10,** 41–53.

Roff, D.A. (1997) *Evolutionary quantitative genetics*. Chapman and Hall, New York, USA.

Rogers, G.W., Banos, G. & Sander-Nielsen, U. (1999) Genetic correlations among protein yield, productive life, and type traits from the United States and diseases other than mastitis from Denmark and Sweden. *Journal of Dairy Science*, **82,** 1331–8.

Romanov, M.N. (1999) Goose production efficiency as influenced by genotype, nutrition and production systems. *World's Poultry Science Journal*, **55,** 281–94.

Romanov, M.N. & Weigend, S. (2001) Analysis of genetic relationships between various populations of domestic and jungle fowl using microsatellite markers. *Poultry Science*, **80,** 1057–63.

Romanov, M.N., Wezyk, S., Cywabenko, K., *et al.* (1996) Poultry genetic resources in the countries of eastern Europe – history and current state. *Poultry and Avian Biology Reviews*, **7,** 1–29.

Roth, E.A. (1996) Traditional pastoral strategies in a modern world: an example from northern Kenya. *Human Organization*, **55,** 219–24.

Rothschild, M.F., Skow, L. & Lamont, S.J. (2000) The major histocompatibility complex and its role in disease resistance and immune responsiveness. *Breeding for disease resistance in farm animals*. 2nd edn. (Ed. by R.F.E. Axford, S.C. Bishop, F.W. Nicholas & J.B. Owen), pp. 73–105. CABI Publishing, Wallingford, UK.

Roughsedge, T., Thompson, R., Villanueva, B., *et al.* (2001) Synthesis of direct and maternal genetic components of economically important traits from beef breed-cross evaluations. *Journal of Animal Science*, **79,** 2307–19.

Roy, B.A. & Kirchner, J.W. (2000) Evolutionary dynamics of pathogen resistance and tolerance. *Evolution*, **54,** 51–63.

Royal, M.D., Darwash, A.O., Flint, A.P.F., *et al.* (2000) Declining fertility in dairy cattle: changes in traditional and endocrine pameters of fertility. *Animal Science*, **70,** 487–501.

Royal Society of Edinburgh (2002) *Inquiry into foot and mouth disease in Scotland. July 2002*. Royal Society of Edinburgh, Edinburgh, UK.

Ruane, J. (1999) A critical review of the value of genetic distance studies in breed conservation. *Journal of Animal Breeding and Genetics*, **116,** 317–23.

Ruane, J. (2000) A framework for prioritizing domestic animal breeds for conservation purposes at the national level: a Norwegian case study. *Conservation Biology*, **14,** 1385–93.

Rudge, M.R. (1970) Mother and kid behaviour in feral goats (*Capra hircus* L.). *Zeitschrift für Tierpsychologie*, **27,** 687–92.

Rudge, M.R. (1986a) The decline and increase of feral sheep on Campbell Island. *New Zealand Journal of Ecology*, **9,** 89–100.

Rudge, M.R. (1986b) A role for feral mammals in conserving the genetic diversity of livestock. *Animal Genetic Resources Information*, **6,** 7–17.

Russel, A.J.F. (1991) Fibre production from sheep and goats. In: *Progress in sheep and goat research* (Ed. by A.W. Speedy), pp. 235–56. CAB International, Wallingford, UK.

Russel, A.J.F., Gunn, R.G. & Doney, J.M. (1968) Components of weight loss in pregnant hill ewes during winter. *Animal Production*, **10,** 43–51.

Rutledge, J.J. (2001) Greek temples, tropical kine and recombination load. *Livestock Production Science*, **68,** 171–9.

Ryan, K., Munene, K., Munyiri, S.K., *et al.* (1991) Cattle naming: the persistence of a traditional practice in modern Maasailand. *MASCA Research Papers in Science and Archaeology*, **8,** supplement, 91–6.

Satibekova, N., Galliard, C., Obexer-Ruff, G., *et al.* (1999) Genetic diversity in Swiss goat breeds based on microsatellite analysis. *Animal Genetics*, **30**, 36–41.

Sage, R.D., Atchley, W.R. & Capanna, E. (1993) House mice as models in systematic biology. *Systematic Biology*, **42,** 523–61.

Sage, R.D., Heyneman, D., Lim, K.-C., *et al.* (1986) Wormy mice in a hybrid zone. *Nature*, **324,** 60–3.

Samson, D.E. & Slee, J. (1981) Factors affecting resistance to induced body cooling in newborn lambs of 10 breeds. *Animal Production*, **33,** 59–65.

Sangaré, M. & Pandey, V.S. (2000) Food intake, milk production and growth of kids of local, multipurpose goats grazing on dry season natural Sahelian rangeland in Mali. *Animal Science*, **71,** 165–73.

Sanudo, C., Nute, G.R., Campo, M.M., *et al.* (1998) Assessment of commercial lamb meat quality by British and Spanish taste panels. *Meat Science*, **48,** 91–100.

Sauter, C.M. & Morris, R.S. (1995). Dominance hierarchies in cattle and red deer (*Cervus elaphus*): their possible relationship to the transmission of bovine tuberculosis. *New Zealand Veterinary Journal*, **43,** 301–5

Scarpa, R., Drucker, A. G., Anderson, S. *et al.* (2003). Valuing genetic resources in peasant communities: the case of 'hairless' creole pigs in Yucatan. *Ecological Economics*, **45,** 427–43.

Scherf, B.D. (1995) *World watch list for domestic animal diversity*. 2nd edn. Food and Agricultural Organization of the United Nations, Rome, Italy.

Scherf, B.D. (2000) *World watch list for domestic animal diversity*. 3rd edn. Food and Agricultural Organization of the United Nations, Rome, Italy.

Scherner, E.R. (2000) Genealogy, distribution and frequency of the *immutabilis* allele in the mute swan (*Cygnus olor*). *Zeitschrift für Jagdwissenschaft*, **46,** 213–35.

Schild, H. & Rammensee, H.-G. (2000) Perfect use of imperfection. *Nature*, **404,** 709–10.

Schlecht, E., Blümmel, M. & Becker, K. (1999) The influence of the environment on feed intake of cattle in semi-arid Africa. In: *Regulation of feed intake* (Ed. by D. van der Heide, E.A. Huisman, E. Kanis, *et al.*), pp. 167–85. CABI Publishing, Wallingford, UK.

Schmid, M., Saitbekova, N., Gaillard, C., *et al.* (1999) Genetic diversity in Swiss cattle breeds. *Journal of Animal Breeding and Genetics*, **116,** 1–8.

Schmutz, S.M., Marquess, F.L.S., Berryere, T.G. *et al.* (1995) DNA marker-assisted selection of the polled condition in Charolais cattle. *Mammalian Genome*, **6,** 710–3.

Schork, N.J., Cardon, L.R. & Xu, X. 1998 The future of genetic epidemiology. *Trends in Genetics*, **14,** 266–72.

Schuurman, T., de Jong, F.H., Wichers Schreur, M.J.M., *et al.* (1999). Effects of domestication on foraging behavior: a study in domesticated and crossbred (Landrace x wild boar) pigs. In: *Regulation of feed intake* (Ed. by D. van der Heide, E.A. Huisman, E. Kanis, *et al.*), pp. 103–8. CABI Publishing, Wallingford, UK.

Schwabe, A.E. & Hall, S.J.G. (1989) Dystocia in nine British breeds of cattle and its relationship to the dimensions of the dam and calf. *Veterinary Record*, **125,** 636–9.

Scobie, D.R., Bray, A.R. & O'Connell, D. (1999) A breeding goal to improve the welfare of sheep. *Animal Welfare*, **8,** 391–406.

Scoones, I. (1995) Exploiting heterogeneity: habitat use by cattle in dryland Zimbabwe. *Journal of Arid Environments*, **29,** 221–37.

Shaw, A.P.M. & Hoste, C.H. (1987) *Trypanotolerant cattle and livestock development in west and central Africa vol 1. The international supply and demand for breeding stock. FAO Animal Production and Health Paper 67/1.* Food and Agricultural Organization of the United Nations, Rome, Italy.

Sheridan, A.K. (1981) Crossbreeding and heterosis. *Animal Breeding Abstracts*, **49,** 131–44.

Shuster, D.E., Kehrli, M.E., Ackermann, M.R., *et al.* (1992) Identification and prevalence of a genetic defect that causes leukocyte adhesion deficiency in Holstein cattle. *Proceedings of the National Academy of Sciences of the USA*, **89,** 9225–9.

Siemes, H. (2001) Export of dairy cattle dropped. *Veepro Magazine for Cattle Improvement*, no. **43**, 2.

Silanikove, N. (1992) Effects of water scarcity and hot enviroment on appetite and digestion in ruminants: a review. *Livestock Production Science*, **30,** 175–94.

Silva, J.R. & Noakes, D.E. (1984) Pelvic dimensions, bodyweight and parturition in rare breeds of sheep. *Veterinary Record*, **115,** 242–5.

Silver, L.M. (1995) *Mouse genetics. Concepts and applications.* Oxford University Press, Oxford, UK.

Simm, G. (1998) *Genetic improvement of cattle and sheep.* Farming Press, Ipswich, UK.

Simm, G., Conington, J., Bishop, S.C., *et al.* (1996) Genetic selection for extensive conditions. *Applied Animal Behaviour Science*, **49,** 47–59.

Simm, G., Villanueva, B., Sinclair, K.D., *et al.* (Eds) (2004) *Farm animal genetic resources. BSAS publication 30.* Nottingham University Press and British Society of Animal Science, Nottingham, UK.

Simon, D.L. & Buchenauer, D. (1993) *Genetic diversity of European livestock breeds. EAAP Publication 66.* Wageningen Pers, Wageningen, The Netherlands.

Slee, J. (1984). Improving lamb survival. *ABRO Report*, 1984, 11–6.

Slee, J., Simpson, S.P. & Wilson, S.B. (1987) Comparative methods for inducing and measuring non-shivering thermogenesis in newborn lambs. *Animal Production*, **45,** 61–7.

Slee, J. & Springbett, A. (1986) Early post-natal behaviour in lambs of ten breeds. *Applied Animal Behaviour Science*, **15,** 229–40.

Slee, J. & Stott, A.W. (1986) Genetic selection for cold resistance in Scottish Blackface lambs. *Animal Production*, **43,** 397–404.

Sleiman, F.T. & Saab, S.A. (1995) Influence of environment on respiration, heart-rate and body temperature of filial crosses compared to local Awassi sheep. *Small Ruminant Research*, **16,** 49–53.

Slingerland, M.A. (1999) Feeding behaviour on natural pasture, a complex of human-livestock-environment interactions. In: *Regulation of feed intake* (Ed. by D. van der Heide, E.A. Huisman, E. Kanis, *et al.*), pp. 187–96. CABI Publishing, Wallingford, UK.

Smil, V. (1999) Crop residues: Agriculture's largest harvest. *Bioscience*, **49,** 299–308.

Smith, C. (1984) Economic benefits of conserving animal genetic resources. *Animal Genetic Resources Information*, **3,** 10–4.

Smith, C. (1985) Scope for selecting many breeding stocks of possible economic value in the future. *Animal Production*, **41,** 403–12.

Smith, L.A., Cassell, B.G. & Pearson, R.E. (1998) The effects of inbreeding on the lifetime performance of dairy cattle. *Journal of Dairy Science*, **81,** 2729–37.

Smith, L.C., El Obeid, A.E. & Jensen, H.H. (2000) The geography and causes of food insecurity in developing countries. *Agricultural Economics*, **22,** 199–215.

Sölkner, J., Filipcic, L. & Hampshire, N. (1998) Genetic variability of populations and similarity of subpopulations in Austrian cattle breeds determined by analysis of pedigrees. *Animal Science*, **67,** 249–56.

Spencer, S. (1897) *Pigs. Breeds and management.* Vinton, London.

Solti, L., Crichton, E.G., Loskutoff, N.M., *et al.* (2000) Economical and ecological importance of indigenous livestock and the application of assisted reproduction to their preservation. *Theriogenology*, **53,** 149–62.

Spielman, D. & Frankham, R. (1992) Modeling problems in conservation genetics using captive *Drosophila* populations: improvement of reproductive fitness due to immigration of one individual into small partially inbred populations. *Zoo Biology*, **11,** 343–51.

Spindler, F. (1996) [Beef breeds: a result of sharecropping. Infuence of sharecropping on the development of husbandry in France]. *Ethnozootechnie*, **57,** 15–24.

Spitz, F. (1986) Current state of knowledge of wild boar biology. *Pig News and Information*, **7,** 171–5.

Sprinkle, J.E., Holloway, J.W., Warrington, B.G., *et al.* (2000) Digesta kinetics, energy intake, grazing behavior, and body temperature of grazing beef cattle differing in adaptation to heat. *Journal of Animal Science*, **78,** 1608–24.

Spruijt, B.M., van Hooff, J.A.R.A.M. & Gispen, W.H. (1992) Ethology and neurobiology of grooming behaviour. *Physiological Reviews*, **72,** 825–52.

Squires, E.L., Carnevale, E.M., McCue, P.M., *et al.* (2003) Embryo technologies in the horse. *Theriogenology*, **59,** 151–70.

Stanhope, M.J., Lupas, A., Italia, M.J. *et al.* (2001) Phylogenetic analyses do not support horizontal gene transfers from bacteria to vertebrates. *Nature*, **411,** 940–4.

Steane, D E. (2002) Experience from Asia. *Landbauforschung Völkenrode*, special issue 228, pp. 45–53.

Stear, M.J., Bairden, K., Bishop, S.C., *et al.* (1997a) The genetic basis of resistance to *Ostertagia circumcincta* in lambs. *Veterinary Journal*, **154,** 111–9.

Stear, M.J., Bairden, K., Duncan, J.L., *et al.* (1997b) How hosts control worms. *Nature*, **389,** 27.

Steinfeld, H. & Mack, S. (1995) Livestock development strategies. *World Review of Animal Production*, **84,** 18–24.

Stennett, A. (2000) An old breed looks to the new millennium. *Ark*, **27,** 146.

Stookey, J.M. & Goonewardene, L.A. (1996) A comparison of production traits and welfare implications between horned and polled beef bulls. *Canadian Journal of Animal Science*, **76,** 1–5.

Sullivan, P.G., Wilton, J.W. & van Doormaal, B.J. (1988) Production traits of red and white relative to black and white Holsteins. *Canadian Journal of Animal Science*, **68,** 299–303.

Swanson, T.M. (Ed.) (1995) *The economics and ecology of biodiversity decline. The forces driving global change*. Cambridge University Press, Cambridge, UK.

Sykes, A.R. (1994) Parasitism and production in farm animals. *Animal Production*, **59,** 155–72.

Syrstad, O. & Ruane, J. (1998) Prospects and strategies for genetic improvement of the dairy potential of tropical cattle by selection. *Tropical Animal Health and Production*, **30,** 257–68.

Tanabe, Y. (1991) The origin of Japanese dogs and their association with Japanese people. *Zoological Science*, **8,** 639–51.

Tarricone, B.J., Hingtgen, J.N., Belknap, J.K., *et al.* (1995) Quantitative trait loci associated with the behavioral response of B x D recombinant inbred mice to restraint stress: a preliminary communication. *Behavior Genetics*, **25,** 489–95.

Taylor, St.C.S. (1976) Multibreed designs 1. Variation within breeds. *Animal Production*, **23,** 133–44.

Taylor, St.C.S., Murray, J.I. & Thonney, M.L. (1989) Breed and sex differences among equally mature sheep and goats. 5. Lipid in dry tissue. *Animal Production*, **49,** 411–22.

Teale, A., Agaba, M., Clapcott, S., *et al.* (1999) Resistance to trypanosomosis: of markers, genes and mechanisms. *Archiv für Tierzucht*, **42,** 36–41.

Templeton, A.R. & Read, B. (1994) Inbreeding: one word, several meanings, much confusion. In *Conservation genetics* (Ed. by V. Loeschcke, J. Tomiuk & S.K. Jain), pp. 91–105. Birkhäuser, Basel, Switzerland.

Thaon d'Arnoldi, C., Foulley, J.-L. & Ollivier, L. (1998). An overview of the Weitzman approach to diversity. *Genetics Selection Evolution*, **30,** 149–161.

Thielscher, H.H. (1987) The pig's heart – a problem of pathophysiology. *Pro Veterinario*, 12.

Thiessen, R.B., Hnizdo, E., Maxwell, D.A.G., *et al.* (1984) Multibreed comparisons of British cattle. Variation in body weight, growth rate and food intake. *Animal Production*, **38,** 323–40.

Thiessen, R.B., Taylor, St.C.S. & Murray, J. (1985) Multibreed comparisons of British cattle. Variation in relative growth rate, relative food intake and food conversion efficiency. *Animal Production*, **41,** 193–9.

Thompson, E.A. (1995) Genetic importance and genomic descent. *Population management for survival & recovery. Analytical methods and strategies in small population conservation* (Ed. by J.D. Ballou, M. Gilpin & T.J. Foose), pp. 112–23. Columbia University Press, New York, USA.

Thompson, J.R., Everett, R.W. & Hammerschmidt, N.L. (2000a) Effects of inbreeding on production and survival in Holsteins. *Journal of Dairy Science*, **83,** 1856–64.

Thompson, J.R., Everett, R.W. & Wolfe, C.W. (2000b) Effects of inbreeding on production and survival in Jerseys. *Journal of Dairy Science*, **83,** 2131–8.

Tolhurst, S., Oates, M. (2001) *The breed profiles handbook. A guide to the selection of livestock breeds for grazing wildlife sites*. English Nature, Peterborough, UK.

Toro, M., Barragán, C., Óvilo, C., *et al.*, (2002) Estimation of coancestry in Iberian pigs using molecular markers. *Conservation Genetics*, **3,** 309–20.

Toro, M.A., Rodriganez, J., Silio, L., *et al.* (2000) Genealogical analysis of a closed herd of black hairless Iberian pigs. *Conservation Biology*, **14,** 1843–51.

Towne, E.G. (2000) Prairie vegetation and soil nutrient responses to ungulate carcasses. *Oecologia*, **122,** 232–9.

Townsend, S.J. (2000) *Genetic diversity and domestication in sheep* (Ovis). PhD thesis, University of East Anglia, UK.

Townsend, S.J. (2003a) Scrapie news. *Ark*, **31,** 28.

Townsend, S.J. (2003b) GENIPED launched. *Ark*, **31,** 29.

Townsend, S.J., Warner, R. & Dawson, M. (in press) *PrP* genotypes of rare sheep breeds in Great Britain. *Veterinary Record.*

Townsend, S.J., Warren, S. & Wilson, H. (2002) FMD – how the disease has impacted on different sectors of the rare breeds' world. *Ark*, **30,** 65–7.

Trimeche, A., Anton, M., Renard, P., *et al.* (1997) Quail egg yolk: a novel cryoprotectant for the freeze preservation of Poitou jackass sperm. *Cryobiology*, **34,** 385–93.

Trinderup, M., Jørgensen, J.N. & Hansen, M. (1999a) Inbreeding and loss of founder alleles in four variations of a conservation programme using circular mating, for Danish Shorthorn cattle. *Animal Genetic Resources Information*, **25,** 99–106.

Trinderup, M., Jørgensen, J.N. & Hansen, M. (1999b) Conservation considerations on Danish Shorthorn cattle using pedigree analysis. *Animal Genetic Resources Information*, **26,** 27–33.

Troy, C.S., MacIlugh, D.E., Bailey, J.F., *et al.* (2001) Genetic evidence for Near-Eastern origins of European cattle. *Nature*, **410,** 1088–91.

Tselutin, K., Narubina, L., Mavrodina, T., *et al.* (1995) Cryopreservation of poultry semen. *British Poultry Science*, **36,** 805–11.

Tubbs, C.R. (1997) The ecology of pastoralism in the New Forest. *British Wildlife*, **9,** 7–16.

Tucker, H.A. (1982) Seasonality in cattle. *Theriogenology*, **17,** 53–9.

Vagenas, D., Jackson, F., Russel, A.J.F., *et al.* (2002) Genetic control of resistance to gastro-intestinal parasites in crossbred cashmere-producing goats: responses to selection, genetic parameters and relationships with production traits. *Animal Science*, **74,** 199–208.

Vajta, G., Holm, P., Kuwayama, M., *et al.* (1998) Open pulled straw (OPS) vitrification: a new way to reduce cryoinjuries of bovine ova and embryos. *Molecular Reproduction and Development*, **51,** 53–8.

van Arendonk, J.A.M., Nieuwhof, G.J., Vos, H., *et al.* (1991) Genetic aspects of feed intake and efficiency in lactating dairy heifers. *Livestock Production Science*, **29,** 263–75.

van der Lende, T., Komen, J. & Brascamp, E.W. (1998) The role of reproductive techniques in genetic improvement and dissemination of genetic gain in livestock production and aquaculture. *Acta Agriculturae Scandinavica*, **29,** supplement, 90–7.

van der Waaij, E.H. & van Arendonk, J.A.M. (2000) Introgression of genes responsible for disease resistance in a cattle population selected for production: genetic and economic consequences. *Animal Science*, **70,** 207–20.

Van Laere, A.-S., Nguyen, M., Braunschweig, M. *et al.* (2003) A regulatory mutation in *IGF2* causes a major QTL effect on muscle growth in the pig. *Nature*, **425,** 832–6.

van Liere, D. (1999) *Large ungulates in self-regulating natural sites: the Dutch Society for the Protection of Animals' views, including reactions to the paper 'Goede tijden, slechte tijden' by Keulartz et al., 1998.* Floris, Den Haag, The Netherlands.

Van Vleck, L.D. (1999) Implications of cloning for breed improvement strategies: are traditional methods of animal improvement obsolete? *Journal of Animal Science*, **77,** supplement, 2, 111–21.

van Wieren, S.E. (1995) The potential role of large herbivores in nature conservation and extensive land use in Europe. *Biological Journal of the Linnaean Society*, **56,** supplement, 11–23.

van Wyk, J.A., Stenson, M.O., van der Merwe, J.S., *et al.* (1999) Anthelmintic resistance in South Africa: surveys indicate an extremely serious situation in sheep and goat farming. *Onderstepoort Journal of Veterinary Research*, **66,** 273–84.

Velarde, F.L., Espinosa, D., Monge, C., *et al.* (1991) [A genetic response to high altitude hypoxia - high haemoglobin-oxygen affinity in chicken (*Gallus gallus*) from the Peruvian Andes]. *Comptes Rendus de l'Académie des Sciences Série III. Sciences de la Vie (Paris),* **313,** 401–6.

Vera, F.W.M. (2002) *Grazing ecology and forest history*. CABI Publishing, Wallingford, UK.

Verbeke, W., Van Oeckel, M.J., Warnants, N., *et al.* (2000) Consumer perception, facts and possibilities to improve acceptability of health and sensory characteristics of pork. *Meat Science*, **53,** 77–99.

Verbeke, W.A.J. & Viaene, J. (2000) Beliefs, attitude and behaviour towards fresh meat consumption in Belgium: empirical evidence from a consumer survey. *Food Quality and Preference*, **10,** 437–45.

Vercoe, J.E. (1996) Pastures for prosperity 5. The role of animal science. *Tropical Grasslands*, **30,** 58–72.

Verrier, E. & Bouffartigue, B. (1993) [AOC products as a way to preserve regional cattle breeds? French North Alps as an example]. *Ethnozootechnie*, **52,** 33–42.

Via, S. & Lande, R. (1987) Evolution of genetic variability in a spatially heterogenous environment: effects of genotype-environment interaction. *Genetical Research, Cambridge*, **49,** 147–56.

Vilà, C., Leonard, J.A., Gotherstrom, A., *et al.* (2001) Widespread origins of domestic horse lineages. *Science*, **291,** 474–7.

Vilà, C., Maldonado, J.E. & Wayne, R.K. (1999) Phylogenetic relationships, evolution, and genetic diversity of the domestic dog. *Journal of Heredity*, **90,** 70–7.

Visscher, P.M., Smith, D., Hall, S.J.G., *et al.* (2001) A viable herd of genetically uniform cattle. *Nature*, **409,** 303.

Vu Tien, J., Verrier, E. & de Rochambeau, H. (1994) Some results on the change in genetic variability in four livestock populations of small size. *Proceedings of the 5th World Congress on Genetics applied to Livestock Production. Guelph, Ontario, Canada*, **21,** 540–3.

Wade-Martins, P. (1993) *Black faces. A history of East Anglian sheep breeds.* Norfolk Museums Service/Geerings of Ashford, Ashford, Kent, UK.

Wakelin, D. (2000) Rodent models of genetic resistance to parasitic infections. *Breeding for disease resistance in farm animals.* 2nd edn. (Ed. by R.F.E. Axford, S.C. Bishop, F.W. Nicholas & J.B. Owen), pp. 107–26. CABI Publishing, Wallingford, UK.

Wall, R.J. (1997) A new lease of life for transgenic livestock. *Nature Biotechnology*, **15,** 416–7.

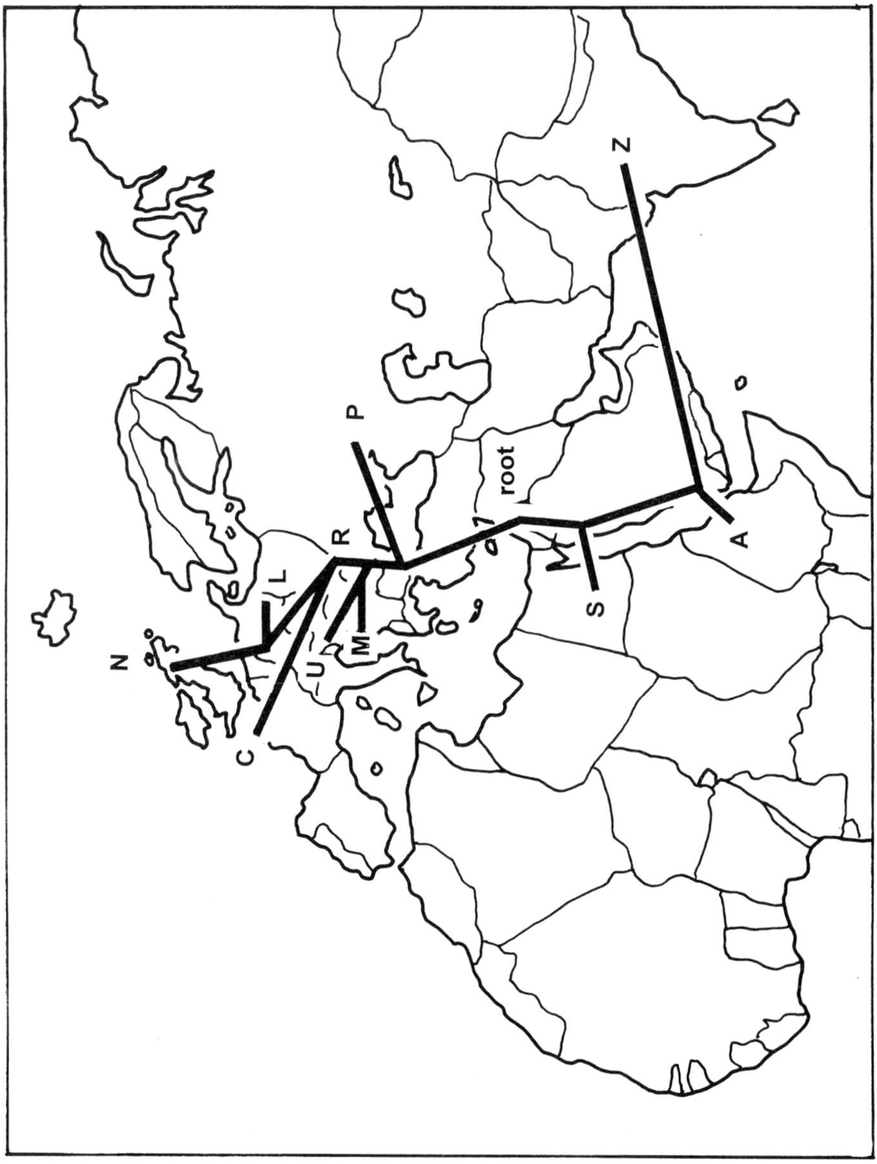

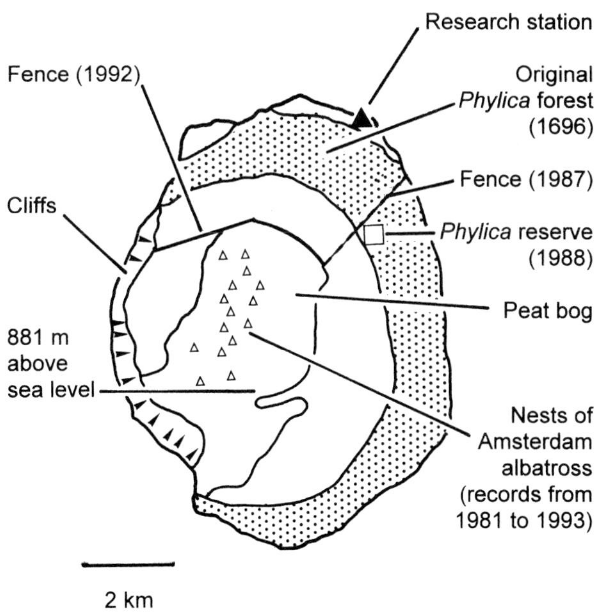

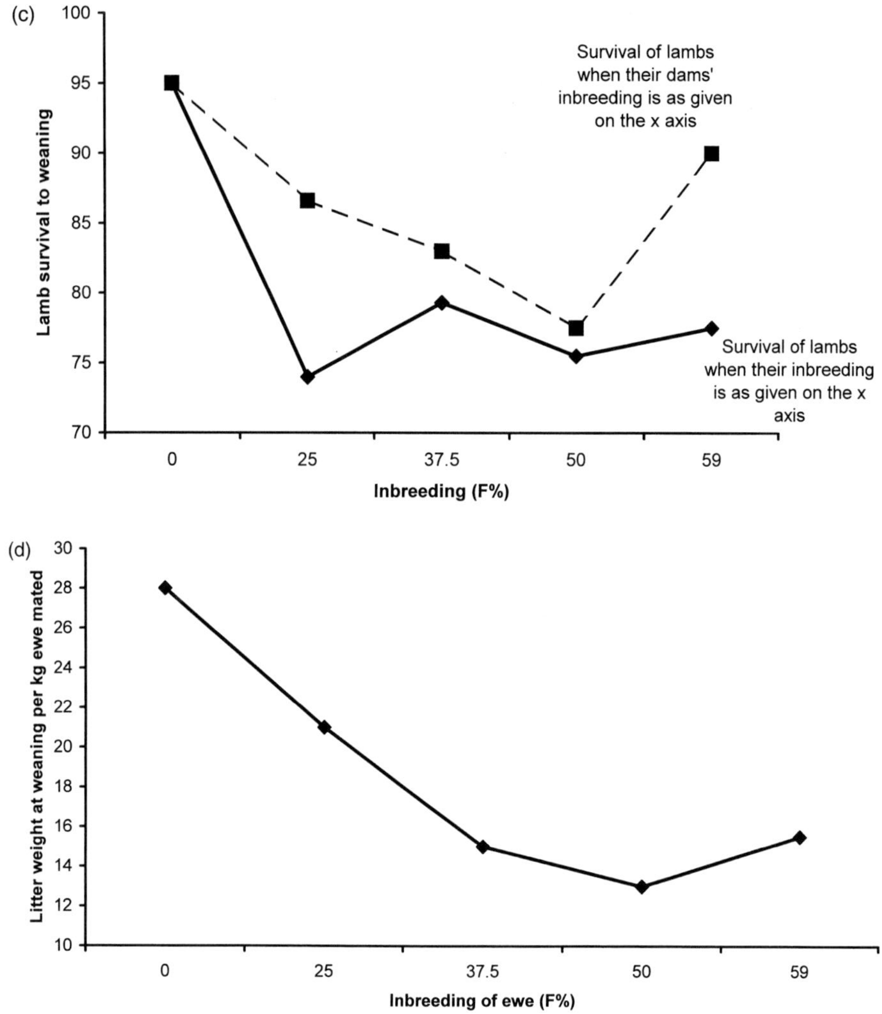

Wallis de Vries, M.F. (1994) *Foraging in a landscape mosaic. Diet selection and performance of free-ranging cattle in heathland and riverine grassland.* PhD thesis, Agricultural University, Wageningen, The Netherlands.

Wallner, B., Bren, G., Müller, M. *et al.* (2003) Fixed nucleotide differences on the Y chromosome indicate clear divergence between *Equus przewalskii* and *Equus caballus. Animal Genetics*, **34**, 453–6.

Wallis de Vries, M.F. (1995) Large herbivores and the design of large-scale nature reserves in western Europe. *Conservation Biology*, **9,** 25–33.

Wang, J.L., Hill, W.G., Charlesworth, D., *et al.* (1999) Dynamics of inbreeding depression due to deleterious mutations in small populations: mutation parameters and inbreeding rate. *Genetical Research*, **74,** 165.

Wang, R.-L., Stec, A., Hey, J., *et al.* (1999) The limits of selection during maize domestication. *Nature*, **398,** 236–9.

Wang, S., Liu, Y., Holyoak, G.R., *et al.* (1998) A protocol for *in vitro* maturation and fertilization of sheep oocytes. *Small Ruminant Research*, **29,** 83–8.

Wanner, J.M., Rogers, G.W., Kehrli, M.E., *et al.* (1998) Intramammary infections in primiparous Holsteins: heritabilities and comparisons of bovine leukocyte adhesion deficiency carriers and noncarriers. *Journal of Dairy Science*, **81,** 3293–9.

Ward, S. (2003) Dolly creators begin mass slaughter. *The Scotsman*, 15 July 2003 (www.thescotsman.co.uk/index.cfm?id = 768342003).

Wareing, S., Hartaningsih, N., Wilcox, G.E., *et al.* (1999) Evidence for immunosuppression associated with Jembrana disease virus infection of cattle. *Veterinary Microbiology*, **68,** 179–85.

Warriss, P.D. (2000) *Meat Science. An introductory text.* CABI Publishing, Wallingford, UK.

Warriss, P.D., Brown, S.N., Franklin, J.G., *et al.* (1990) The thickness and quality of backfat in various pig breeds and their relationship to intramuscular fat and the setting of joints from the carcass. *Meat Science*, **28,** 21–9.

Warriss, P.D., Kestin, S.C., Brown, S.N., *et al.* (1996) The quality of pork from traditional pig breeds. *Meat Focus International*, May/June, 179–82.

Wassmuth, R., Wallbaum, F. & Langholz, H.J. (1999) Outdoor wintering of suckler cows in low mountain ranges. *Livestock Production Science*, **61,** 193–200.

Watanabe, T., Hayashi, Y., Kimura, J., *et al.* (1986) Pig mitochondrial DNA: polymorphism, restriction map orientation and sequence data. *Biochemical Genetics*, **45,** 385–96.

Watve, M.G. & Sukumar, R. (1995) Parasite abundance and diversity in mammals: correlates with host ecology. *Proceedings of the National Academy of Sciences of the USA*, **92,** 8945–9.

Webb, R., Gosden, R.G., Telfer, E.E., *et al.* (1999) Factors affecting folliculogensis in ruminants. *Animal Science*, **68,** 257–84.

Weigend, S. & Romanov, M.N. (2002) The world watch list for domestic animal diversity in the context of conservation and utilisation of poultry biodiversity. *World's Poultry Science Journal*, **58,** 411–25.

Weigend, S., Vef, E., Wesch, G., *et al.* (1995) Conception for conserving genetic resources in poultry in Germany. *Archiv für Geflugelkunde*, **59,** 327–34.

Weiler, U., Appell, H.-J., Kremser, M., *et al.* (1995) Consequences of selection on muscle composition. A comparative study on gracilis muscle in wild and domestic pigs. *Anatomia Histologia Embryologia*, **24,** 77–80.

Wells, D.N., Misica, P.M., Tervit, H.R., *et al.* (1998) Adult somatic cell nuclear transfer is used to preserve the last surviving cow of the Enderby Island cattle breed. *Reproduction Fertility and Development*, **10,** 369–78.

Whitlock, M.C. & Fowler, K. (1999) The changes in genetic and environmental variance with inbreeding in *Drosophila melanogaster*. *Genetics*, **152,** 345–53.

Wiener, G. (1994) *The tropical agriculturalist. Animal breeding.* Macmillan and Technical Centre for Agricultural and Rural Cooperation, London, UK and Wageningen, The Netherlands.

Wiener, G., Lee, G.J. & Woolliams, J.A. (1992a) Effects of rapid inbreeding and of crossing of inbred lines on the body weight growth of sheep. *Animal Production*, **55,** 89–99.

Wiener, G., Lee, G.J. & Woolliams, J.A. (1992b) Effects of rapid inbreeding and of crossing of inbred lines on conception rate, prolificacy and ewe survival in sheep. *Animal Production*, **55,** 115–21.

Wiener, G., Suttle, N.F., Field, A.C., *et al.* (1978) Breed differences in copper metabolism in sheep. *Journal of Agricultural Science, Cambridge*, **91,** 433–41.

Willham, R.L. (1982) Genetic improvement of beef cattle in the United States: cattle, people and their interaction. *Journal of Animal Science*, **54,** 659–66.

Williams, O.B., Wells, T.C.E. & Wells, D.A. (1974) Grazing management of Woodwalton Fen: seasonal changes in the diet of cattle and rabbits. *Journal of Applied Ecology*, **11,** 499–516.

Wilson, R.T. (1995) Livestock production and animal genetic resources in Croatia. *Animal Genetic Resources Information*, **16,** 41–50.

Wimmers, K., Ponsuksili, S., Schmoll, F., *et al.* (1999) Application of microsatellite analysis to group chicken according to their genetic similarity. *Archiv für Tierzuchtung, Dummerstorf*, **42,** 629–39.

Winder, J.A., Walker, D.A. & Bailey, C.A. (1995) Genetic aspects of diet selection in the Chihuahuan desert. *Journal of Range Management*, **48,** 549–53.

Wolf, J.B., Brodie, E.D. & Wade, M.J. (2000) *Epistasis and the evolutionary process.* Oxford University Press, Oxford, UK.

Womack, J.E. (1992) Molecular genetics arrives on the farm. *Nature*, **360,** 108–9.

Wood, A. (in press) Could Africa be like America? *Proceedings of the Annual Bank Conference on Development Economics April–May 2002*. World Bank, Washington, DC, USA.

Woods, G.L., White, K.L., Vanderwall, D.K., *et al.* (2003) A mule cloned from fetal cells by nuclear transfer. (www.sciencemag.org/cgi/doi/10.1126/science.1086743)

Woolhouse, M.E.J., Webster, J.P., Domingo, E., *et al.* (2002) Biological and biomedical implications of the co-evolution of pathogens and their hosts. *Nature Genetics*, **32,** 569–77.

Woolliams, J.A. & Mäantysaari, E.A. (1995) Genetic contributions of Finnish Ayrshire bulls over four generations. *Animal Science*, **61,** 177–87.

Woolliams, J.A. & Meuwissen, T.H.E. (1999) Development of an expert system for conservation. In: *Genebanks and the conservation of farm animal genetic resources* (Ed. by J.K. Oldenbroek), pp. 113–9. DLO Institute for Animal Science and Health, Lelystad, The Netherlands.

Wright, I.A., Dalziel, A.J.I. & Ellis, R.P. (2002) *The status of traditional Scottish animal breeds and plant varieties and the implications for biodiversity*. Scottish Executive Social Research, Edinburgh.

Wright, I.A., Jones, J.R., Maxwell, T.J., *et al.* (1994) The effect of genotype x environment interactions on biological efficiency in beef cows. *Animal Production*, **58,** 197–207.

Wright, S. (1923) Mendelian analysis of the pure breeds of livestock. 2. The Duchess family of Shorthorns as bred by Thomas Bates. *Journal of Heredity*, **14,** 405–422.

Wright, S. (1977) *Evolution and the genetics of populations. Volume 3. Experimental results and evolutionary deductions*. University of Chicago Press, Chicago, IL, USA.

Wu, N. (2000) Yak breeding programmes in China. *Developing breeding strategies for lower input animal production environments. ICAR Technical Series no. 3* (Ed. by S. Galal, J. Boyazoglu & K. Hammond), pp. 409–27. International Committee for Animal Recording, Rome, Italy.

Wu, R.L. (2000) Partitioning of population genetic variance under multiplicative-epistatic gene action. *Theoretical Population Biology*, **100,** 743–9.

Wu, X.Q., Ding, B. & Zhang, Y.-P. (1999) Genetic diversity of Chinese native pigs inferred from protein electrophoresis. *Biochemical Genetics*, **37,** 159–73.

Yazdi, M.H., Rydhmer, L., Ringmar-Cederberg, E., *et al.* (2000) Genetic study of longevity in Swedish Landrace sows. *Livestock Production Science*, **63,** 255–64.

Yu, Y., Nie, L., He, Z.Q., *et al.* (1999) Mitochondrial DNA variation in cattle of South China: origin and introgression. *Animal Genetics*, **30,** 245–50.

Zajchowski, L.D. & Etches, R.J. (2000) Transgenic chickens: past, present, and future. *Avian and Poultry Biology Reviews*, **11,** 63–80.

Zepeda Sein, C. (1989) Prospects of replacing the pig population in Haiti. *Veterinaria-Mexico*, **20,** 65–70.

Zinsstag, J., Ankers, P., Njie, M., *et al.* (2000) Heritability of gastrointestinal nematode faecal egg counts in west African village N'Dama cattle and its relation to age. *Veterinary Parasitology*, **89,** 71–8.

Zuckerman, S. (1994) Historical address. *Ark*, **21,** 168–81.

Subject index

Country & breed index

Countries

Breeds, camel

Breeds, cattle

Breeds, donkey

Breeds, goat

Breeds, horse

Breeds, pig

Breeds, poultry

Breeds, sheep